INSTALLATION, SERVICING AND MAINTENANCE

S.N. BHATTACHARYA

M.E. (MECH), M.I.A

S. CHAND
PUBLISHING

S Chand And Company Limited

(ISO 9001 Certified Company)

S. CHAND
PUBLISHING

S Chand And Company Limited

(ISO 9001 Certified Company)

Head Office: Block B-1, House No. D-1, Ground Floor, Mohan Co-operative Industrial Estate, New Delhi – 110 044 | Phone: 011-66672000
Registered Office: A-27, 2nd Floor, Mohan Co-operative Industrial Estate, New Delhi – 110 044
Phone: 011-49731800
www.**schandpublishing.com**; e-mail: **info@schandpublishing.com**

Branches

Ahmedabad	:	Ph: 27542369, 27541965; ahmedabad@schandpublishing.com
Bengaluru	:	Ph: 22354008, 22268048; bangalore@schandpublishing.com
Bhopal	:	Ph: 4274723, 4209587; bhopal@schandpublishing.com
Bhubaneshwar	:	Ph: 2951580; bhubaneshwar@schandpublishing.com
Chennai	:	Ph: 23632120; chennai@schandpublishing.com
Guwahati	:	Ph: 2738811, 2735640; guwahati@schandpublishing.com
Hyderabad	:	Ph: 40186018; hyderabad@schandpublishing.com
Jaipur	:	Ph: 2291317, 2291318; jaipur@schandpublishing.com
Jalandhar	:	Ph: 4645630; jalandhar@schandpublishing.com
Kochi	:	Ph: 2576207, 2576208; cochin@schandpublishing.com
Kolkata	:	Ph: 23357458, 23353914; kolkata@schandpublishing.com
Lucknow	:	Ph: 4003633; lucknow@schandpublishing.com
Mumbai	:	Ph: 25000297; mumbai@schandpublishing.com
Nagpur	:	Ph: 2250230; nagpur@schandpublishing.com
Patna	:	Ph: 2260011; patna@schandpublishing.com
Ranchi	:	Ph: 2361178; ranchi@schandpublishing.com
Sahibabad	:	Ph: 2771238; info@schandpublishing.com

First Edition 1995
Second Edition 2001; Reprints 2005, 2007, 2008, 2009, 2010 (Twice), 2012, 2013, 2015
Reprint 2017

ISBN: 978-81-219-0831-3 **Product Code:** H3ISM62PAQC10ENAB01O

PRINTED IN INDIA

By Vikas Publishing House Private Limited, Plot 20/4, Site-IV, Industrial Area Sahibabad, Ghaziabad – 201 010 and Published by S Chand And Company Limited, A-27, 2nd Floor, Mohan Co-operative Industrial Estate, New Delhi – 110 044

Dedicated

to

My respectfully beloved

Late father and mother

FOREWORD

We are at the threshold of a new technological era. The tools, the equipment, the machines are getting modernised everyday. Even the maintenance techniques, planning are being improved continuously to achieve quality-accuracy-precision and to create a non-interrupted production flow. The carefulness during installation, the procedural steps on servicing of machines with utter skill and experience and a proper maintenance schedule are the vital criteria to improve the productivity and to extract optimum output from the existing infrastructure.

A good number of books regarding maintenance of some individual machines are there in the market. But Prof. S.N. Bhattacharya has done an excellent job covering a wide range of instruments and machines in this book.

Specially it will be of immense help to the students to get acquainted with most of the diversified but interrelated information in a single book. The technical and engineering students were also badly in need of such a book for a long period of time.

Prof. Bhattacharya took much pain to write the book with the background of an excellent teaching experience of more than twenty years and also with rich practical experience in various capacities of a number of related concerns.

I take this opportunity to congratulate Prof. Bhattacharya for his praiseworthy, need-based performance of such a useful job, both for the academic and shop-floor interest.

I hope this book will certainly be accepted by all concerned.

Dr. S.M. Chatterjee
Director of Technical Education
West Bengal

PREFACE TO THE REVISED SECOND EDITION

With a great pleasure I am placing this revised second edition before you. A few additions have been made in this edition, as per the valuable suggestions of a number of readers of the book. I am really indebted to them.

The 'Maintenance and Work Simplification' will certainly enrich the book regarding the maintenance planning. It has been added in the chapter –1. The topic 'PINS' has been inserted in the chapter–6. The chapter –9 has been made complete by the introduction of the topic 'CUTTING FLUIDS' and the topics 'Cranes', 'Crane Hooks' and 'Slings' have been made available in the chapter–14, considering the need of the students.

It may again be mentioned that I will be happy and grateful to receive further suggestions to enrich the book for the academic interest and they will be incorporated in the following editions on careful scrutiny.

S.N. BHATTACHARYA

PREFACE TO THE FIRST EDITION

Installation of machineries, right maintenance work at right time and servicing at scheduled interval are the essential features of the industrial world. But an academic discussion on such topics involving at least the common devices in a single book is unfortunately rarely available.

It is needless to mention that such a subject has literally got no range. However, I have tried to design the contents considering the urgent need of the Engineering Diploma and Degree students. The syllabus on Diploma Course in Mechanical engineering has been almost totally covered.

A major emphasis has been given at every step to furnish figures which may be easily understandable and reproducible by the students. The language also has been intentionally made as simple as possible.

I have got hesitation to confess that I consulted a lot of authentic books both of Indian and foreign authors while writing this book and at the same time I have received positive help and cooperation from number of persons, institutions and and organisations.

I am grateful to the Publisher S. Chand & Co. Ltd. New Delhi, not only for accepting this book for publication but also for permitting me to reproduce a number of figures and information for topics "repairing of guide surface" and "cracks on C.I. body" from the book 'Industrial maintenance', written by respectable H.P. Garg.

I am thankful to Mr. N. Ramanuja, General Manager, Machine Tool Marketing, H.M.T. Ltd. Bangalore for encouraging me to write this book and allowing me to reproduce certain figures, test charts etc. from their manuals. I extend my gratitutde to the Engineers and Officers of their Eastern Regional Office at Calcutta for their cooperation in this regard. My gratitude is also due to Dr. R. Mathur, Director (Publications) of Bureau of Indian Standards along with the BIS as a whole for providing me permission to reproduce a few extracts from their publications.

Mr. R.P.Banerjee and Sri. P. Barat put their honest effort and extensive care for typing the manuscript and sketching the figures respectively. My colleague Prof. A. Biswas and Prof. R.M. Dutta of J.C. Polytechnic, West Bengal extended their valuable help to formulate the contents etc. and Dr. S. Bhattacharya took every pain to perform the tedious job of comnparing the manuscript. My Principal Prof. S. Bandyopadhyay, Smt. S. Chakrovorty, K. Bhattacharya, Mr. A.K. Bhattacharya, Mr. S. Roy, Proprietor, Bharti Industries Berhampur also extended their hands of cooperation at various stages. My deepest thanks and gratitude go to all of them.

I am indebted to Dr. S.M. Chatterjee, Director of Technical Education, Govt. of West Bengal, as he showed keen interest in this work and kindly agreed to write a foreword for the book.

I am also thankful to the Department of Education, Govt. of West Bengal for providing me necessary permission to publish the book.

This book was written over a long period of time, encroaching upon the time that had to be spent towards the stern reality of the family job. But my better half encouragingly bore with it and my appreciation therefore, is also due to her.

The book contains fifteen chapters. Besides the general discussion on maintenance and its planning, a number of devices like pump, compressor, heat engines, boilers etc. starting from the major handtools of day-to-day use have been taken into consideration.

The pipings, dust collectors, etc. which have been discussed in the 14th chapter are not really miscellaneous equipments or devices in true sense. however, such a title has been put, as a number of different things have been included in a single chapter.

The maiden venture of this author will be cosidered to be successful, if the book serves the interests of the students effectively fulfilling their long standing necessity. I will honour each and every constructive suggestion of any concerned reader of any level for the improvement of this book, if their comments are communicated and I will remain grateful to them.

Author

CONTENTS

1

MAINTENANCE AND MAINTENANCE PLANNING

Introduction

The basic principle of maintenance is to direct the efforts to upkeep the production facilities for their trouble-free operation throughout its economic life. A concern may be said to be successful over the years, when it runs uninterrupted, maintains a smooth production flow and continuously belongs to optimum productivity level. This is not possible only with the installation of fully automated and sophisticated machineries. It is an established phenomenon, that no plant, irrespective of its type of facilities, can achieve the productivity, at least up to a satisfactory level, without proper maintenance work.

Maintenance is the only way, on effective application of which, the gap between the achieved and achievable productivity can be minimised. The function of maintenance is composite in nature. A wide range of activities are involved in it. But it is an established truth that an efficient maintenance not only reduces the probability of breakage of machine elements and shut down of machines, hampering the scheduled production, but at the same time, such a function enhances the efficiency and accuracy of the production machines, lengthening their span of life with usual reliability.

Of course, no definite all-purpose maintenance procedure can be re-commended or prescribed in general. It varies from plant to plant, depending on the size, policies, technology and prevailing condition of an individual concern. But the major objectives of maintenance are always same.

Objectives of Maintenance

The plant maintenance is usually having a number of objectives of various types. But the ultimate aim is to have maximum profitability and to achieve optimum productivity. The objectives may be summarised as below:

(*i*) To minimise breakage;

(*ii*) To maximise plant availability;
(*iii*) To enable the facilities to function at maximum efficiency;
(*iv*) To ensure longer service life of the equipment, machineries and other facilities;
(*v*) To ensure safety to the workers;
(*vi*) To establish a satisfactory working condition.

The other cost-related objectives are obviously to minimise the maintenance cost and to maximise the profitability.

Maintenance Policies

The management is always after attaining the scheduled production. The ultimate aim in broad sense is to spend less and earn more, without disturbing the designed accuracy, quality and other parameters of a product.

Naturally, policies to achieve the objectives of maintenance will be framed by the managerial people, considering the availability of personnel, instruments, materials etc. attaching proper weightage to the cost of maintenance.

The maintenance department on the other hand will have to be supplied a well-defined maintenance policy. What they will do? Up to what extent they will do? What will be the basis of selection of maintenance procedure among the various alternatives?

It is an usual practice to make an annual budget on maintenance and to formulate a maintenance policy. It is preferable to have such policies in written form and should be known to all concerned departments.

The policies may be framed, covering the following points:

(*i*) Maintenance budgets
(*ii*) Functional scope of maintenance
(*iii*) The level of maintenance work
(*iv*) Systems to be followed and procedures to be adopted
(*v*) Inventory control planning
(*vi*) Purchase procedure for maintenance system
(*vii*) New investments, if any
(*viii*) Modification of projects in near future
(*ix*) Personnel policies
(*x*) Indigenisation and standardisation
(*xi*) Facility for research and development
(*xii*) Maintenance Organisation

The maintenance policies, once framed, may be reviewed continuously and changes, if required to achieve the objectives, are to be adopted at proper time.

Types of Maintenance

Primarily, maintenance may be categorised as either 'planned' or 'unplanned'.

The unplanned procedures permit failure to take place before any maintenance work is done. It may be termed as an "Emergency Maintenance". It involves less direct maintenance cost but enhances the probabilities of down time of machineries. Generally an unplanned system may be employed in small scale industries, where a planned maintenance is not economically viable. But for a large scale industry, an unplanned system has been proved to be most unproductive.

Conversely, a planned maintenance system incorporates all functions like planning, scheduling, dispatching, recording, analysis and controlling. Then a planned work cycle may be represented as shown in Fig. 1.1.

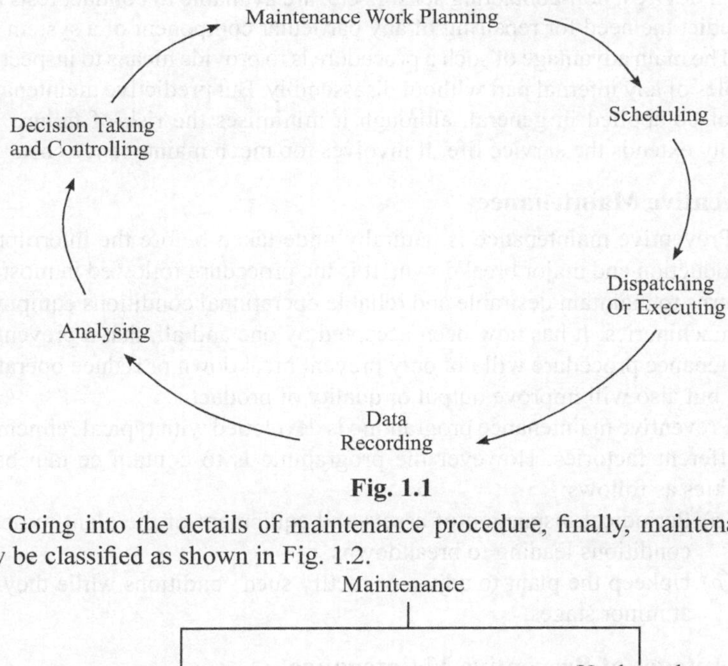

Fig. 1.1

Going into the details of maintenance procedure, finally, maintenance may be classified as shown in Fig. 1.2.

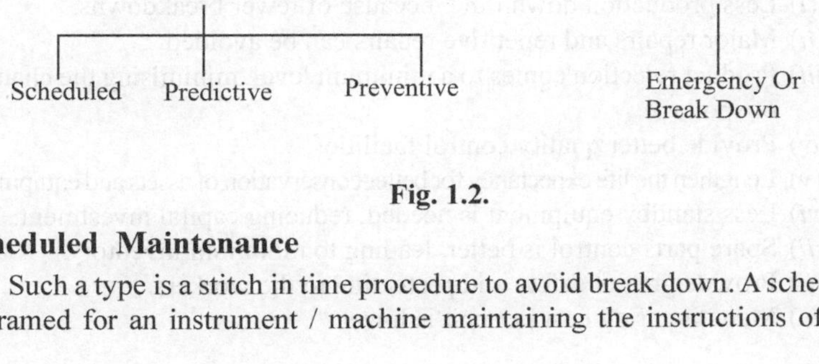

Fig. 1.2.

Scheduled Maintenance

Such a type is a stitch in time procedure to avoid break down. A schedule is framed for an instrument / machine maintaining the instructions of the manufacturer and applying the concept gained through the experience incorporating inspection, lubrication, repair and overhaul. Generally overhauling of machineries, cleaning of water tank etc. are some of the examples scheduled maintenance procedures are followed.

Predictive Maintenance

It is comparatively newer term, now coming into general use. Sensing, measuring or monitoring techniques are followed normally to recognize the cause of any change of physical conditions of a system. Today, a large number of commercial instruments such as vibration analyser, axial displacer monitors, optical devices, non-contacting sensors etc. are available to conduct tests and to predict the need for repairing of any particular component of a system.

The main advantage of such a procedure is to provide means to inspect the troubles of any internal part without disassembly. But predictive maintenance cannot be applied in general, although it minimises the risk of failure and reliably extends the service life. It involves too much maintenance cost.

Preventive Maintenance

Preventive maintenance is naturally undertaken before the interruption of production and major breakdown. It is the procedure followed in most the concerns to maintain desirable and reliable operational conditions equipment and machineries. It has now been accepted by one and all, that a preventive maintenance procedure will not only prevent breakdown or reduce operating costs but also will improve output or quality of product.

Preventive maintenance programme is developed with typical refinement in different factories. However the programme is to contain certain basic activities as follows:

(*a*) Periodical inspection of assets and equipment of the plant to detect conditions leading to breakdowns.

(*b*) Upkeep the plant to adjust or rectify such conditions while they are at minor stages.

Advantages of Preventive Maintenance

(*i*) Less production down time, because of fewer breakdowns.

(*ii*) Major repairs and repetitive repairs can be avoided.

(*iii*) Product rejection comes to a minimum level, minimising the chances of spoilage.

(*iv*) Provide better quality control facilities.

(*v*) Lengthen the life expectancy for better conservation of assets and equipment.

(*vi*) Less standby equipment is needed, reducing capital investment.

(*vii*) Spare parts control is better, leading to minimum inventory.

(*viii*) Provide greater safety and protection to the workers.

(*ix*) Manufacturing cost per unit is lower.

Breakdown Maintenance

Breakdown maintenance can only be suitable for small scale industries, as stated earlier, because of its less direct cost involvement. But the indirect cost, eventually becomes more and it does not provide any guarantee of reliability to the smooth production running.

The maintenance work, such as repairing, replacements etc. are performed only when breakdown occurs. This procedure can actually be taken as justified when the cost of preparation for a preventive maintenance required for a breakdown, along with the preventive inspections, is same as its repair cost.

But when the failure may create an alarming damage to the equipment or the downtime due to the failure results severe monetary loss, it is better to avoid breakdown maintenance. A scheduled instruction procedure on the equipment will produce better result.

Organisational set-up and maintenance departments

The success of a concern mostly depends on the allocation of work to proper personnel, which is on the rightly designed infrastructure of the organisation.

Similarly, for an effective functioning of the maintenance department a suitable organization structure is to be designed, befitting to the need of the industry.

For small scale industries, a separate maintenance department may not be economical. The usual practice is to employ a few fitters and mechanics for all types of maintenance work. However, for large scale industries or even for medium scale industries, an organisational set up of maintenance personnel is necessary, clearly defining their responsibilities and lines of authorities.

Position of maintenance department in a typical organisational structure is shown in Fig. 1.3.

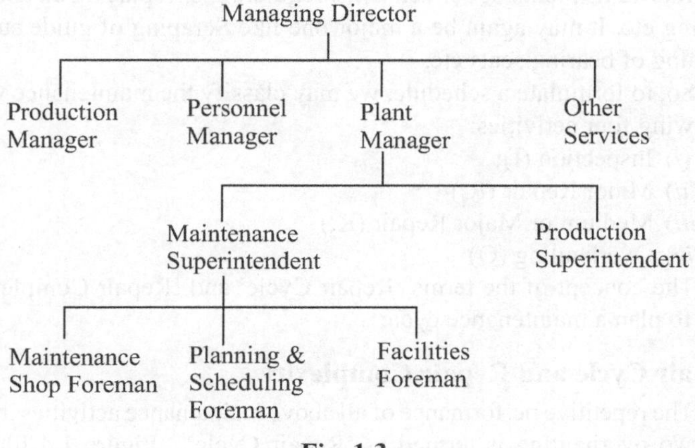

Fig. 1.3.

The basic structure may differ with respect to the difference of departmentalisation pattern, or type of production and size of the plant. The number of maintenance personnel to be provided for different branches are also calculated on the basis of workload, considering the past data.

It may be noted that concept of line and staff organisation is not being felt proper in designing the structure. The maintenance departments are directly under the control of Plant Manager, as the maintenance work is considered, today, to be associated with plant engineering activities.

Maintenance Planning

A successful maintenance, specially a preventive maintenance, is logically dependent on the skill and efficiency of the maintenance personnel. But technically, it is having a complete dependence on the following basic features.

(A) Planning and Scheduling

(B) Controlling: (*i*) Corrective measure
(*ii*) Monitoring

(A) Planning and Scheduling

Planning means formulation of work in advance and scheduling is the time phase of the sequential activities.

''What should be done" and ''when it should be done" - are the major criteria of successful maintenance. Before an elaborate discussion, the following terms may be discussed in brief:

Repair Cycle

Straightway it can be stated that maintenance schedule generally includes Inspection, Repair and Overhauling. Again the term "Repair" does not reflect any proper meaning where time-duration is a major factor. It may be a minor one, like adjustments of belt tensions, regulating or replacing an anti-friction bearing etc. It may again be a major one like Scraping of guide surfaces or cleaning of bearing seats etc.

So, to formulate a schedule, we may classify the maintenance work into following four activities:

(*i*) Inspection (I);

(*ii*) Minor Repair (R_1)

(*iii*) Medium or Major Repair (R_2)

(*iv*) Overhauling (O)

The concept of the terms 'Repair Cycle' and 'Repair Complexity' will help to plan a maintenance cycle.

Repair Cycle and Repair Complexity

The repetitive performance of all above maintenance activities in between the two overhauling is termed as 'Repair Cycle'. Figure 1.4 illustrates a repair cycle. From the point of overhauling point (0), the activity is taken to be started in a clockwise direction. It is seen that at first an inspection is scheduled, then a minor and a major repair is planned. Inspection is again done and a minor repair is performed. After the third inspection a major repairing takes place and then comes the time of next overhauling, completing one repair cycle. So the typical repair cycle (Fig. 1.4) suggests to have three inspections, two minor and two major repairing in between the time-span of two consecutive overhauling. This may simply be represented as $O_1 - I_1 - R_{11} - R_{21} - I_2 - R_{12} - I_3 - R_{22} - O_2$.

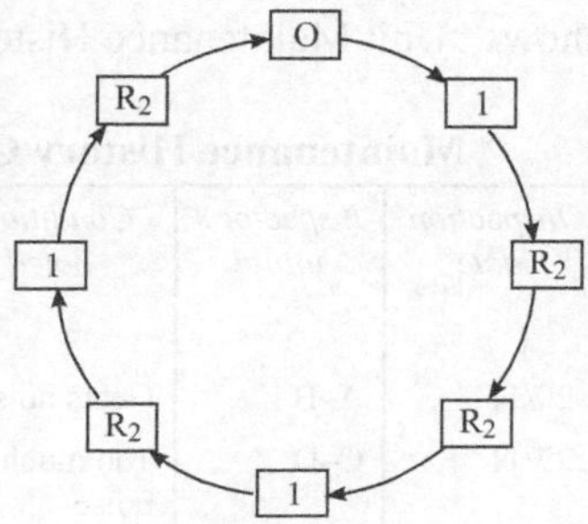

Fig. 1.4.

Now it is evident that the time period for a repair cycle is dependent on the time-intervals in between two activities. If it is six months, then time duration of above referred cycle is four years.

The planning of time intervals may be scheduled, considering the past data of the machine, following manufacturing instructions and as per the complexities of the instruments or machine tools. An index is generally used to communicate a comparative complexity of different machines. Such an index has been finalized, taking into account the power transmission devices, hydraulic units, guide surfaces, intricate mechanisms associated with the functioning of the machine. This index is technically known as ''Repair complexity number". For instance, repair complexity number for a boiler may be 12, for an air compressor it may be 8 or for a turbine it may be 14.

The repair complexity number is helpful to the maintenance department to finalize the 'maintenance staff-size', to design the 'inventory required or to forecast 'maintenance costs' etc.

It may be remembered that higher the repair complexity index, higher number of activities (inspection - major or minor repairing) are involved in a repair cycle.

Equipment History Card

Maintenance planning may be effectively performed, if the past data regarding inspection or repairing etc. are available in the department. The best way is to have a "history card" for each machine or instrument. Analysis of the past data can only enable the management to build a confident work-planning structure.

Two types of history cards are shown in the following charts. Chart 1-1 shows "Unit Maintenance History" for a Geared Turbine Generator.]

Chart 1-1

Maintenance History Card, Inspection frequency : Weekly

Inspection date	*Inspector's initial*	*Conditions noted*	*Corrective action taken*	*Cost*		
				Labour	*Material*	*Total*
20/11	A–B	Gears noisy	None	–	–	–
27/11	C–D	Too much noise	Gears removed, machined true & replaced (Downtime-6 days)	–	–	–
4/12	C–D	Negative	–	–	–	–

Chart 1-2 is shown for a production machine. Both the charts may also be combined together. The design of history card format may be designed by the maintenance department to store data for the future planning.

Chart 1-2

Equipment History Card

Date	*Unit No.*	*M/C*	*Hours lost*	*Production lost*	*Details of Breakdown & action taken*	*Comments*
20/11	1102	AA	1/2 hrs.	B	Vertical shaft slipped, as set screws were loose, Screwed and realigned shaft.	'Tightening of Allen screw during each inspection' may be added in the check list.

Such a format (chart 1-2) provides suggestions to modify the inspection check-list along with other information.

Master Schedule Card

In case of a Preventive Maintenance, the total planning should be reflected in a master schedule card. The annual record of the total programme, as is denoted in the master schedule, may be broken into part-work step by step and then necessary work-order may be issued.

Chart 1-3 will provide an idea of the master schedule, which is planned involving all the maintenance activities for the total plant. Chart 1-3 shows only a part of the total schedule, designed for a power plant.

Chart 1-3

Master Maintenance Schedule (I–Inspection; O–Overhaul)

Units	*Ist quarter*	*2nd quarter*
Turbine generator (Electrical]	I I I I I I I I I I I I I	O O O O
Turbine generator [Mechanical]	I I I I I I I I I I I I I	O O O I
Condenser	I I I I I I I I I I I I I	O O O I
Air Preheater	I I I I	O O O I

After preparing 'master maintenance schedule', weekly schedule may also be planned. But while planning, the activities required to be performed for an effective maintenance are to be identified.

Chart 1-4 has been tabulated to identify the activities involved in different types of maintenance.

Chart 1-4

Work activities

Types of Maintenance	*Work activities in brief*
Preventive	(*i*) Cleaning, (*ii*) Inspection, (*iii*) Lubrication, (*iv*) Repairing – minor, medium or major, (*v*) Minor adjustment etc.
Scheduled	Overhauling, Reconditioning, Replacing, Lubricating etc.
Predictive	Monitoring of vibration, Temperature, Analysing used lubricants etc.
Breakdown	Repairing as required, Replacing if required, Resetting etc.

Of course, work like cleaning, lubricating etc. are associated with all types of maintenance.

Work-Order

Once scheduling is completed, the next phase is dispatching, that is to issue order to start work.

It may be noted that a well-defined assignment to any maintenance team or individual is always necessary to avoid conflicting instructions.

So, work-order should be issued either directly from the plant engineer or by some competent authority.

A work-order format is shown in chart 1-5.

Chart 1-5

Work-Order

Work-Order No.	Date ....
Equipment	No.
Shop	
Details of defects	Work required
______________________	
______________________	
______________________	
Planned time ..	Work by ....
Actual time ..	
The equipment has been repaired.	Signature

Before issuing the work-order, the authority should check:

(*a*) Assurance on the availability of the equipment.

(*b*) The availability of the required tools and spares, if necessary.

(*c*) If any necessary drawing is to be attached with the work-order for proper understanding of the job.

Work-Report

Success of maintenance cannot be evaluated in absence of any feedback system. A work schedule directs 'what to do' and there should be a provision of work report to communicate 'what has been done'. Analysis of such reports will enable the management to have a direct control on maintenance and to review the 'maintenance planning' which is extremely necessary for any large-scale production centre.

So, work-report should be a vital part of the paper work associated with the maintenance work. The reporting format may be designed by the management as per their convenience. But it should be a simple one and should consume minimum time to communicate maximum information. This may be a part of work-order itself. For the activity of inspection or checking, it may enlist the machine parts to be examined and particular characteristics which require special checking. A typical work report *proforma* is shown in chart 1-6 for a metal cutting machine. A column has been provided for an evaluation of each part as to whether it is in good, fair or poor condition.

Chart 1-6

Maintenance Work-Report

Machine	No. shop	By			Date Time
Machine Part	Check For	Good	Fair	Poor	Work done
Bed Foundation	Good condition				
Pulleys	Alignment				
Gears	Smooth transmission				
Spindle bearing	Good condition				
Belts	Cleanliness				
Guide surface	Cleanliness & Flatness				
Lubricating system	Good condition				
Note :					

If any special work is done that can also be recorded at the bottom as a special note.

Controlling

The accomplishment of 'controlling part' literally completes the planning cycle. For the activities of maintenance also a centralised effective control should exist to enjoy the benefits of planning.

Whenever the plan deviates or does not occur as per schedule, a corrective measure is to be followed. But monitoring of the total project actually integrates the phases of control and it provides information, if any change of action is to be made in course of work.

A plant maintenance control flow-chart may be suggested, as shown in Fig. 1.5. The maintenance work-order and work-report may be checked and compared in course of work, whether the work has already been done, or work needed is waiting for any spare, or is facing any type of administrative or technical difficulties, are always to be monitored. The both way information line should exist with the superior personnel. If the work has been satisfactorily done the details of the work may be recorded in the equipment history-card as shown in the flow-chart (Fig. 1.5).

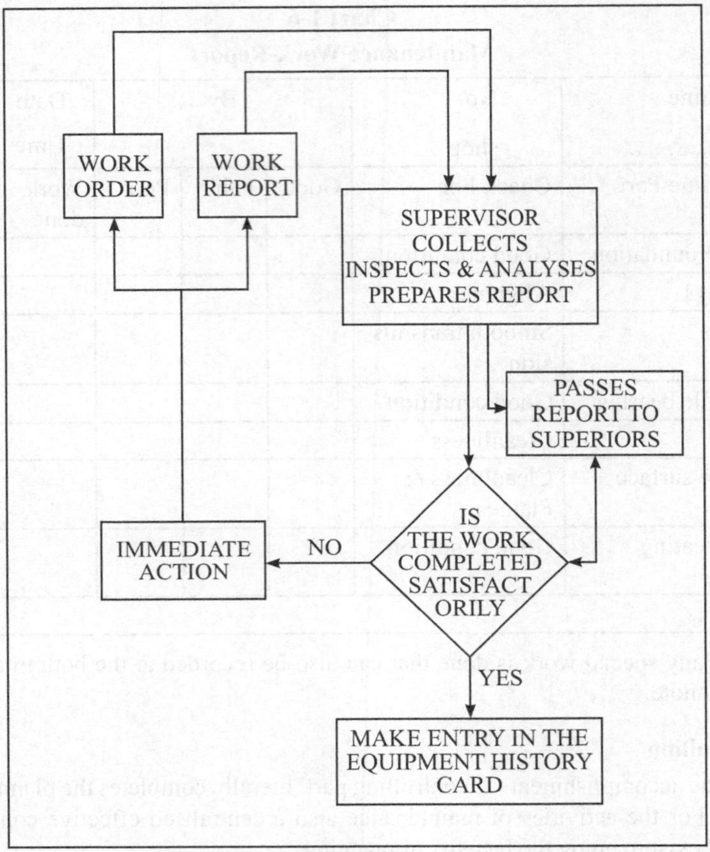

MAINTENANCE CONTROL FLOW-CHART

Fig. 1.5.

The functional requirements of the control process may briefly be outlined as follows:

(*i*) The environmental and internal conditions should be sensed and monitored.

(*ii*) Provision and procedure to receive, process and store external and internal information should exist.

(*iii*) Stored information (plans, goals, schedules etc.) should be continuously compared with the incoming information.

(*iv*) Decision making should be followed according to available infrastructure, technical necessities and cost criteria.

(*v*) Communicating procedure should be effective to transmit instructions.

(*vi*) The decision based action should be implemented.

Maintenance and Reliability

Reliability may be defined in several ways. A well-defined statement to understand the concept of reliability is as follows:

Reliability is ''the probability of performing, without failure, a specified function under given conditions for a specified period of time." For a complex equipment, a large amount of money is spent for its upkeep and maintenance. The total reliability of an equipment is actually a function of design, maintenance and operational reliability.

Let us concentrate only on the maintenance support in this part. The most significant maintenance support factors are the criteria of adequate performance, manuals and technical order, test equipment and facilities and spares provisioning.

The operator as well as the maintenance personnel, should have the skill and knowledge for assessment of situation and should have a thorough technical know-how about the performance of the equipment. Training is an aid to make the personnel conversant on the nature of the equipment and simple, sequential, understandable maintenance instructions should be available, as has been discussed earlier also.

To replace spares till the trouble disappears is a bad practice, though it is unfortunately being followed in many places. Detailed step-by-step procedures of locating the trouble is mandatory and replacement of spares are made after proper identification.

But 'Spares Provisioning' is an important criterion in this regard.

It may be mentioned that replacements of spares are needed for marginal testing, for primary failure (parts damaged during operation of maintenance) and also for secondary failure, which may be caused as a result of the primary failure. So while planning for storage of spare parts, not only the failure rates but all types of replacements are also to be taken into account.

Spare Parts Planning

For an effective planning of spare parts, the following factors must be followed:

(*i*) Codification of Spares

Codification or to employ a definite code for a definite spare, simplifies the work of identification. Otherwise, you may purchase the materials, you are already having in the store.

Moreover, codification of spares ensures issue of proper items, as is requisitioned. However, codes may be of different types.

Say for instance, for a code number, first two digits may indicate a main group, a later part may indicate its chief characteristics. A few examples are given below:

M 20 – Ball bearing

M 201111 – Radial ball bearing of size 20/52 × 15

H 99 – Taper roller bearing

H 9921134 – Taper roller bearing of size 70 × 125 × 31

M 208110 – Thrust ball bearing

Of course, it is also having limitations. To meet the varieties, the number of digits may increase too much.

So generally an accepted method is to limit the codes within eight digits.

(ii) Preservation of Past Data

Each and every replacement of spares should be recorded in a card for an individual machine or instrument.

The cause breakdown, the spare replaced, the cost of spare etc. should be recorded clearly.

Such data card will ultimately provide information to quantify the average requirement of spares.

(iii) Practice to use standard spares

It is always recommended to purchase standard spares. They are generally marked with the standard code numbers. Such a practice assures the reliability of fitting and operation and on the other hand easy procurement is also ensured.

(iv) Effective purchase procedures

Purchasing department should have the control on inventories. They should be in a position to detect the manufacturer or supplier of spares instantly, they should know the lead time of procurement etc. to maintain the right spares in sufficient quantity and of equal and better quality.

Classification of spare parts

The spare parts may be needed throughout the year are generally classified to have a clear idea on the nature of spares. They may be as follows:

(*i*) Spare parts to be manufactured in the concern itself

(*ii*) The wearing criteria of parts

(*iii*) Spares used in numbers in similar machines

(*iv*) Spares which are difficult to procure from regular market without special order etc.

Such classification will ultimately provide information to finalize the amount of spares to be procured at a time or the interval of procurement.

Utilisation of Human Resources in Maintenance

The maintenance planning succeeds often with the perfect utilisation of human resources. The total maintenance activity is a complex, dynamic and goal oriented system. So it needs prior planning and proper control. At the same time, to select the 'right man at the right job' takes an effective role in maintenance management.

A term "Operator Maintenance" or "Maintenance by production" is very common in todays' industrial world with regard to the manpower utilisation. This is the practice of maintenance, carried out by the operators themselves, while engaged in production. But this practice is only suitable on small scale industries. However, for the large scale industries also, a part of this concept can be employed. Such as, daily cleaning, lubrication may be done by the machine operators. This will ultimately minimise the work load of the maintenance department.

However, in a large plant, besides 'scheduling', 'man loading' etc., the organisation should positively have the means to motivate the workers and to train them properly.

The planning may be effective to extract the skill of individuals, only when the personnel engaged in work, do have proper depth of knowledge on the required field of disciplines and activities.

The training objective plays a vital role in this purpose. A concern may have highly qualified personnel and they will have general people with some specific and average background also.

A three dimensional graph, as depicted in figure 1.6, is hereby suggested to study the situation and to formulate the objectives of the training programme.

It may be seen that small rises in the 'x - y' and 'y - z' planes (curves marked 'a') show the depth of knowledge of the available human resources in the plant on individual activity or disciplines. At the same tune, the curve for a qualified person is shown through curve 'b'. In each case the peak belongs to any single activity or discipline.

But the utilisation of the people, associated with maintenance management system, can only be effective (if not optimum), if the objective of the training is laid to approach the curve 'c', for each set of people, whether belong to curve 'a' or 'b'.

The course content of such a training programme should depend on the diversities of 'fields of work', necessary to accomplish the task of an individual firm.

Such a principle will equip the available human resources to tackle the situation confidently and successfully within the maintenance system, followed up in the factory.

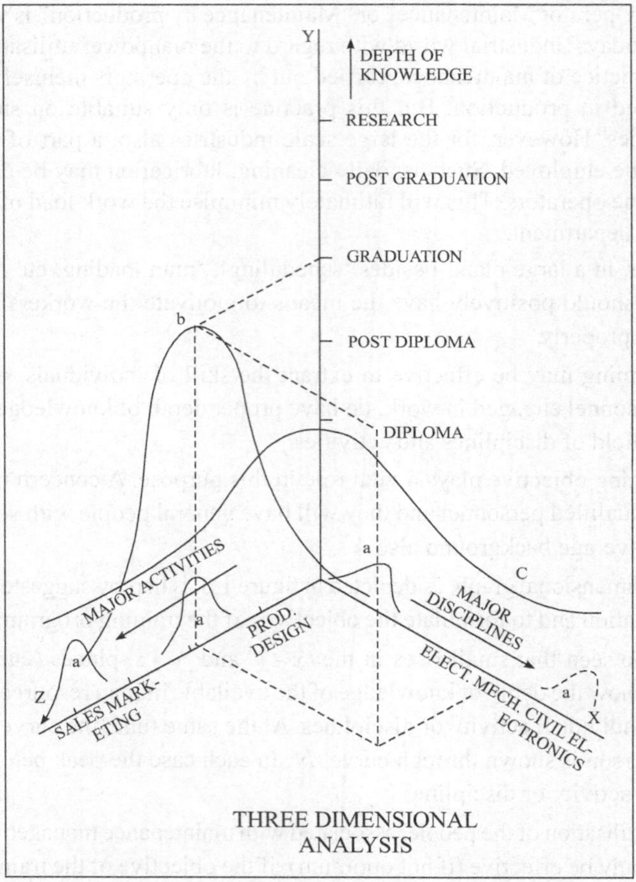

Fig. 1.6. Three Dimensinal Analysis

Maintenance-Work Simplification:

Maintenance can be made more fruitful and justified when it is performed by right person through right methods and within a right span of time. Maintenance management is in charge of controlling the work. However, the simplification of work plays an important role for the effective and desired quality of maintenance. To perform this job, the maintenance department should be active to analyse its own activity through method study and time study or work measurement. It should simplify and standardise its work.

The 'simplification of work' is nothing but a technique of searching the optimum improvement of methods. The method study technique may be cussed in brief.

It is to be remembered that such studies involve time, money and manpower and so, one is to select the method worth studying. It is better to deal with the problems on priority basis. The scope of application of method study can be sorted from the situations as listed below:

(*i*) The maintenance procedure that causes considerable down time
(*ii*) Hazardous or dangerous work
(*iii*) Work that involves high material cost or manual labour
(*iv*) Periodic work that is being performed repeatedly at certain time interval such as inspection job, overhauling etc.

The steps to be followed under method study are:

(*a*) Select the work to be studied
(*b*) Finalize the objective of the study
(*c*) Record the relevant useful data
(*d*) Analyse the data
(*e*) Develop new or modified method
(*f*) Install the same to achieve the objective
(*g*) Maintain the new procedure and review the effectiveness through routine inspection and check up

(*a*) Select the work

Maintenance job may range from a simple cleaning or replacing an item to a major shut-down problem. But one is to judiciously select a job for the method study. The main factors to be emphasized in this regard is techno-economic feasibility. The problem may be in respect of a job or a section, or the total plant it is always to be kept in mind that the selection of work should yield maximum benefit after the improvement or modification of the methods.

(*b*) Finalize objective

The objective of the work undertaken should always be well defined. It may be either to minimize cost involvement or to reduce fatigue or to remove bottlenecks of any process. The pre-settled objective will lead us to move to a positive direction. Otherwise all the efforts may be futile.

(*c*) Record data

On studying the existing method, one is to record all the relevant facts and information regarding it. The data collection should be scientific and perfect. A number of scientific procedures have already been established to record the data. They may be as follows:

	....	Out line process chart
Process Charts	....	Flow process chart
	....	Two handed process chart
Diagrams	....	Flow diagtram
	....	String diagram

Outline process chart:

The operations and inspections are sequentially shown in this chart to provide an overall broad outline of a process.

An entire process can be readily understood through such charts and the elimination of some steps or to combine some activities may be decided by studying an outline process chart. The standard time for each activity should also be marked on the chart itself.

Two handed process chart:

Two handed process chart records the activities of the hands of workers. Naturally such jobs are of small duration. This chart is very much useful for a repetitive nature of work. The job is broken into small elements like grip, hold, transport, inspect etc. and are entered into the chart sequentially, with proper symbols while observing any work.

Two handed process charts are not much useful in the maintenance work although they have got wide applications in production line.

Diagrams:

Flow diagram is generally used to show the pattern of movement of human being, materials etc. within a definite work area.

Sometimes strings are used to show the path of movement. The optimal movement involved in a task is clearly shown in the flow or string diagram.

(*d*) Analysis of Data:

The data available through the above procedures are to be critically examined. One is to put different questions to be assured the utility of each activity.

The questioning and style of searching alternative may be examined as follows:

Who is doing the work?	Who else can do it?
Why it is being done?	What else can be done?
Where is it done?	Where else it can be done?
How it is done?	In what other way it can be done?
What is being done?	What else can be done?
Is it necessary?	Would something else provide better results?

This questioning attitude offers scope to analyse the detailed procedure step by step and to develop improved style of functioning.

The alternative procedure, if found more effective at any step, for any activity, may be suggested for improving performance.

(*e*) Develop new ideas:

The developed method is then charted with proper symbols to be sure that the proposed modification either saves time of some activity of provides information for low cost involvement etc.

There may be a number of alternatives for some specific improvement. The best method is then to be identified through comparison and proper evaluation and to be recommended for implementation.

(*f*) Install to achieve:

After the finalisation of improved method, it is to be implemented. But before implementing, it is to be noted that all the related workers, supervisors are well aware of the changes. If it is needed, a short training may also be arranged for the concerned people for some complicated process(es).

(*g*) Maintain the new process:

The newly suggested methods of work performances are to be finally maintained properly to achieve the planned improvement. Monitoring is a must after the installation of a modified method to check the desired results. Further scope of improvement of the process may also be explored, if it is needed.

Working Measurement

Work measurement plays an important role for planning and control which are the keys of effective management. The measurement of work is equally important in the production, as well as in maintenance planning. The time standards, framed through work measurement, balances the resources requirement with the resources available.

So, when the standard methods are suggested through method studies, it becomes necessary to finalize the time standards to perform a job or segments of a job and the workers may be trained in specific methods. Whenever the methods are changed or modified, the time standards should also be reviewed accordingly.

Procedures of Work Measurement in Maintenance Work

This may be noted that most of the maintenance works are not repetitive in nature, as are mostly selected for time study in production sphere. There fore, the normal practice of work measurement may not always be efficient in case of maintenance jobs. However, the following techniques may be applied for maintenance work:

(*a*) Job estimating, (*b*) Time study technique, (*c*) Statistical analysis, (*d*) Work Sampling, (*e*) Analytical procedure, (*f*) Methods Time measurement (MTM) etc.

(*a*) Job estimating

This is a crude form of estimating time for a definite piece of work through experience and mental approximation. It is not at all perfect, but still is sometimes used to make some schedule, knowing its limitations fully well, for its simplicity. However, the man in charge of estimating the standard should be an experienced and knowledgeable person.

(*b*) Time Study

Time study is a classical technique of estimating time standard. The first step is to break the maintenance job into a number of elements like pick-up, grasp, reach, move, release etc. Time is to be recorded for each element separately with the help of a stop watch. Observations are repeated several times to avoid variability of performance of an individual worker.

Moreover, to get the 'Normal time', the average observed time is multiplied by a factor known as Performance Rating Factor (PRF), which takes care of the variations of speed/skill efficiency of different workers.

$$\text{Performance rating factor (PRF)} = \frac{\text{Performance level of the worker}}{\text{Standard performance level expected}}$$

The allowable allowances are then added with the normal time for a particular job.

(*c*) Statistical analysis:

It is a simple statistical technique to calculate time standards for various types of jobs. The total time spent for a definite type of maintenance work, over a period of time is divided by the number of times the work has been done. The average time, thus calculated, is taken as standard.

For such analysis, the jobs are categorically classified. They may be some repair jobs. It may be repetitive or non-repetitive type it also may be a work order job. Time standard for each classification is calculated separately.

The past data may be segregated for different classification and the number of occasions, it has been done is also found out to get the average time taken per occasion or per job.

This method, although an approximate one, is sometimes used in the industry for its simplicity.

(*d*) Work sampling

This is a technique based on statistical principle, to measure the work content in any job. It was first invented by L.H.C. Tippet in 1934 at Great Britain and R.L. Morrow implemented this technique at America: in around 1945, and named it 'Ratio- Delay analysis'. In this technique a large number of observations are made over a span of time and the individual observation records what is happening at what instant. The average recorded value tends to represent the time spent by a man or machine in the state of job. Say for instance, out of some 50 observations on a particular job, it is seen that an operator is working on the machine for 25 times, for 20 times – he was setting up the machine, and for 5 - he was doing nothing.

From the above study, it may be calculated that actual working time for that job is 50%, whereas the setting up time is 40% and the remaining 5% time is the idle time.

The job of long duration cannot generally be timed with stop watches and so the work sampling procedure is found to the most appropriate in such cases. Although, this technique does not provide the element details of a job and does not also reflect the efficiency of the worker, hut still, it is seemed to be most economical for timing of group activities and the activities of long duration.

(*e*) Analytical Procedure

This method determines the time values of some elements from the knowledge and experience of the elements concerned.

Here also, the job is broken up into a number of elements. The time requirement for an element may either be available from the synthetic data, as observed by time study technique, or it may be estimated from the past experience by the estimator.

(*f*) Methods Time measurement (MTM)

Methods Time Measurement technique shows the basic motion - time values for small elements of work. All the jobs are made of a set of body and limb movements of workers. The job may be broken down into small elements, as it is done in the time study technique and then the time may be observed for the motions involved in each element separately. The summation of time standards of each element may represent the time for a job.

To start the work, one is to be thoroughly conversant to the detailed information regarding the work methods, equipment, requirement of materials etc. It is although simple, but it is a time consuming procedure and it is to be conducted by an experienced person.

Advantages of Work Measurement:

The advantages of work measurement is primarily to have a clear knowledge regarding all sorts of work and it provides information to control over work methods and job allocation. Planning and scheduling of work and man loading can also be effectively controlled. Maintenance work measurement specially provides effective means for planning and control of 'routine', resulting efficient planned and preventive maintenance, reducing the cost of overall maintenance.

REVIEW QUESTIONS

1. What is maintenance? What may be its objectives and policies?

2. What are different types of maintenance? Why and at what situation preventive maintenance is preferred over breakdown maintenance?

3. What is maintenance planning? Discuss sequential steps of maintenance planning and control.

4. Write short notes on:

(*i*) Equipment history card
(*ii*) Repair cycle
(*iii*) Master maintenance schedule
(*iv*) Repair complexity
(*v*) Spare-parts planning

5. Suggest an efficient position of maintenance department in an organisational structure and justify your suggestion.

6. Identify the correct and false statements and rewrite the false statements correctly:

(*a*) More standby equipment is needed in case of preventive maintenance.
(*b*) Breakdown maintenance is suitable for large scale industries.
(*c*) Predictive maintenance is a type of unplanned maintenance.
(*d*) Maintenance department should be under the control of plant manager.
(*e*) Repair complexity number is helpful to forecast the maintenance cost.
(*f*) Codification of spare may complicate its identification.
(*g*) Codification is generally limited to eight digits.

2

GENERAL HAND TOOLS

Introduction

Installation and maintenance work necessitates the use of a large variety of measuring tools, gauges and instruments. We will discuss only a few of them, which are very common in daily work schedule.

1. Spirit Level (Fig. 2.1)

It is a common instrument to test or adjust horizontal surfaces. It is essential in all erection work or in repair works. It is simply a glass tube, 200 to 300 mm long (l) almost filled with a liquid. An air bubble (2) remains at the top-most position of the tube. This glass tube is placed within a large radius (3) having a flat base at the bottom (4). When the spirit level is moved, the air bubble will move left or right, depending on the amount of tilting of its base. The angle of tilting can be quantified from the calibrated graduations. The work will be taken as levelled, when the air bubble is exactly centred on the centre mark. The level may be moved in both horizontal or vertical directions.

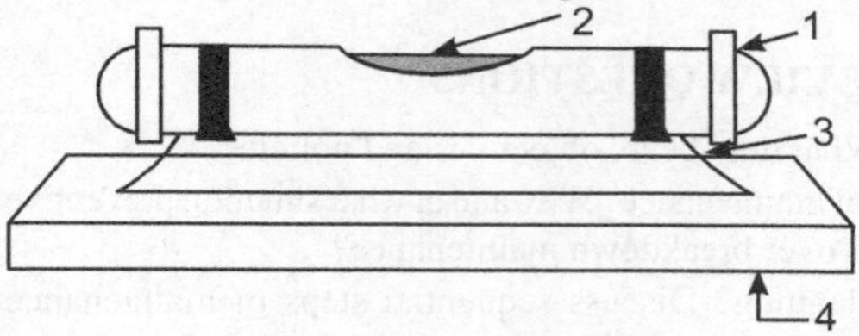

Fig. 2.1. Spirit level

Accuracy of instruments like spirit level is an essential criterion to offer accuracy to the machine tools. So the spirit levels should be checked at least once a year on a standard surface plate.

2. Surface Plat (Fig. 2.2)

Surface plates are used to mark out work of small sizes and to test the flatness of work. For the marking of

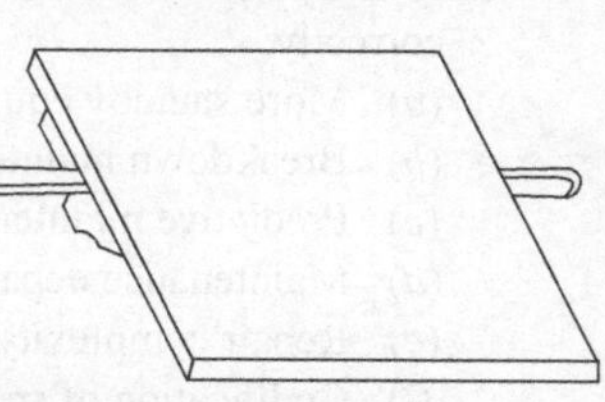

Fig. 2.2. Surface Plate

large work marking-off tables are used. The surface plates are of rigid castings of grey cast iron. The surface is hand scraped and is of high accuracy. While using a surface plate, great care should be taken not to scratch or make it defaced in any way.

It should always be kept clean and must be protected from rusting by a smear of oil or grease. It should also be provided with a wooden cover, the underside of which should be covered with felt cloth, to protect the surface when not in use.

The accuracy of a surface plate may be expressed in the following way as class 'II', 'I' and '0' accuracy in points/sq. inch.

1. 'Class II' accuracy — 16 to 20 points/sq. inch
2. 'Class I' accuracy — 20 to 25 points/sq. inch
3. 'Class '0' accuracy — 25 to 35 points/sq. inch

3. Straight Edge (Fig. 2.3)

It is a tool, used for testing whether or not a plane surface is straight or flat. It is commonly a rectangular metal piece. Sometimes the edge is made bevelled and is called a knife edge. It may be laid down in different positions on a surface, which is to be tested. If the surface is perfectly flat, no light will pass through the junction of the surface and the straight edge. If the light passes through any point, it indicates the inaccuracy at that point. Flatness may be measured with a straight edge upto an accuracy of 3 to 5 microns.

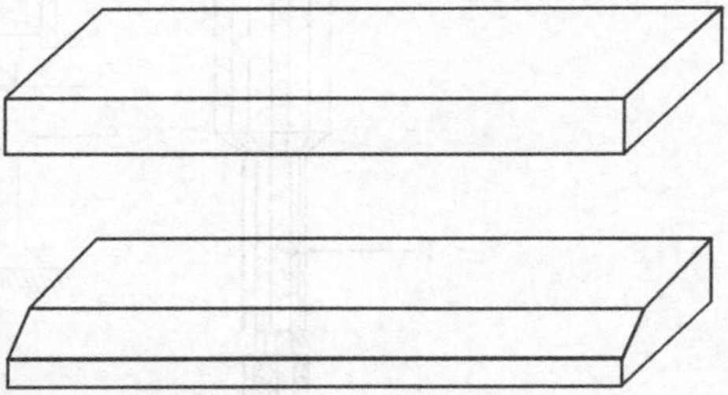

Fig. 2.3. Straight edge

4. Dial Indicator (Fig. 2.4)

The dial gauges (Fig. 2.4*a*) are used to check the irregularities of flat surfaces (Fig. 2.4*b*), to check the trueness of running of shaft, parallelism of work piece and for other similar purposes. It is having a feeler pin (1) below the dial. Slightest pressure in it is transmitted to the pointer (2) through a gear-train drive in such a ratio that the exact reading is shown on the dial. The dial gauge is frequently used together with block gauges, (Fig. 2.5) to measure the difference between actual dimensions with the designed dimensions (3 is '+ve' and 4 is '—ve' error) of the work piece (2). While using the dial indicator, care is to be taken to take the zero graduation exactly under the pointer by turning the dial before the testing work is started. One complete revolution of the pointer indicates movement of 1 mm and as the dial is divided into 100 divisions, it gives a reading with an accuracy of 0.01 mm.

The dial gauges are available with a range from zero to 3, 5 and 10 mm.

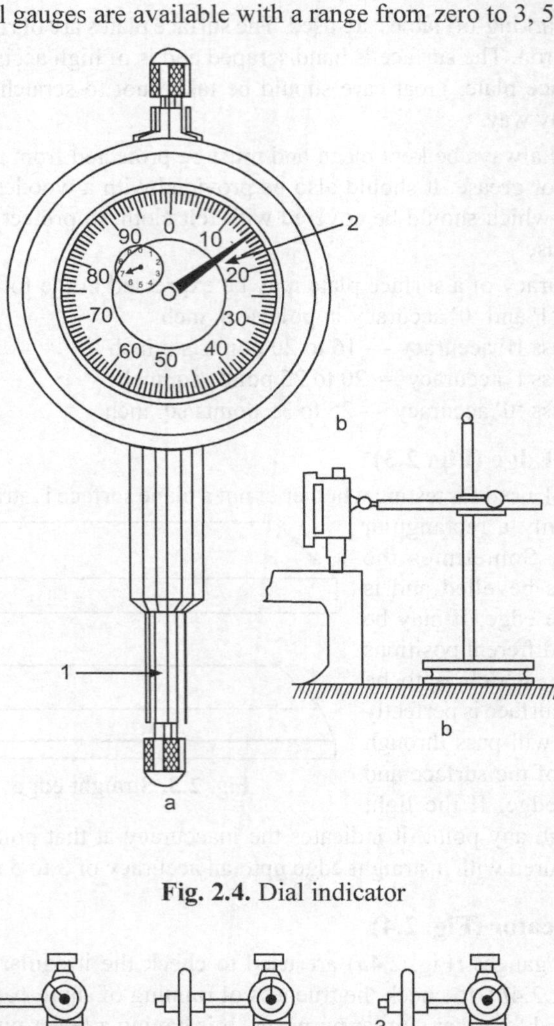

Fig. 2.4. Dial indicator

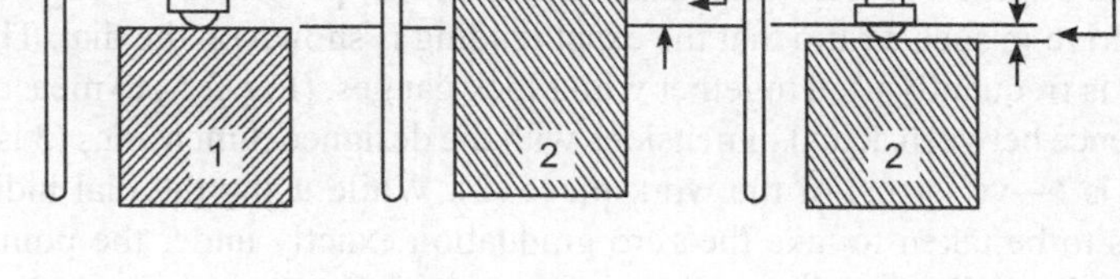

Fig. 2.5. Checking by dial gauge

5. Feeler Gauges (Fig. 2.6)

The gap or clearance between the mating surfaces can better be checked with feeler gauges. This may also be used to reach the actual size in conjunction with other dimension gauges. They are simply precision steel blades(1), the thickness of which are 0.03 to 1 mm pivotted together. The thickness of the gauges are cleanly marked on it. As per 'ISI standards', seven sets (No. 1 to No. 7) are available in different ranges or thickness. While using the feeler gauge, a single blade or as the number necessary to get the required size may be used at a time. But it should never be inserted by force, which may disfigure the blades.

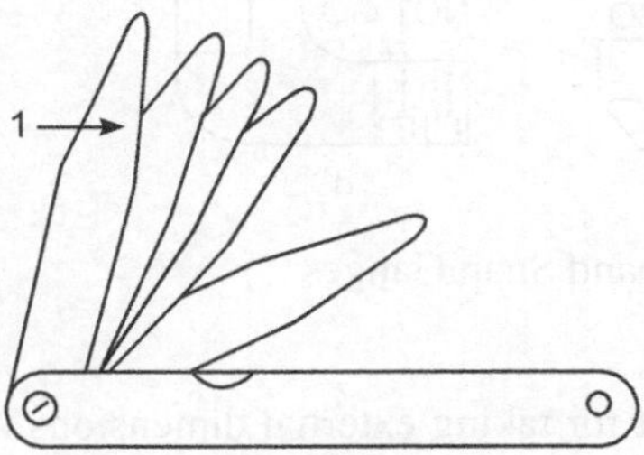

Fig. 2.6. Feeler Gauges

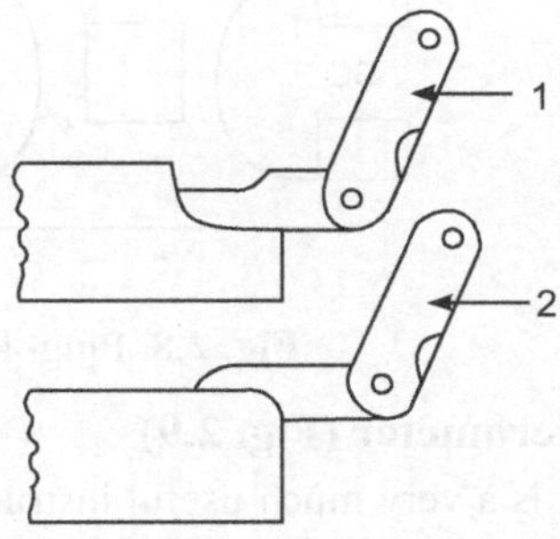

Fig. 2.7. Radius gauge

6. Radius Gauge (Fig. 2.7)

The radius of the curves, fillets etc. are tested by radius gauges. It may be of concave (1) or convex (2) radius types. Radius gauges are available in sets over a range of 1 to 25 mm. Each set contains 16 blades either of concave or of convex designs.

7. Plug-Ring and Snap Gauges (Fig. 2.8)

Plug and ring gauges are usually very common in regular use. Holes of different shapes and sizes are tested by a plug gauge (Fig. 2.8*a*) and ring gauges (Fig. 2.8*b*) are used for external shaft diameters. They are available in various standard sizes. Snap gauges are also called 'limit gauges'. They are used for testing external dimensions. The two sides provide minimum and maximum permissible sizes. 'Go' and 'Not go' ends simplify the identification of accuracy during testing. They may be a solid type (Fig. 2.8*c*) or an adjustable type (Fig. 2.8*d*).

All the gauges should be handled carefully. They should not fall down and strike with some other instruments.

The user should be careful always to avoid any distortion of the gauges.

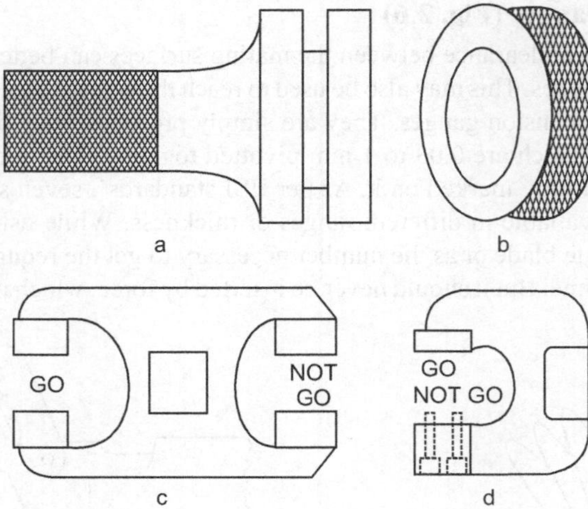

Fig. 2.8. Plug–Ring and Snap Gauges

8. Micrometer (Fig. 2.9)

It is a very much useful instrument for taking external dimensions upto an accuracy of 0.01 mm. The spindle (2) is attached to the thimble or sleeve (5) and it is having thread for the portion which is inserted in the barrel or hub (4) and sleeve (5). So the spindle can be moved by rotating the thimble. The

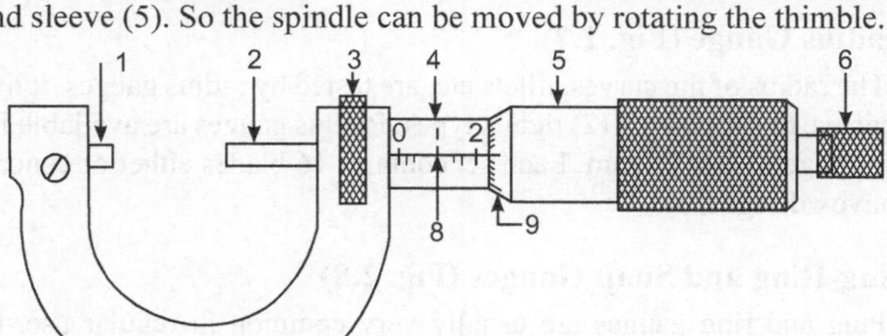

Fig. 2.9. Micrometer

ratchet (6) is a small extension of the thimble. It is also known as friction stop and is used for final tightening when the spindle comes closer to the job, placed against the anvil (1). The frame (7) is stationary. Once the anvil and the spindle touch the job, properly held, the clamp nut (3) may be tightened and the micrometer may be taken out of the job. The dimensions now can be read primarily from the linear scale (8) on the barrel which possesses 1 mm and 0.5 mm divisions and then the circular reading (9) is to be noted. For each turn of the circular scale (having 50 divisions) the thimble moves one smallest division on the linear scale, that is 0.5 mm. So a single division on the circular scale is equivalent to (0.5 x 1/50) or 0.01 mm. Both linear and circular readings are to be added to get the final reading. Micrometers are available in different ranges (0 to 25 mm, 25 to 50 mm, 50 to 75 mm or 75 to 100 mm). While taking readings, special care should be taken to operate the ratchet, at right moment and to withdraw the micrometer after clamping the clamp ring. It should never be used to measure a rotating work. Before taking measurement, the errors if any should be identified and error correction in the reading should be properly made. Micrometer should always be maintained clean and may be kept slightly oiled. It should be kept in a box, while not in use.

9. Depth Gauges (Fig. 2.10)

Depth gauges are used to check the depth of hole, grooves, recesses etc. It is frequently fitted with either a vernier or micrometer attachment. Micrometer depth gauge is generally used to measure the depths of high precision. It should be noted that graduations on thimble are positioned in inverse order to facilitate taking readings. The measuring range can be varied by taking the help of inserts.

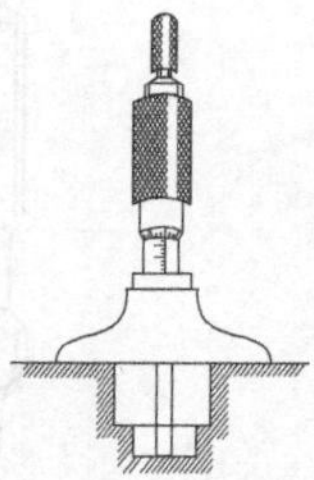

Fig. 2.10. Depth gauges

10. Try Square (Fig. 2.11)

Generally a try square is used to measure the perpendicularity of two adjacent surfaces (Fig. 2.11), when either of the two faces are true surface. It is a simple square having a beam (1) and a blade (2). While using a try square, both the measuring surface and the try square should be cleaned and the burrs if any, should be removed. Try square is used for laying out also. But before testing, the trueness of try square itself should be checked. It can be checked by placing it first against a board which is having a perfectly straight edge. A line is to be drawn against the blade. The square is then to be turned to check

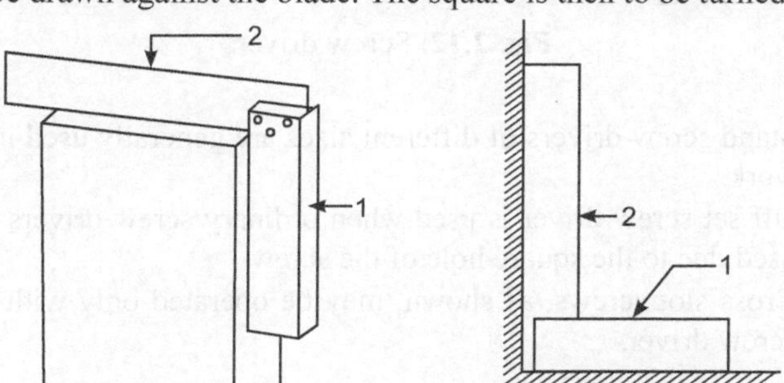

Fig. 2.11. Try Square

whether the line already drawn, coincides with the blade edge or not. If not, the square is not true.

A combination square is also available to check angles of 45° and 90° and also for marking off lines.

11. Screw driver (Fig. 2.12)

It is a common tool for maintenance and assembly work. Various types of screw drivers are available today. They may be as follows in respect to the Figure 2.12.

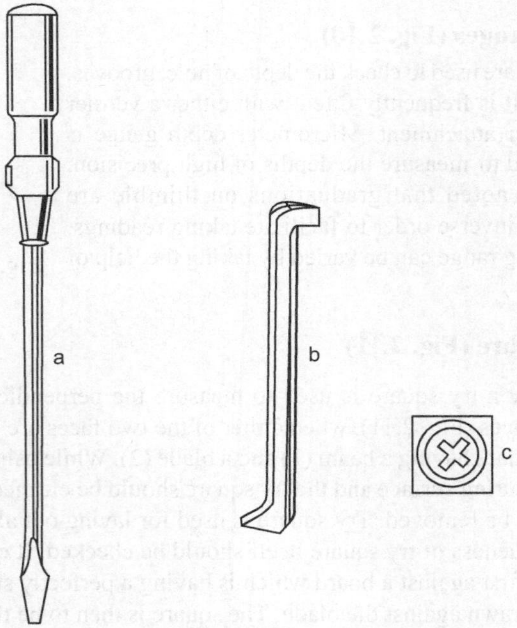

Fig. 2.12. Screw drivers

(*a*) Stand screw drivers of different sizes are generally used in ordinary work.

(*b*) Off set screw driver is used when ordinary screw drivers cannot be used due to the square hole of the screw.

(*c*) Cross slot screws, as shown, may be operated only with a Phillips screw driver.

Care to use screw drivers

1. The shape of the blade is to be maintained properly. The uniform thickness of the tip and its squareness with the broad surface and to keep the centre of the tip on the centre line of the shank indicates proper shape of screw driver blade.
2. The end should be cleaned from oil or grease while using.
3. The fit of the screw driver with the slot of the screw head should always be perfect. In the Figure 2.13, '1' is shown as incorrect fit, while '2' is correct fit.
4. The centre of the screw and that of the screw driver should be in line.

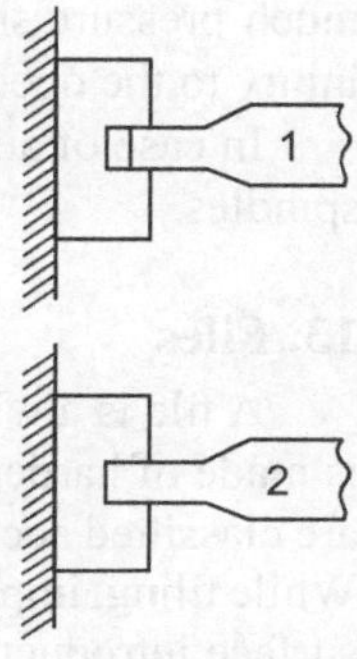

Fig. 2.13

It may finally be remembered that wrong selection of screw driver, application of undue pressure may spoil both the screw head and the screw driver tip.

12. Wrenches (Fig. 2.14)

It is a very common tool in erection or maintenance service. It is a lever-tool made of forged steel and is used to tighten or loosen the nuts and bolts. It may be a single ended (Fig. 2.14*a*) or a double ended (Fig. 2.14*b*), spanner. It may also be closed end spanner (Fig. 2.14*c*) where one of its end is closed completely to form a hexagon or square. This removes the possibility of slippage. To operate any inaccessible nuts, *e.g.*, in engines and boilers etc. T. Socket wrench or Box Spanner (Fig. 2.14*d*) may be used. Such a spanner may be an off set socket wrench (Fig. 2.14*e*) also. It provides a greater leverage. Moreover, it may be adjustable one (Fig. 2.14*f*) Pipe wrenches should also be included in this series. Pipe wrenches (Fig. 2.14*g*) are used for connecting or disconnecting rigid conduit pipes.

As the wrenches are generally made of cast steel, they are having less shock resistance capacity. So they should not fall down while working and they should never be used as hammers.

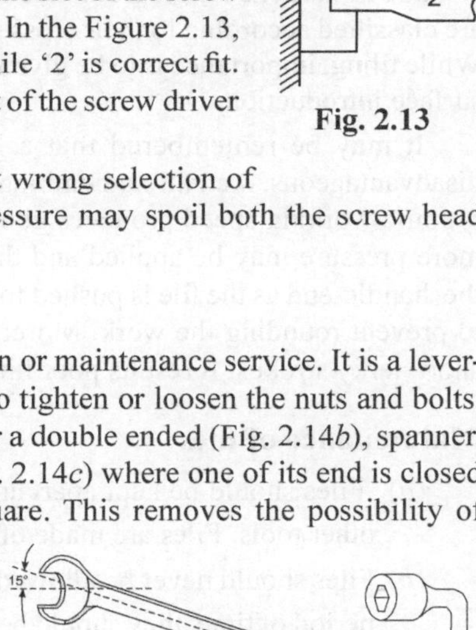

Fig. 2.14. Wrenches

Moreover while in use, holding should be firmly made. Application of too much pressure should always be avoided. The slippage of wrenches causes injury to the operator in addition to blunting the nuts or bolts edges.

In case of adjustable wrenches, proper grease should be applied on jaw-spindles.

13. Files

A file is a widely used hand tool for production or maintenance work. It is made of hardened high grade steel. There are different types of files. They are classified according to their cross section, length, types of cut and quality. While filing, importance is to be given on its technique for speedy and smooth surface introduction.

It may be remembered that a heavy pressure on the file is usually disadvantageous. Actually a combination of steady continuous pressure and rythmic working speed produces good result. At the beginning of the stroke more pressure may be applied and the pressure may gradually be shifted to the handle-end as the file is pushed forward to achieve a desired file level and to prevent rounding the work. Moreover, none should use unnecessary fast, short, jerky strokes. It results poor finishing and makes the file dull rapidly.

Maintenance of files

(*a*) Files should be kept apart and they should not rub or knock against other tools. Files are made of cast steel and are extremely brittle.

(*b*) Files should never be allowed to go rusty. If they are to set aside for a period of time, they should be greased or may be coated with machine oil. But before re-using, they should be washed with petrol or carbon tetrachloride or caustic soda.

(*c*) While using files, they may be clogged with 'small particles of metal. This happens generally when filing a soft metal or dirty, greasy pieces. Such small particles of metal will often tear out and push ahead of the file teeth and scratch the work. The teeth may be cleaned by vigorous rubbing with fine wire brush or file card. The file surfaces are generally rubbed with chalk to prevent pinning.

(*d*) Worn out file does not function properly. They may be made usable by dipping the file in hydrochloric acid. The time to keep the file in the solution depends on the formation of teeth profile upto desired extent. The file is then washed with caustic soda or in fresh water and is to be dried properly. It is a general practice to rub the file firm against a piece of cloth and then to apply saw dust to make it dry.

14. Chisel (Fig. 2.15)

Chisels are the most common metal cutting tool used in shop floor. They may be broadly classified into four categories:

(*a*) ***Flat chisel:*** They are used for cutting off and surfacing.

(*b*) ***Round nose chisel***: They are mainly used for cutting grooves and channels.

(*c*) ***Cross cut chisel***: They are also called as cape chisel. It is used for roughing and grooving on large surfaces. Key ways are also cut with it.

(*d*) ***Diamond point chisel***: They are used to cut grooves, square small holes and to clean corners.

Side and half round chisels are also available for chipping surplus metal in slots etc. and for cutting small grooves like oil holes in bearings etc. respectively.

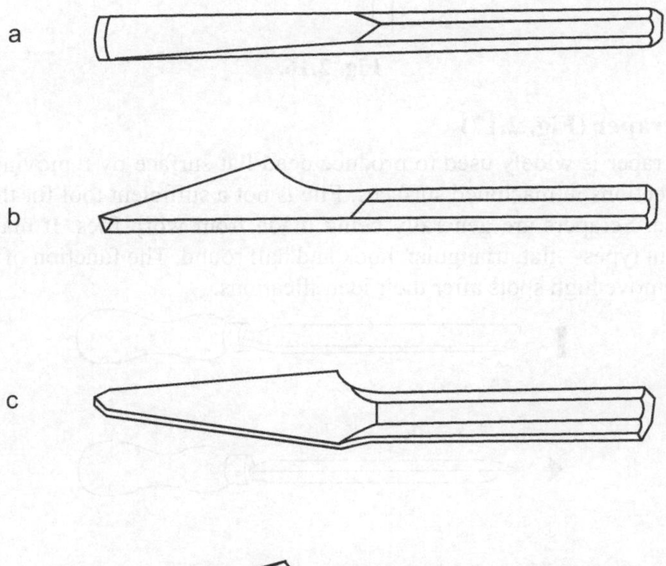

Fig. 2.15. Different types of chisels

Care on use of chisel

The angle at which the chisel is to be held depends on the amount of thickness of chipping (Fig. 2.16). When a large surface is to be worked upon, normal procedure is to cut a number of narrow grooves of equal depth. Then the uprise metal in between the groovs is chipped away. For final surface finish, spirit level, straight edge etc. are to be used for checking.

One more precaution should be undertaken regarding the disfigured shape of the chisel head. A ragged top produced by repeated blows may create accident. Rounding off through grinding whenever necessary is a general practice.

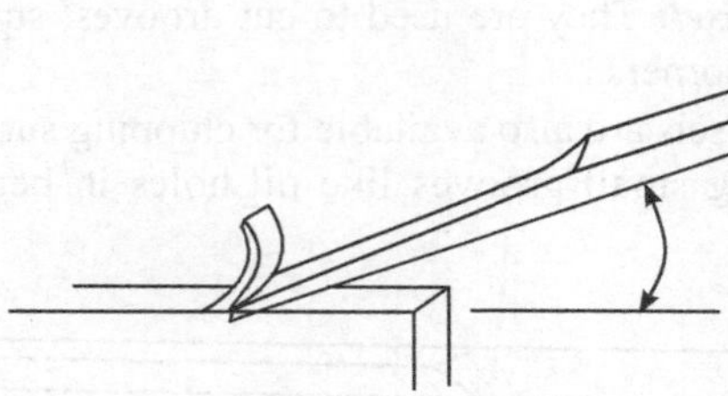

Fig. 2.16.

15. Scraper (Fig. 2.17)

Scraper is widely used to produce dead-flat surface by removing slight imperfections in machined surfaces. File is not a sufficient tool for the same purpose. Scrapers are generally being made from worn files. It may be of different types—flat, triangular, hook and half round. The function of scraper is to remove high spots after their identifications.

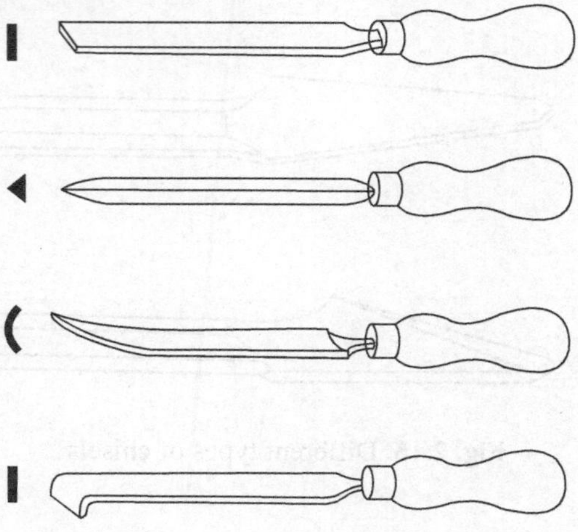

Fig. 2.17. Different types of Scraper

Before using scrapers grinding may be necessary for their sharpening. But care is to be taken before use to rub them against oil stone to eliminate the marks of grinding. While scraping, none should scrap at a place too deeply. The total surface is to be inspected over and over again and repeated scraping will finally produce the required degree of accuracy.

More accurately, scraping may be defined as a process of removal of small particles from identified localised places to produce certain degree of surface finish and proper fit for mating parts.

It is a fine art and proficiency can only be attained through considerable practice.

16. Hacksaw

Cutting metal by hand is generally performed with a hacksaw. Hacksaw frames available, are of two types solid and adjustable and the blades used, may also be of two types. (1) All hard and (2) flexible. All hard blades are having the whole blade tempered. They are suitable for the harder metals and where the job can be held securely. But the flexible blades have only the teeth hardened without hardening the back. They are generally used when the sawing is to be done in awkward positions or the job cannot be held securely and for cutting flexible materials. It may be noted that wrong selection of saws with respect to the pitch of the teeth is generally responsible for its breakage. The hacksaw blades may be classified as coarse, medium or fine teeth with a pitch of 1.8 mm, 1.4 mm and 1.0 mm respectively. Coarse and medium teeth are used for general purpose. The blade with fine pitch is mostly suitable for sawing thin metal or hollow pipes.

Care for Using Hacksaw

(1) Blade is to be fixed in the frame with proper strain. While using a new blade, it will get stretched after a few strokes. It needs retightening.

(2) Selection of blade is to be done properly considering the material to be cut and its position.

(3) Sawing should continue with long, even and steady strokes. Strokes should not be too fast. 40 to 50 strokes per minute are generally better.

(4) During forward stroke, the pressure is to be applied and it is to be released in the backward stroke when the blade is better to be lifted slightly.

(5) For proper control two hands must always be used.

(6) The sawing should start with a backward stroke.

(7) In case a blade gets broken, the new blade should be started working at a new place, possibly on the opposite side of the workpiece. Otherwise, the new blade would bind and there would be a chance of breakage.

REVIEW QUESTIONS

1. Refer a few hand tools which are mostly needed for maintenance work. Describe them with sketches.

2. "Accuracy of instruments like spirit level is an essential criterion to offer accuracy to the machine tools." Justify the statement.

3. Mention in details the special cares to be taken in case of hacksaw and chisel.

4. State the procedure of filing a job. What are the major maintenance procedures of files. Explain how worn files are reconditioned.

5. What is the utility of a Try Square? How its accuracy can readily be checked?

6. What tools are used for measuring the following:

(*a*) The straightness of a flat surface;

(*b*) Flatness of a job;

(*c*) Perpendicularity of two adjacent surfaces;

(*d*) Measuring the depth of a groove;

(*e*) Clearance between mating surface;

(*f*) Diameter of a shaft upto an accuracy of 0.01 mm.

(*g*) Accuracy of a taper hole.

7. Select the correct word and rewrite the following sentences :

(*a*) A heavy pressure on the file is (advantageous/disadvantageous) for efficient filing.

(*b*) Class I-type surface plate is (inferior/superior) to Class II-type, regarding accuracy.

(*c*) (Sometimes/Always) the edge of a 'straight edge' is made bevelled.

(*d*) Socket wrench provides (lesser/greater) leverage.

(*e*) A (thick/thin) layer of metal may be removed by chipping.

(*f*) A (thick/thin) layer of metal may be removed by scraping.

(*g*) Ring gauges are used to check (internal/external) measurements.

(*h*) (Seven/eight) sets of feeler gauges are available in different ranges of thickness.

3

FITS-TOLERANCE AND SURFACE TOPOGRAPHY

Introduction

The machining of a work piece cannot be exactly performed to its nominal dimensions. The dimensions differ, however small it may be, even among the work-pieces, produced under identical conditions. The degree of precision depends on the skill of the worker, machine tools, materials etc. Whatever may be the situation, the deviation on nominal size is generally inevitable. More is the deviation allowed, more is the flexibility available during production. But the extent of deviation should have a permissible limit to meet the objectives of the work-pieces, produced. Moreover, if a definite limit of deviation for a particular work-piece is maintained, it will enable the production engineer to produce parts for an assembly with a freedom of interchangeability, that is, the mating parts can be assembled randomly within the limits of designed tightness or looseness to meet any definite purpose.

Thus, maintaining the limits of deviation is a pre-condition of mass production. The 'interchangeable system' is also termed as a 'limit system' or 'system of limits and fits'.

Terms associated with an assembly

Basic size and Nominal size. Nominal size is the dimensions, shown in the drawing for general identification and Basic size is the size with reference to which all other variations are determined.

The nominal size of an assembly of a shaft and a hole may be 60 mm, whereas the basic size of the shaft may be 59.5 mm and that of the hole may be 60 mm.

Actual size and the Limits

Actual size is the dimension of a component which results after the final machining.

The extent of deviation, allowed in the design and drawing is termed as 'limit'. The limits are permissible on both the higher and lower sides of the nominal size, denoting 'Higher' and 'Lower' limit of the shaft or the hole. It is obvious that the higher and lower limit of a shaft represents the 'maximum' and 'minimum' size of the shaft respectively, whereas in case of a hole it will be reverse.

Tolerance

This may be defined as the difference between the maximum and minimum limits or sizes.

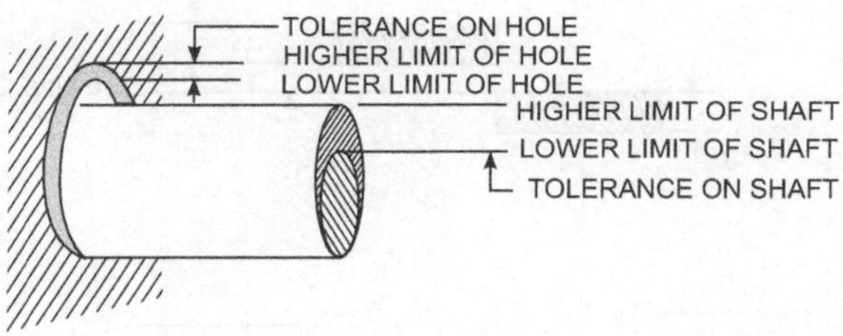

Fig. 3.1. Limits and Tolerance.

The higher and lower limits are always termed as the limiting sizes with respect to the nominal size. Considering the limiting dimensions the toler ance may be represented as a zone in between the higher and lower limits.

Thus, the dimension denoted by $50^{+0.01}_{-0.01}$ indicates that the higher limit is 50.1 mm while the lower limit is 49.99 mm. So, the amount of tolerance attached to this particular example is (50.01 mm - 49.99 mm) 0.02 mm.

If in the deviation on both the above (+ve) and below (–ve) the nominal size is allowed, it is called the 'bilateral system', "as shown in the above example". If the total tolerance is specified in one side only, it is said to be the 'unilateral system of tolerance', e.g. $50^{+0.02}_{-0.00}$ or $50^{+0.00}_{-0.02}$.

Such unilateral tolerances are very much common in interchangeable systems, specially when a precision fit is in question. On the other hand, bilateral tolerance is commonly used in mass production, as it simplifies the machine setting.

Deviation and Zero line

From the above discussion it is now clear that 'deviation' is the measure of boundaries of a tolerance zone with respect to the 'basic size'.

And 'zero line' is a straight line to which the deviations are referred in a graphical representation of limits and fits. The zero line is the line of zero deviation and represents the basic size. When a zero line is drawn horizon tally, the positive and negative deviations are shown above and below it, respectively .

The 'Maximum and Minimum' or the 'upper and lower' deviations are the differences between the maximum and minimum limits with the basic size respectively.

'Fundamental deviation' is a term which may be applicable to either of the two deviations which is conventionally chosen to define the position of the tolerance zone in relation to the zero line, as shown in Fig. 3.2.

The Fig. 3.2 explains the deviations of shafts or holes and the fundamental deviation with respect to the zero line.

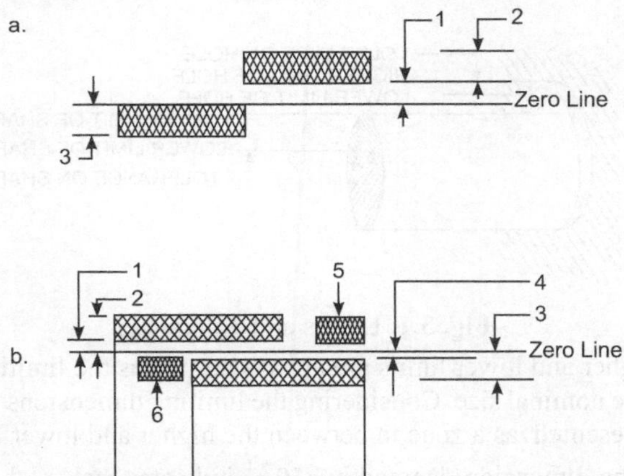

Fig. 3.2. Deviation of shaft and hole .

(*a*) 1. Fundamental deviation. 2. Tolerance zone for a hole. 3. Tolerance zone for shafts.

(*b*) 1. Lower deviation of a hole. 2. Upper deviation of a hole. 3. Lower deviation of a shaft. 4. Upper deviation of a shaft. 5. Tolerance zone of a hole. 6. Tolerance zone of a shaft.

Allowance and Clearance

These two terms are having a sharp difference. The difference between a hole diameter and a shaft diameter is called 'allowance'. It may be '+ve' when the shaft diameter is less than a tole diameter, and when the shaft diameter is more it is '–ve' allowance.

But, a 'clearance' is always the '+ve' difference between a hole and a shaft diameter. It may be noted that,

Max. allowance = Largest hole size - Smallest shaft size.

Min. allowance = Smallest hole size - Largest shaft size.

The positive differences of the above relations indicate the maximum or minimum clearances.

Fits

The importance of tolerance for the manufacturing processes have been discussed. But, the main objective, is to maintain the desired level of accuracy while two mating parts are fitted together.

In an assembly, there is one 'enveloped surface', which is termed as a 'shaft' and another 'enveloping surface', termed as a 'hole'. These two surfaces are to be assembled together, sometimes to move freely or sometimes, more or less tightly, as may be reserved by the assembly to function properly.

The degree of looseness or tightness with respect to the mating parts is called the 'fit'.

The classification of fit is shown in the chart 3.1.

Chart 3.1

Classification of Fits

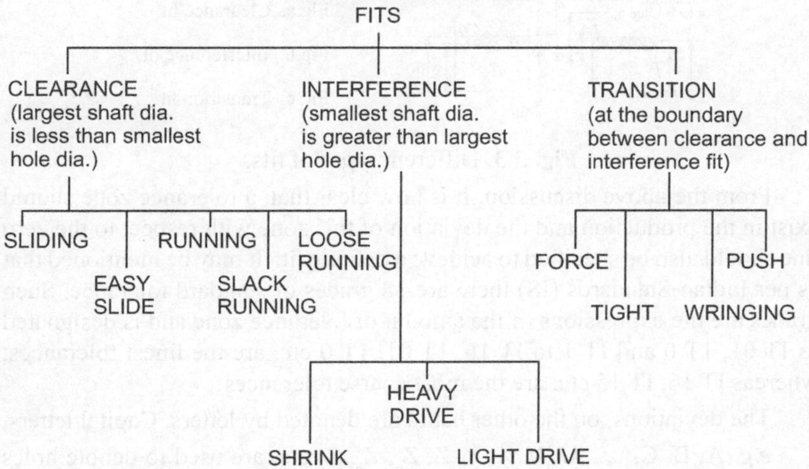

The conditions to be satisfied to achieve a definite class of a fit is also mentioned in the above chart. The terms associated with the sub classifications are self explanatory, as the concerned name indicates. This may only be mentioned that the positive allowance in the mating parts produces 'clearance fit'; negative allowance produces 'interference fit', that is, the 'tolerance zone' of the 'hole' is entirely above the 'tolerance zone' of the shaft in case of a "clearance fit", and is entirely below that of the shaft in case of an "interference fit". But, for a "transition fit", both the 'tolerance zones' of the hole and shaft overlap.

The different types of fits may be represented by sketches showing the degree of clearance or interference (Fig. 3.3).

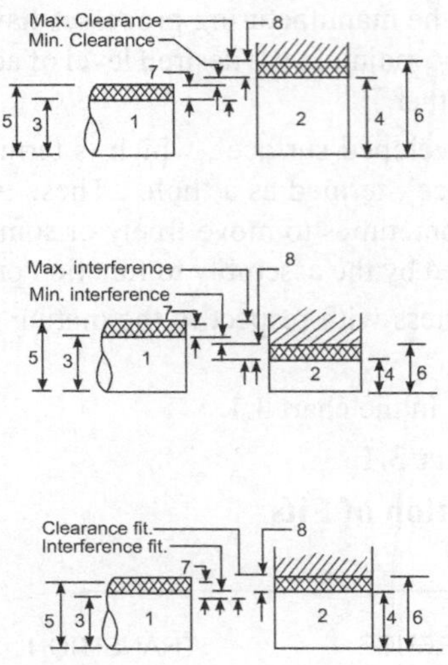

1. Shaft
2. Hole.
3. Minimum size of a shaft.
4. Minimum size of a hole.
5. Maximum size of a shaft
6. Maximum size of a hole.
7. Tolerance on a shaft.
8. Tolerance on a hole.

Fig.a. Clearance fit.

Fig.b. Interference fit.

Fig.c. Transition fit.

Fig. 3.3. Different types of fits.

From the above discussion, it is now clear that, a tolerance zone should exist in the production and the deviation of this zone with respect to the zero line should also be specified to achieve a definite fit. It may be mentioned that as per Indian Standards (IS) there are 18 grades of standard tolerance. Such grades are the expressions of the amount of tolerance zone and is designated as IT 01, 1T 0 and IT 1 to IT 16. 1T 01, 1T 0 etc. are the finest tolerances, whereas IT 16, IT 15 etc. are the most coarse tolerances.

The deviations, on the other hand, are denoted by letters. Capital letters, *e.g.* A, B, C, ...U, V, X, Y, Z, Z_a, Z_b, Z_c etc. are used to denote holes and small letters, *e.g. a, b, c,* ....*u, v, x, y, z,* z_a, z_b, Z_c etc. are used for shafts.

In total, 25 types of fundamental deviations are available as per the Indian Standard System. An example is given in Fig. 3.4, showing the position of the tolerance zone with respect to the zero line. It is indicated by a letter (by two letters in some cases) symbol. Capital letters have been used to denote holes and small letters for shafts, as usual.

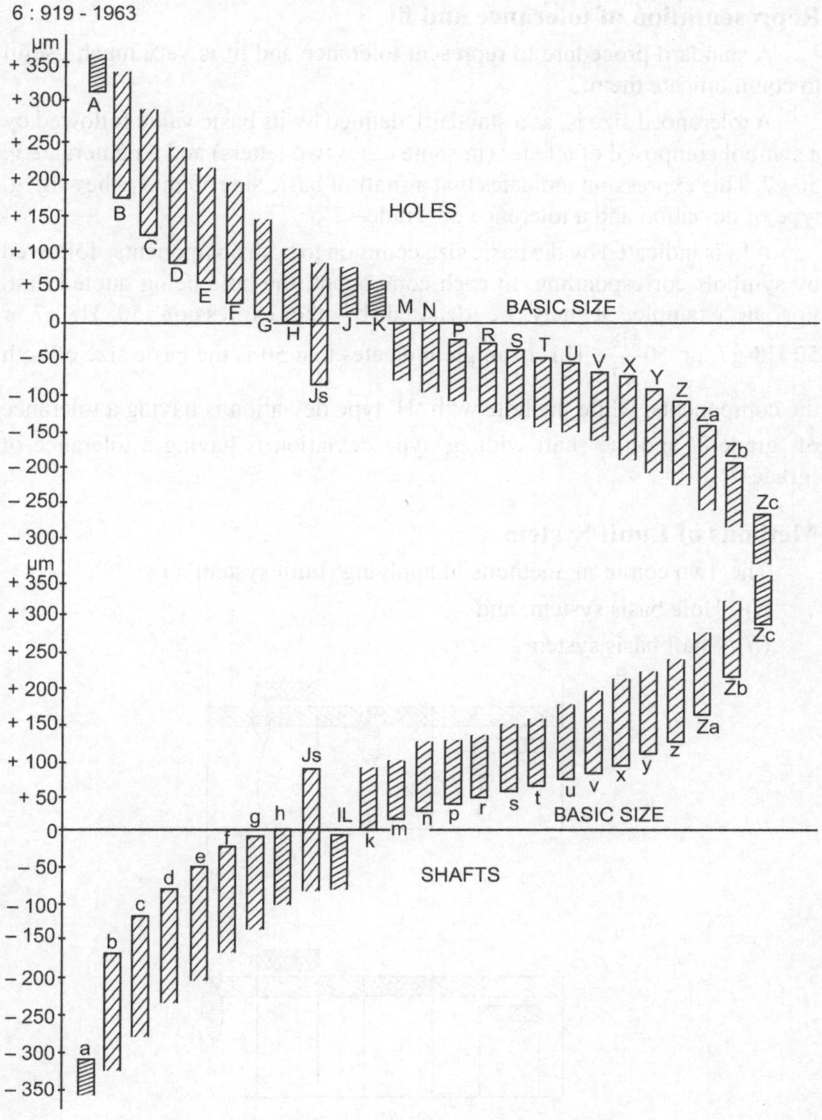

Fig. 3.4. Letter Symbols for Tolerances.
(Examples for Diameters 30 to 40 mm)

Representation of tolerance and fit

A standard procedure to represent tolerance and fit is very much useful to communicate them.

A toleranced size is, as a standard, defined by its basic value followed by a symbol composed of a letter (in some cases two letters) and a numeral, *e.g.* 50 g7. This expression indicates that a shaft of basic size 50 mm is having 'g' type of deviation and a tolerance of 'grade - 7'.

A fit is indicated by the basic size common to both components, followed by symbols corresponding. to each component, the hole being quoted first. For an example, it may be denoted by the expression 50 H8/g7 or 50 H8-g7 or $50\frac{H8}{g7}$. This example indicates that 50 is the basic size of both the components, while the hole with 'H' type deviation is having a tolerance of 'grade-8' and the shaft with 'g' type deviation is having a tolerance of 'grade-7'.

Methods of Limit System

The two common methods of applying 'limit system' are :

(*i*) Hole basis system; and

(*ii*) Shaft basis system.

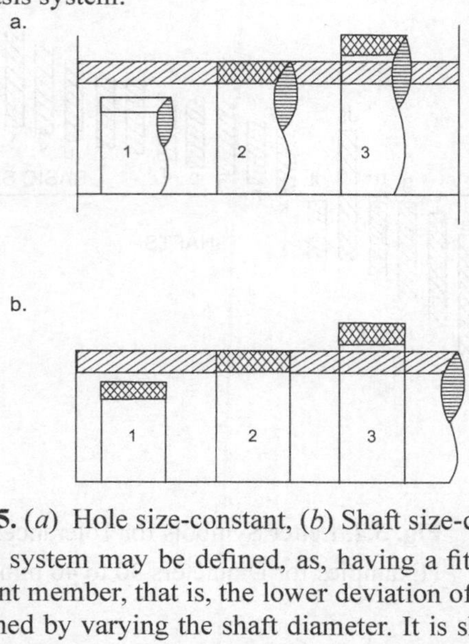

Fig. 3.5. (*a*) Hole size-constant, (*b*) Shaft size-constant.

A hole basis system may be defined, as, having a fit, when the hole is taken as a constant member, that is, the lower deviation of a hole is zero and the fits are obtained by varying the shaft diameter. It is shown in Fig. 3.5a, drawing three shafts 1, 2 and 3 in a single hole and indicating (1) clearance, (2) clearance or interference and (3) interference fits.

The reverse is the shaft basis system. The shaft is taken as a constant member, that is, its upper deviation is zero. Different fits are obtainable by varying the hole sizes as shown in Fig. 3.5*b*. Three holes 1, 2 and 3 are shown in combination with a single shaft. The available fits can also be illustrated in the same way, as stated above. The hole basis and shaft basis system can be shown in other way (Fig. 3.5*c*) with respect to the zero line.

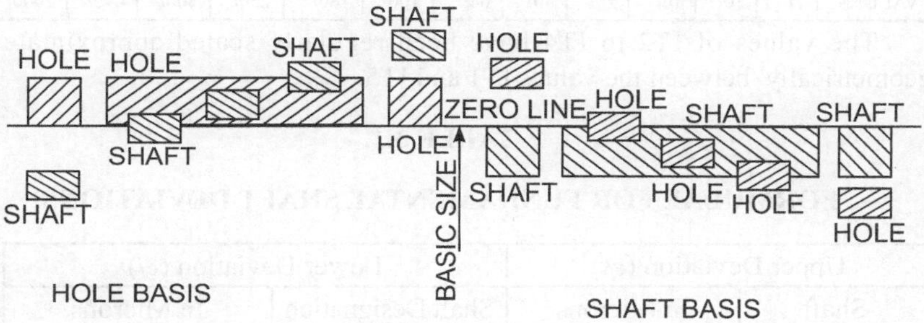

Fig. 3.5. (*c*) Hole Basis and Shaft Basis System.

Out of these two systems a hole basis system is always preferred. Uniformity in the size of holes is the beginning of any interchangeable work. Holes are usually produced and finished by standard tooling like drill, reamers etc. To make a variety of classes of holes, varieties in tooling are essential and essentially the cost involvement becomes too high. Practically, a large varieties of such tools are not also commercially avail able.

Conversely, it is easily in practice to vary the size of the shaft which is to go into the hole.

So, most economic and accepted system is the 'hole basis system', though shaft basis system is followed in some cases, when it is deemed to be advantageous.

The Indian Standard System: (IS 919 - 1963)

The Indian Standard System provides standard information in detail regarding the tolerance and deviation for particular diameter (D).

TOLERANCE

Grades of tolerance from IT 01, IT 0 and IT 1 are given below:

Values in microns	IT 01	IT 0	IT 1
for D in millimetres	0.3 + 0.008D	0.5 + 0.012D	0.8 + 0.020 D

The values for grades 5 to 16 can be calculated in terms of a standard tolerance '*i*', where

$$i \text{ (microns)} = 0.45\sqrt[3]{D} + 0.001D \text{ (D in mm)}$$

The relative magnitude of each grade is as given below:

GRADE	IT5	IT6	IT7	IT8	IT9	IT10	IT11	IT12	IT13	IT14	IT15	IT16
VALUES	7*i*	10*i*	16*i*	25*i*	40*i*	64*i*	100*i*	160*i*	250*i*	400*i*	640*i*	2000*i*

The values of IT2 to IT4 have been regularly scaled approximate geometrically between the values IT1 and IT5.

TABLE 3.1.

FORMULAE FOR FUNDAMENTAL SHAFT DEVIATIONS

Upper Deviation (*es*)		Lower Deviation (*ej*)	
Shaft Designation	In Microns (*for D in mm*)	Shaft Designation	In Microns (*for D in mm*)
a.	$= -(265 + 1.3\,D)$ for $D \le 120$	5 to j8	No formula
		k4 to k7	$= +0.6\sqrt[3]{D}$
	$= -3.5\,D$ for $D > 120$	k for grades ≤ 3 and ≥ 8	$= 0$
b.	$\cong -(140 + 0.85\,D)$ for $D \le 160$	m	$= +(IT\,7 - IT\,6)$
		n	$= +5\,D^{0.34}$
	$\cong -(1.8\,D$ for $D > 160$	P	$= +IT + 0$ to 5
c.	$= -52\,D^{0.2}$ for $D \le 40$	r	= geometric mean of values el for p and s
	$= -(95 + 0.8\,D)$ for $D > 40$		$= +IT\,8 + 1$ to 4 for $D \le 50$
d.	$= -16\,D^{0.44}$	s	$= +IT\,7 + 0.4\,D$ for $D > 50$
e.	$= -11\,D^{0.41}$	t	$= IT\,7 + 0.63\,D$
f.	$= -5.5\,D^{0.41}$	u	$= +IT\,7 + D$

Contd. Table 3.1.

g.	$= -2.5^{0.34}$	v	= +IT 7 + 1.25D
h.	= 0	x	= +IT 7 + 1.6D
		y	= +IT 7 + 2 D
		z	= +IT 7 + 2.5 D
		za	= +IT 8 + 3.15 D
		zb	= +IT 9 + 4 D
		zc	= +IT 10 + 5D
For js : The two deviations are equal to $\pm \frac{IT}{2}$			

Fundamental Deviations

Different formulae have been accepted as the ISA system to calculate the fundamental deviations. The following symbols have also been accepted as standard:

Upper deviation of a hole – ES

Lower deviation of a hole – EI

Upper deviation of a shaft – es

Lower deviation of a shaft – ei

Deviation calculation in case of a shaft

For each letter symbol, the magnitude and sign for one of the two deviations, which is known as 'fundamental deviation' (es or ei) have been determined by means of the formulae given in the Table 3.1. The other deviations may be derived directly using the absolute value of the standard tolerance 'IT', by means of the following algebraic relationship:

ei = es – IT, or es = ei + IT

It may be noted that,

(*a*) except in the case of shafts j and js, for which, strictly speaking, there is no fundamental deviation, the value of the deviation shown in the table is independent of grade.

(*b*) the deviation given in the table is that corresponding, in principle, to the limit closest to the zero line; in other words, the upper deviations' for shafts 'a' to 'h' and the lower deviation 'ei' for shafts 'j' to 'zc'.

Deviation calculation in case of a hole

Deviations for holes are derived from those of the corresponding shafts as given in the Table – 3.2.

TABLE – 3.2.

All deviations except those given below:			*General rule*: Hole limits are identical with the shaft limits of the same symbol (letter and grade) but disposed on the other side of the zero line. EI = Upper deviation 'es' of the shaft of the same letter symbol but of opposite sign.
	N	g and coarser grades	ES = O
For sizes above 3mm	'J', 'K', 'M','N' 'P' to 'ZC'	Up to grade 8 inclusive Up to grade 7 inclusive	*Special rule*: ES = Lower deviation 'ei' of the shaft of the same letter symbol but one grade finer and of 'P'to opposite sign increased by the difference 'ZC' between the tolerances of the two grades in question.

In all cases, the other deviation may be derived from the first one using the absolute value of the IT tolerance, by means of the algebraic relationship :

ES = EI + IT, or

EI = ES – IT

The special rule is applicable only to sizes above 3 mm and has been provided in order that in those fine grades two comparable fits, such as H7 – H6, shall have exactly the same clearance or interference.

The principle of the special rule can be explained diagrammatically.

It is as illustrated in Fig. 3.6. The formulae used are as follows:

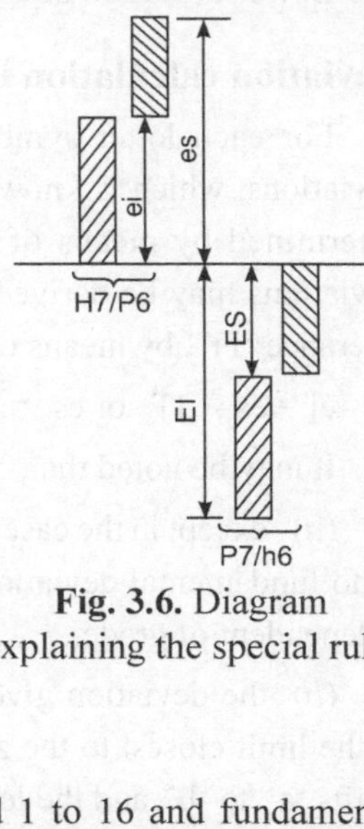

Fig. 3.6. Diagram explaining the special rule.

$EI_n = -\ es\ (n-1)$, and thus for the upper deviation : $ES_n = EI_n + IT_n$

$= -\ es\ (n-1) + IT.$

$= -\ ei\ (n-1) - IT(n-1) + IT_n$

$= -\ ei\ (n-1) + i\ [IT_n - IT(n-1)]$

Fundamental tolerances of grades 01, 0 and 1 to 16 and fundamental deviations for shafts 'a' to 'zc', for diameters up to 500 mm is given here in the following tables :

TABLE 3.3. FUNDAMENTAL TOLERANCES OF GRADES 01, 0 AND 1 TO 16

Diameter steps in mm		Values of Tolerances in Microns																(1 microns = 0.001 mm)	
		Tolerance Grades																	
		01	0	1	2	3	4	5	6	7	8	9	10	11	12	13	14*	15*	16*
To and inc	3	0.3	0.5	0.8	1.2	2	3	4	6	10	14	25	40	60	100	140	250	400	600
Over To and inc	3 6	0.4	0.6	1	1.5	2.5	4	5	8	12	18	30	48	75	120	180	300	480	750
Over To and inc	6 10	0.4	0.6	1	1.5	2.5	4	6	9	15	22	36	58	90	150	220	360	580	900
Over To and inc	10 18	0.5	0.8	1.2	2	3	5	8	11	18	27	43	70	110	180	270	430	700	1100
Over To and inc	18 30	0.6	1	1.5	2.5	4	6	9	13	21	33	52	84	130	210	330	520	840	1300
Over To and inc	30 50	0.6	1	1.5	2.5	4	7	11	16	25	39	62	100	160	250	390	620	1000	1600
Over To and inc	50 80	0.8	1.2	2	3	5	8	13	19	30	46	74	120	190	300	460	740	1200	1900
Over To and inc	80 120	1	1.5	2.5	4	6	10	15	22	35	54	87	140	220	350	540	870	1400	2200
Over To and inc	120 180	1.2	2	3.5	5	8	12	18	25	40	63	100	160	250	400	630	1000	1600	2500
Over To and inc	180 250	2	3	4.5	7	10	14	20	29	46	72	115	185	290	460	720	1150	1850	2900
Over To and inc	250 315	2.5	4	6	8	12	16	23	32	52	81	130	210	320	520	810	1300	2100	3200
Over To and inc	315 400	3	5	7	9	13	18	25	36	57	89	140	230	360	570	890	1400	2300	3600
Over To and inc	400 500	4	6	8	10	15	20	27	40	63	97	155	250	400	630	970	1550	2500	4000

* Up to 1 mm, Grades 14 to 16 are not provided.

TABLE 3.4. FUNDAMENTAL DEVIATIONS FOR SHAFTS A TO ZC

(*a* to *k*)

Fundamental Deviations in Microns (I micron = 0.001 mm)

Diameter Steps in mm		Upper Deviations (es)									Lower Deviations (ei)				
		*a**	*b**	*c*	*d*	*e*	*f*	*g*	*h*	*js*†	*j*			*k**	
Over	To					All Grades					5-6	7	8	4 to 7	≤ 3 > 7
–	3°	–270	–140	60	–20	–14	–6	–2	0	Deviations = $\pm \frac{IT}{2}$	–2	–4	–6	0	0
3	6	–270	–140	–70	–30	–20	–10	–4	0		–2	–4	–	0 +1	0
6	10	–280	–150	–80	–40	–25	–13	–5	0		–2	–5	–	+1	0
10	14														
14	18	–290	–150	–95	–50	–32	–16	–6	0		–3	–6	–	+1	0
18	24														
24	30	–300	–160	–110	–65	–40	–20	–7	0		–4	–8	–	+2	0
30	40	–310	–170	–120											
40	50	–320	–180	–130	–80	–50	–25	–9	0		–5	–10	–	+2	0
50	65	–340	–190	–140											
65	80	–360	–200	–150	–100	–60	–30	–10	0		–7	–12	–	+2	0
80	100	–380	–220	–170											
100	120	–410	–240	–180	–120	–72	–36	–12	0		–9	–15	–	+3	0
120	140	–460	–260	–200											
140	160	–520	–280	–210	–145	–85	–43	–14	0		–11	–18	–	+3	0
160	180	–580	–310	–230											
180	200	–660	–340	–240											
200	225	–740	–380	–260	–170	–100	–50	–15	0		–13	–21	–	+4	0
225	250	–820	–420	–280											
250	280	–920	–480	–300											
280	315	–1050	–540	–330	–190	–110	–56	–17	0		–16	–26	–	+4	0
315	355	–1200	–600	–360											
355	400	–1350	–680	–400	–210	–125	–62	–18	0		–18	–28	–	+4	0
400	450	–1500	–760	–440											
450	500	–1650	–840	–480	–230	–135	–68	–20	0		–20	–32	–	+5	0

* The Deviations *a* and *b* are not provided for diameters up to 1 mm.

† For js in the particular Grades 7 to 11, the two symmetrical deviations $\pm \frac{IT}{2}$ may possibly be rounded, if the IT value in microns in odd value, by replacing it by the even value immediately below.

TABLE 3.4. FUNDAMENTAL DEVIATIONS FOR SHAFTS A TO ZC (*Contd.*)

k to *zc*

Fundamental Deviations in Microns (1 micron = 0.001 mm)

Diameter Steps in mm		Lower Deviations (ei)													
mm		*m*	*n*	*p*	*r*	*s*	*t*	*u*	*v*	*x*	*y*	*z*	*za*	*zb*	*zc*
Over	To	All Grades													
–	3	+2	+4	+6	+10	+14	–	+18	–	+20		+26	+32	+40	+60
3	6	+4	+8	+12	+15	+19	–	+23	–	+28		+35	+42	+50	+80
6	10	+6	+10	+15	+19	+23	–	+28	–	+34		+42	+52	+67	+97
10	14								–	+40		+50	+64	+90	+130
14	18	+7	+12	+18	+23	+28	–	+33	+39	+45		+60	+77	+108	+150
18	24						–	+41	+47	+54	+63	+73	+98	+136	+188
24	30	+8	+15	+22	+28	+35	+41	+48	+55	+64	+75	+88	+118	+160	+218
30	40						+48	+60	+68	+80	+94	+112	+148	+200	+274
40	50	+9	+17	+26	+34	+43	+54	+70	+81	+97	+114	+136	+180	+242	+325
50	65				+41	+53	+66	+87	+102	+122	+144	+172	+226	+300	+405
65	80	+11	+20	+32	+43	+59	+75	+102	+120	+144	+174	+210	+274	+360	+480
80	100				+51	+71	+91	+124	+146	+178	+214	+258	+335	+445	+585
100	120	+13	+23	+37	+54	+79	+104	+144	+172	+210	+254	+310	+400	+525	+690
120	140				+63	+92	+122	+170	+202	+248	+300	+365	+470	+620	+800
140	160	+15	+27	+43	+65	+100	+134	+190	+228	+280	+340	+415	+535	+700	+900
160	180				+68	+108	+146	+210	+252	+310	+380	+465	+600	+780	+1000
180	200				+77	+122	+166	+236	+284	+350	+425	+520	+670	+880	+1150
200	225	+17	+31	+50	+80	+130	+180	+258	+310	+385	+470	+575	+740	+960	+1250
225	250				+84	+140	+196	+284	+340	+425	+520	+640	+820	+1050	+1350
250	280	+20	+34	+56	+94	+158	+216	+315	+385	+475	+580	+710	+920	+1200	+1550
280	315				+98	+170	+240	+350	+425	+525	+650	+790	+1000	+1300	+1700
315	355				+108	+190	+268	+390	+475	+590	+730	+900	+1150	+1500	+1900
355	400	+21	+37	+62	+114	+208	+294	+435	+530	+660	+820	+1000	+1300	+1650	+2100
400	450				+126	+332	+330	+490	+595	+740	+920	+1100	+1450	+1850	+2400
450	500	+23	+40	+68	+132	+252	+360	+540	+660	+820	+1000	+1250	+1600	+2100	+2600

TABLE 3.5. FUNDAMENTAL DEVIATIONS FOR HOLES A TO ZC (*Contd.*)

Fundamental Deviations in Microns (I micron = 0.001 mm)

Diameter Steps in mm		Upper Deviations (es)												Lower Deviations (ei)						
mm		A*	B*	C	D	E	F	G	H	*js*†	*J*			*K*		M		N		*P to ZC*
Over	To	All Grades									6	7	8	≤8	>8	≤8‡	>8	≤8	>8*	≤7
–	3*	+270	+140	+60	+20	+14	+6	+2	0		+2	+4	+6	0	0	–2	–2	–4	–4	
3	6	+270	+140	+70	+30	+20	+10	+4	0		+5	+6	+10	–1 + Δ	–	–4 + Δ	–4	–8 + Δ	0	
6	10	+280	+150	+80	+40	+25	+13	+5	0		+5	+8	+12	–1 + Δ		–6 + Δ	–6	–10 + Δ	0	
10	14																			
14	18	+290	+150	+95	+50	+32	+16	+6	0		+6	+10	+15	–1 + Δ		–7 + Δ	–7	–12 + Δ	0	
18	24																			
24	30	+300	+160	+110	+65	+40	+20	+7	0		+8	+12	+20	–2 + Δ		–8 + Δ	–8	–15 + Δ	0	
30	40	+310	+170	+120																
40	50	+320	+180	+130	+80	+50	+25	+9	0		+10	+14	+24	–2 + Δ		–9 + Δ	–9	–17 + Δ	0	
50	65	+340	+190	+140																
65	80	+360	+200	+150	+100	+60	+30	+10	0		+13	+18	+28	–2 + Δ		–11 + Δ	–11	–20 + Δ	0	
80	100	+380	+220	+170																
100	120	+410	+240	+180	+120	+72	+36	+12	0	Deviations = $\pm \frac{IT}{2}$	+16	+22	+34	–3 + Δ		–13 + Δ	–13	–23 + Δ	0	
120	140	+460	+260	+200																
140	160	+520	+280	+210	+145	+85	+43	+14	0		+18	+26	+41	–3 + Δ		–15 + Δ	–15	–27 + Δ	0	
160	180	+580	+310	+230																
180	200	+660	+340	+240																
200	225	+740	+380	+260	+170	+100	+50	+15	0		+22	+30	+47	–4 + Δ		–17 + Δ	–17	–31 + Δ	0	
225	250	+820	+420	+280																
250	280	+920	+480	+300																
280	315	+1050	+540	+330	+190	+110	+56	+17	0		+25	+36	+55	–4 + Δ		–20 + Δ	–20	–34 + Δ	0	
315	355	+1200	+600	+360																
355	400	+1350	+680	+400	+210	+125	+62	+18	0		+29	+39	+60	–4 + Δ		–21 + Δ	–21	–37 + Δ	0	
400	450	+1500	+760	+440																
450	500	+1650	+840	+480	+230	+132	+68	+20	0		+33	+43	+66	–5 + Δ		–23 + Δ	–23	–40 + Δ	0	

* the deviations A and B in all grades > 8, are not provided for diameter up to 1 mm.

† For js in the Grades 7 to 11, the two symetrical deviations may possibly be rounded, if the IT value in microns in an odd value, by replacing it by the even value immediately below.

Special case : for M6, ES = – 9 from 250 to 315 (instead of – 11).

TABLE 3.5. FUNDAMENTAL DEVIATIONS FOR HOLES A TO ZC (*Contd.*)

Fundamental Deviations in Microns — N to ZC (1 micron = 0.001 mm)

Diameter Steps in mm		Upper Deviations (ES)												Δ *in microns**					
in mm		P	R	S	T	U	V	X	Y	Z	ZA	ZB	ZC						
Over	To						> 7							3	4	5	6	7	8
–	3	–6	–10	–14	–	–18		–20		–26	–32	–40	–60			Δ = 0			
3	6	–12	–15	–19	–	–23	–	–20	–	–35	–42	–50	–80	1	1.5	1	3	4	6
6	10	–15	–19	–23	–	–28	–	–34	–	–42	–52	–67	–97	1	1.5	2	3	6	7
10	14						–	–40	–	–50	–64	–90	–130						
14	18	–18	–23	–28	–	–33	–39	–45	–	–60	–77	–109	–150	1	2	3	3	7	9
18	24				–	–41	–47	–54	–63	–73	–98	–136	–188						
24	30	–22	–28	–35	–41	–48	–55	–64	–75	–88	–118	–160	–218	1.5	2	3	4	8	12
30	40				–48	–60	–68	–80	–94	–112	–148	–200	–274						
40	50	–26	–34	–43	–54	–70	–81	–97	–114	–136	–180	–242	–325	1.5	3	4	5	9	14
50	65		–41	–53	–66	–87	–102	–122	–144	–172	–226	–300	–405						
65	80	–32	–43	–59	–75	–102	–120	–147	–174	–210	–274	–360	–480	2	3	5	6	11	16
80	100		–54	–71	–91	–124	–146	–178	–214	–258	–335	–445	–585						
100	120	–37	–51	–79	–101	–144	–172	–210	–254	–310	–400	–525	–690	2	4	5	7	13	19
120	140		–63	–92	–122	–170	–220	–248	–300	–365	–470	–620	–800						
140	160	–43	–65	–100	–134	–190	–228	–280	–340	–415	–535	–700	–900	3	4	6	7	15	23
160	180		–68	–108	–146	–210	–252	–310	–380	–465	–600	–780	–1000						
180	200		–77	–122	–166	–236	–284	–350	–425	–520	–670	–880	–1150						
200	225	–50	–80	–130	–180	–258	–310	–385	–470	–575	–740	–960	–1250	3	4	6	9	17	26
225	250		–84	–140	–196	–284	–340	–425	–520	–640	–820	–1050	–1350						
250	280		–94	–158	–218	–315	–385	–475	–580	–710	–920	–1200	–1550						
280	315	–56	–98	–170	–240	–350	–425	–525	–650	–790	–1000	–1300	–1700	4	4	7	9	20	29
315	355		–108	–190	–268	–390	–475	–590	–730	–900	–1150	–1500	–1900						
355	400	–62	–144	–208	–294	–435	–530	–660	–820	–1000	–1300	–1650	–2100	4	5	7	11	21	32
400	450		–162	–232	–330	–490	–595	–740	–920	–1100	–1450	–1850	–2400						
450	500	–68	–132	–252	–360	–540	–660	–820	–1000	–1250	–1600	–2100	–2600	5	5	7	13	23	34

* In determine K, M, N, up to Grade 8 and P to ZC up to Grade 7 (given on P 46) take the Δ values from the columns on the right. Example: For P7 from 18 to 30 Δ = 8 hence ES = – 14.

SURFACE TOPOGRAPHY

Introduction

Manufacturing processes always produce the surface of a 'Work piece Characteristic' pattern of hills and valleys (Fig. 3.7*a*) whatever small in size they may be. They have got their own heights (h_1), spacings (L_1) and shape.

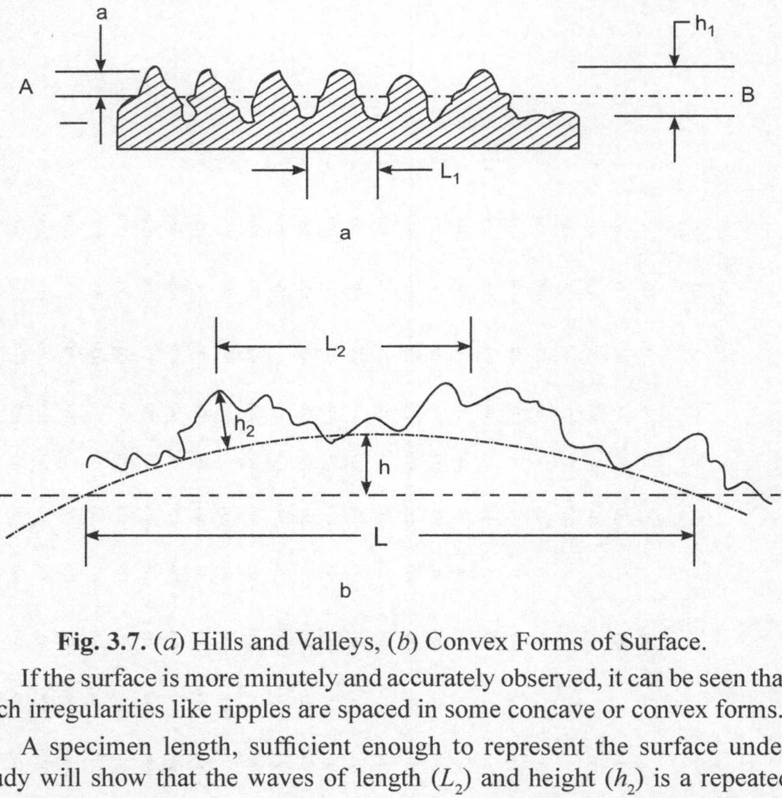

Fig. 3.7. (*a*) Hills and Valleys, (*b*) Convex Forms of Surface.

If the surface is more minutely and accurately observed, it can be seen that such irregularities like ripples are spaced in some concave or convex forms.

A specimen length, sufficient enough to represent the surface under study will show that the waves of length (L_2) and height (h_2) is a repeated phenomenon over the convex (as shown in Fig. 3.7*b*), or concave form of length '*L*' and height '*h*'.

Form of irregularities

To characterise the irregularities, some definite terms, relating to the definite geometrical dimensions, are generally used in practice. They are,

(*a*) **Micro irregularities.** This term represents the 'finely spaced deviations, as shown in Fig. 3.7*a*., which repeats itself throughout the surface. This can be expressed as a ratio of 'L_1' and 'h_1'. It may be longitudinal or transverse and generally occurs during the separation of chips.

(*b*) **Macro irregularities.** These are the irregularities with respect to the longest concave or convex forms on a surface. This may also be expressed as a ratio of its length and height. From Fig. 3.7*b*, it is Uh. Lack of flatness, roundness etc. are due to such faults.

(*c*) **Surface waviness.** These are the regularly repeated wavy forms which occur usually due to machine tool faults. They are generally evenly placed and can be again expressed as a ratio of their pitch (L_1) and height (h_2), as shown in Fig. 3.7*b*. It appears generally in turning, milling, shaping due to tool chatter, feed, etc .

(d) **Surface flaws.** Such irregularities occur at infrequent random intervals or more precisely, it may occur at one or two places, such as a hole or a peak or crack etc. as shown (1) in Fig. 3.8.

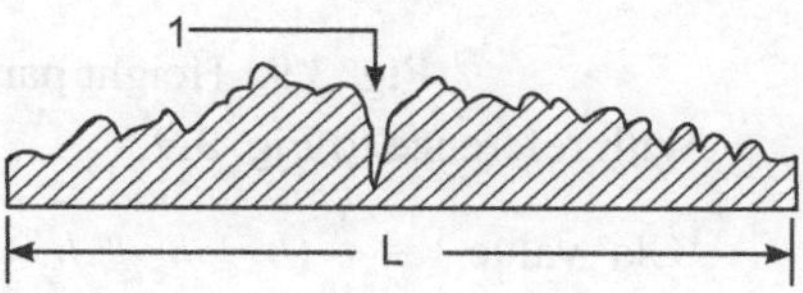

Fig. 3.8. Surface Flaws.

Measurement and Evaluation

There are different techniques of measurement and evaluation of surface roughness.

An ordinary microscope may provide a lot of information regarding the regularities, but it cannot show the nature of heights of the ripples or wave. The scanning electron microscope is also used in this regard, but their application is very much limited to the small specimens. The stylus method is very much popular to trace the profile of a cross-section of surface. Optical interferometers are also helpful to show contours and cross-sections of fine surfaces.

However, the range of heights from peaks to valley for different machining procedures are as follows:

Fine lapping	:	About 0.05 micron
Grinding	:	From 1 to 10 microns
Rough machining	:	up to 50 microns.

The surface roughness discussed above, is needed to be expressed for communication, specially in the drawings. The degree of irregularities can be expressed in several ways.

Such an expression, with respect to their statistical criteria, may be of two types:

(*a*) Root Mean Square Method. The method can be explained with the help of Fig. 3.7*a*. The line AB has been taken as a central reference line. Let the heights and depths of the irregularities be *a*, *b*, *c*,.........., and the number of measurements be '*n*' .

So, The R.M.S. value $= \dfrac{1}{n}\sqrt{a^2 + b^2 + c^2 + \ldots\ldots}$

'Cla' Value or 'Ra' Value. The most commonly used 'Cla' value is the average departures of height parameters from the mean line of the irregularities, on the specimen of the surface, under consideration.

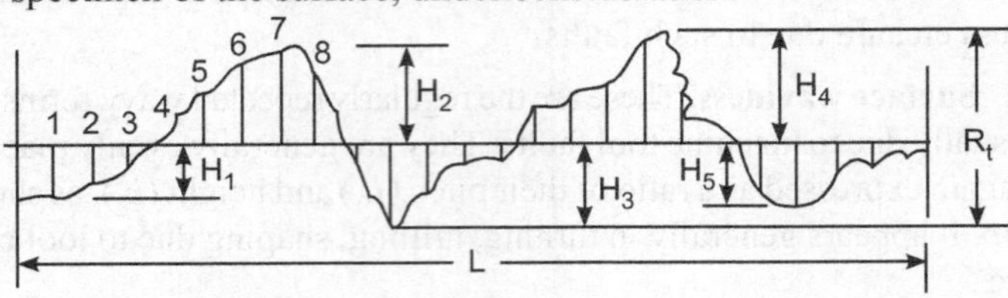

Fig. 3.9. Height parameters for Cla value.

With reference to Fig. 3.9,

$$\text{'Cla' value} = \frac{1}{n}(h_1 + h_2 + h_3 + \ldots\ldots h_n)$$

$$= \frac{\sum h}{n}$$

Again, it may be expressed as:

$$= \frac{H_1 + H_2 + H_3 \ldots\ldots\ldots}{L}$$

$$= \frac{\sum H}{L}$$

The 'Cla' value is commonly known as 'R_a; value.

(*c*) **Ten point height (R_z).** The height from the highest peak to the deepest valley is termed as 'R_t'. As, to recognise such heights or depths are very much chance-dependent and as such expression does not communicate the exact nature of the total surface, the general practice is to take average values of the five highest peaks and five deepest valleys and is termed as 'ten point height'. It is designated as 'R_z'.

Specification

Different values of R_a and R_z, represent different grades of surface finish. 14 classes are generally used as standards.

They are specified as a equilateral triangle '∇'. To express surface finish through 'R_a' value is the most common. ∇9 indicates that the surface should have "R_a" values not more than 0.32 micron, as specified in the available table. If the higher and lower limits, that is the maximum and minimum values of roughness are to be specified, it may be done specifying both the related numbers *e.g.* ∇8-9. So in this case, surface irregularities should be in the limit of 0.32 to 0.63 microns.

The "R_a" values for different classes are shown below:

Class of surface finish	1	2	3	4	5	6	7	8	9	10	11	12	13	14
"R_a" value in microns	80	40	20	10	5	2.5	1.25	0.63	0.32	0.16	0.08	0.04	0.02	0.01

Specification Methods

The surface finish can be specified generally in two methods. In the first method, the following symbols are used:

Symbol	Roughness (in microns)
~	2
∇	1–2
∇∇	0.4– 1
∇∇∇	Up to 0.4

Again the roughness can also be specified in the following method :
a where '*a*' indicates surface roughness in microns.

REVIEW QUESTIONS

1. Explain and illustrate the following terms :
   (*a*) Basic size,
   (*b*) Unilateral and bilateral tolerance,
   (*c*) Upper and lower deviation,
   (*d*) Maximum and Minimum clearance,
   (*e*) Lower limit and upper limit.

2. What are tolerance and fit ? What is an interchangeable system? State its advantages.

3. What is a fit? What are its classifications? Explain with sketches.

4. Explain with neat sketches the difference between 'Basic Hole System' and 'Basic Shaft System' of fits. Why is the Basic Hole System adopted in all modern systems of fits? Explain the symbol 50 H8/g7.

5. What is surface roughness? How can the surface roughness be measured and specified?

6. Fill up the gaps selecting the correct word:
   (*a*) Maximum allowance is the difference between the .... (smallest/largest) hole size and .... (smallest/largest) shaft size.
   (*b*) For an interference fit (smallest/largest) shaft diameter is greater than .... (smallest/largest) hole diameter.
   (*c*) Push fit is a type of (interference/transition) fit.
   (*d*) To denote deviation, (capital/small) letters are used for shafts.
   (*e*) For a hole basis system, different fits are obtained varying the (hole/shaft) diameter.
   (*j*) The zero line represents the .... (basic/actual) size.
   (*g*) The .... (R.M.S. /Cla) values are most commonly used to express surface roughness.

4

SEALS, PACKINGS AND GASKETS

Introduction

Seal is a device used to prevent or minimise leakages of fluids through any mechanical clearance. They are widely used both in static and in dynamic states. Packings or gaskets, the two types of seals are also used for the same purpose. The only difference is that the gaskets are used only to seal the static clearances while packings are used in the case of dynamic clearance also. A few of the wide applications of packings or gaskets may here be mentioned.

(*i*) Piston ring in gas, oil or steam engines

(*ii*) Gland packing on reciprocating shaft of pumps

(*iii*) Gasket below the cylinder head of an I.C. Engine etc.

The seals may be classified, considering their nature of performance, on the basic shape of the surface (cylindrical, spherical etc.) or on the type of motion (rotary, reciprocating etc). However a typical classification is given.

Static Seal (Gasket)

A gasket may be made of soft rubber, where water is being handled or it may even be of thin iron rings to use them under high pressure or at high temperature. Leather and asbestos are popular materials for gaskets. It may be remembered that gaskets of all sizes and shapes may not be available commercially. It is generally made in the plant itself. The hand-cut gasket as is shown in Fig. 4.1 is generally located nearly inside the bolts and is not extended beyond the flange diameter. The compressibility of the gasket material is commonly taken into consideration while selecting a gasket. The load, being offered by the bolts, must seat the gasket to eliminate the effects of flange-surface irregularities. The bolt holes in the gasket serves also the

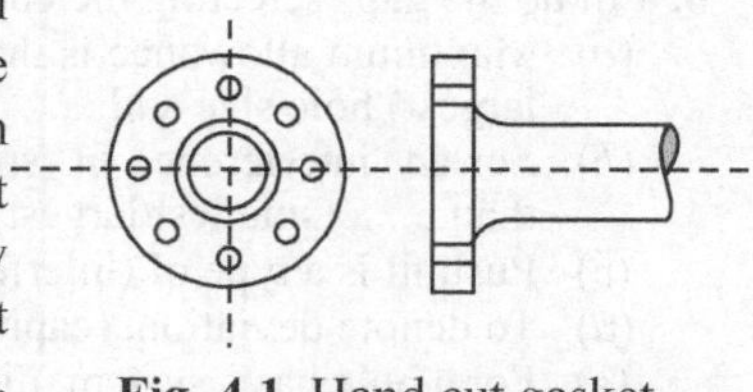

Fig. 4.1. Hand cut gasket.

purpose of locating the same. The best examples of metallic gaskets are the piston rings. The use of an 'O' ring gasket is shown in Fig. 4.2. The rings are extended above the groove surfaces to provide cross-sectional compression. They may be of rectangular or round in cross-section.

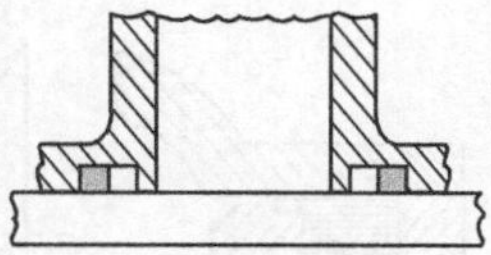

Fig. 4.2. O-ring gasket.

Classification of Seats

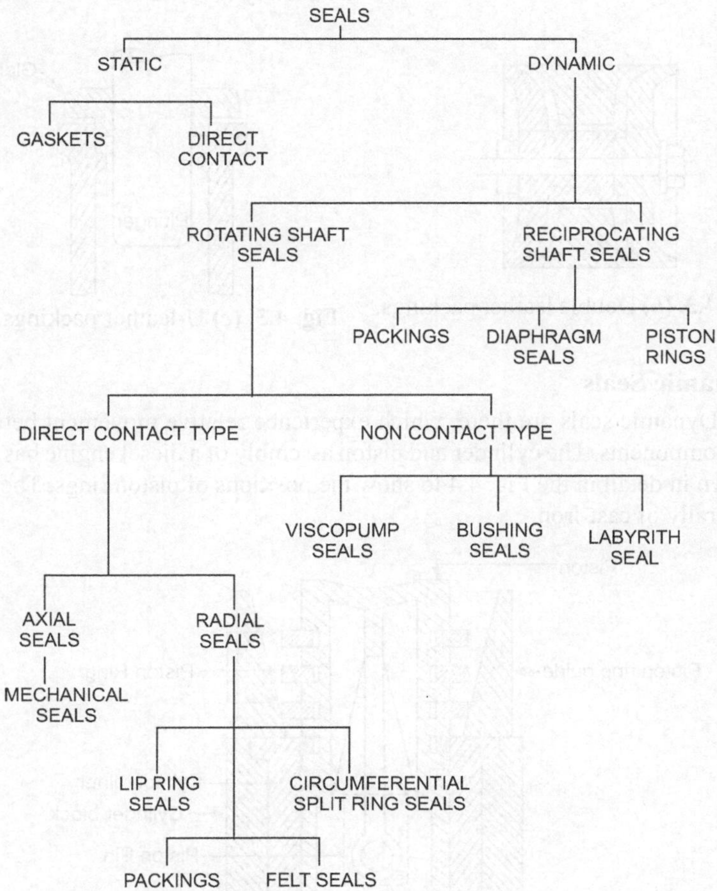

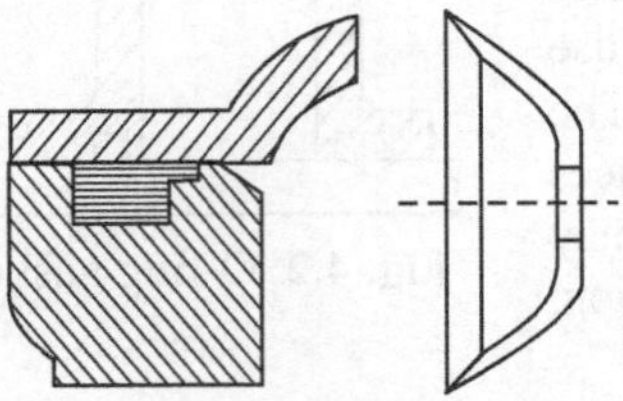

Fig. 4.3. (*a*) Cut gasket.

There are also gaskets, shown partially in the figure 4.3(*a*), which take its shape under pressure being tightened. They may be termed as cut gaskets. For heavy pressure, a double leather packing (Fig. 4.3*b*) is shown for a double acting pump. But for too high pressure 'U-leather' packings are most satisfactory (Fig. 4.3*c*).

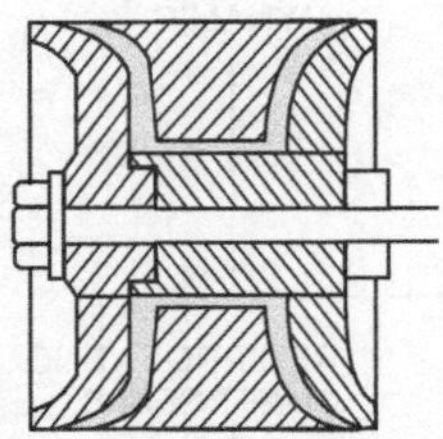

Fig. 4.3. (*b*) Double leather packings.

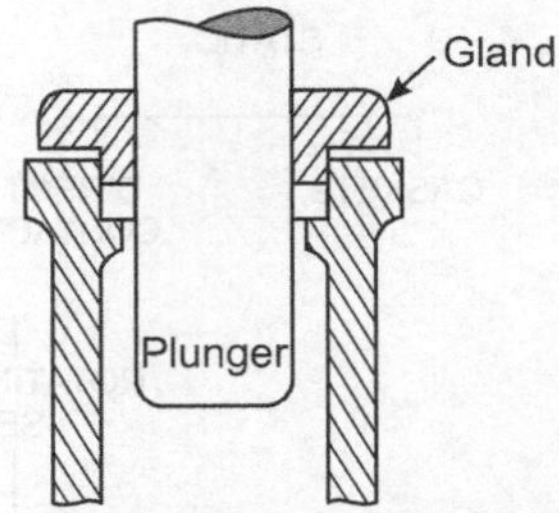

Fig. 4.3. (*c*) U-leather packings.

Dynamic Seals

Dynamic seals are there, which experience relative movement between the components. The cylinder and piston assembly of a diesel engine has been shown in detail in the Fig. 4.4 to show the positions of piston rings. They are generally of cast-iron.

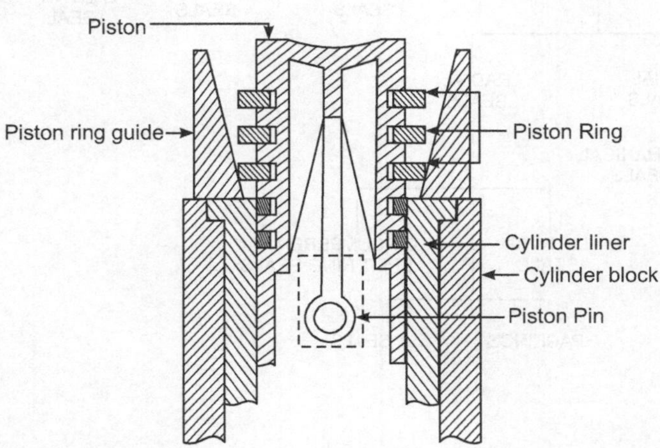

Fig. 4.4. Position of piston rings.

Split piston rings are widely used for gas, oil or steam engines. Dynamic seals carry the hydraulic loads and retain fluid under pressure where pressure is absent or too less, a spring load is applied. They require lubrication as well. Lubricants such as seal oils act as coolants and also serve the purpose of formation of a separating film.

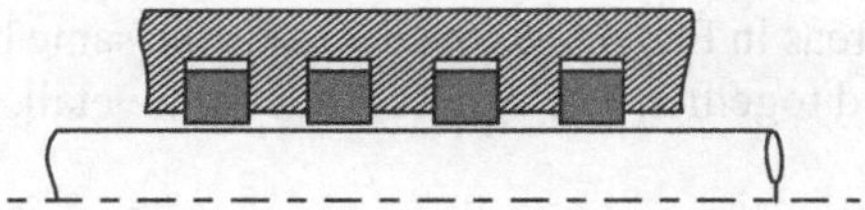

Fig. 4.5. Segmented carbon rings.

For very low pressure, carbon rings are used with a shaft. They are generally segmented carbon rings as shown in Fig. 4.5 (only one-half is shown).

When the component is having an extreme speed, a mechanical seal is used (Fig. 4.6). The seat ring is mounted on the shaft and the seal against the stationery seat, for an absolute control. Such an arrangement makes the real assembly balanced. The balancing is a vital criterion for proper functioning of seals.

Dynamic seal materials may be of different types, such as stainless steel, sprayed ceramic, Stellite, lead, bronze, etc.

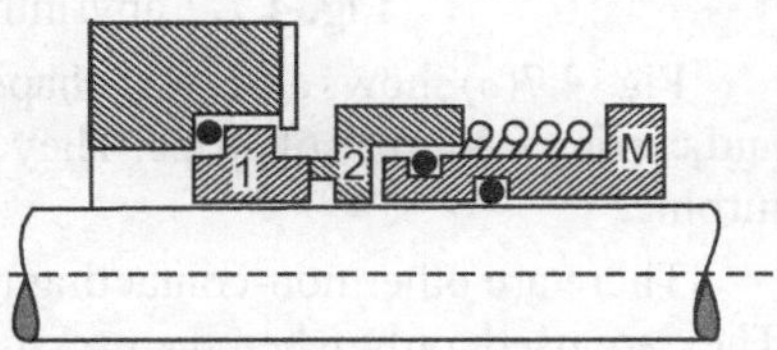

Fig. 4.6. Mechanical seal.

The selection of the face material of packing of a seal should be proper to have a longer life.

The mechanical seal shown in Fig. 4.6 may be explained in brief in this regard. Only one side is shown in the figure where (1) and (2) are rings.

Ring (1) is a 'stationary ring' and rests against the housing, whereas ring (2), a rotating ring is connected to a separate member M, which rotates with a shaft along with ring (2).

Rubbing between the two rings (1) and (2) cannot be avoided. So, suitable materials to be selected depending on the pressure and the diameter of the shaft. The two faces of the rings are to be suitably coated, so that, the body material may be the same, but the rubbing surfaces will be of dissimilar materials.

Labyrinth Seals

A special reference may be made for Labyrinth seal. They are used at extremely high speed but mainly to eliminate rubbing contact. It is widely used in turbines.

Some types of such seals are shown in Fig. 4.7. Fig. 4.7(*a*) shows four strips of brass, copper or steel directly fitted into the grooves in the bore of the diaphragm, whereas in Fig. 4.7(*b*) the design is the same but here two packing strips are caulked together. The ring is made to dovetail.

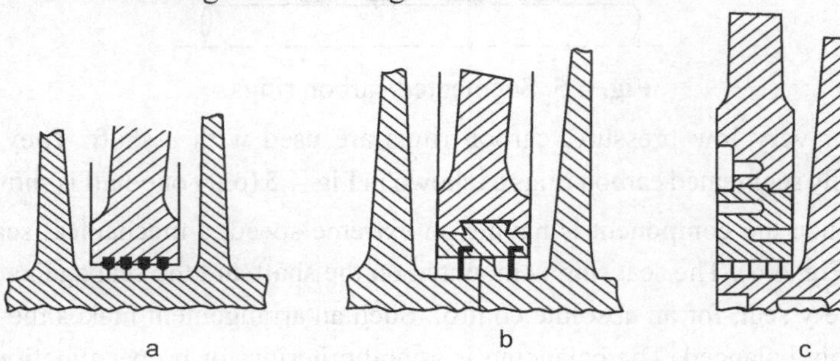

Fig. 4.7. Labyrinth Seals – Contact type

Fig. 4.7(*c*) shows a steel 'S' shaped ring, fitted in the bore of the diaphragm and carries a number of strips. They are mostly used in a multistage impulse turbine.

There are other non-contacting type of seals to provide a controlled gap. They are used, only when the pressure difference is not too much and where a little amount of leakage may be permitted.

Fig. 4.8(*a*) provides radial clearance. Fig. 4.8 (*b*) the axial clearance and the seal shown in Fig. 4.8(*c*) is with both radial and axial clearances

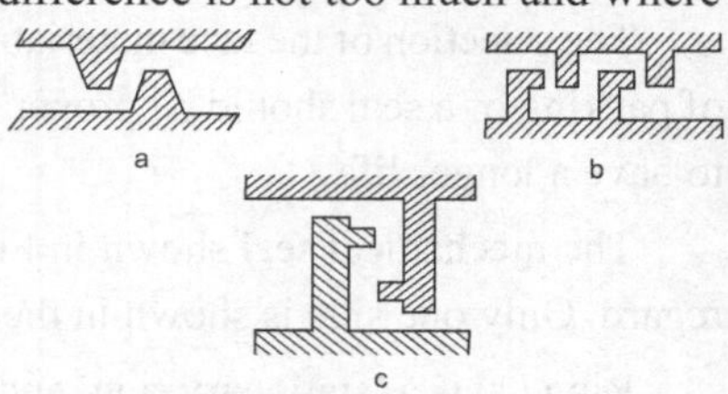

Fig. 4.8 Labyrinth Seals – Non-contacet type

Packing

To discuss packings separately, the usual packing position in a centrifugal pump is shown in the Fig. 4.9 as a typical example. It may be noted that a seal cage-at the end of the packing is mounted just next to an impeller to save the packings from abrasive liquids. On the inside of the stuffing box, in other arrangements, the packing com prises of generally three rings of square cross-section and then a grooved brass ring is used. This is called 'lantern ring'. It is again followed by 2 or 3 rings of square packing held in place by a bronze gland. The packing is subjected to water pressure through the lantern ring. This water also serves the purpose of lubrication. Grease can also be used where water is unsuitable.

Such packing should never be tightened too much. Undue tightening will cause rapid wear of the shaft sleeve. After putting the rings in position, their joints are made staggered 180 degree if the box is having two rings, 120 degree for three rings and 90 degree for four rings as shown in Fig. 4.10. However, a different design of packings may be used in a number of situations. But, always a close attention is to be attached to the packing materials. They should be replaced whenever they will be seemed to be hard.

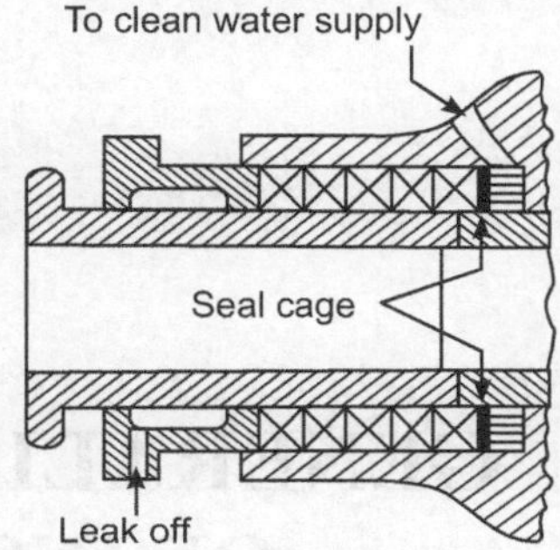

Fig. 4.9. Stuffing Box.

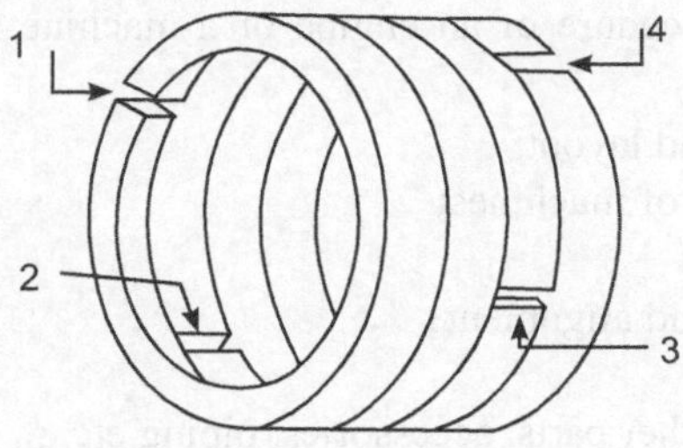

Fig. 4.10. Staggering of rings.

REVIEW QUESTIONS

1. What are seals? How are they classified?

2. What is the difference between the static and dynamic seals? State a few common examples of application of the packings and gaskets.

3. What is a labyrinth seal? Describe an example of the application of such a seal with sketches.

4. Describe the following with a neat sketch:

(*i*) Cut gasket

(*ii*) Split piston ring

(*iii*) U-leather packing

5. Correct the following sentences:

(*i*) Piston rings are rotating shaft seals.

(*ii*) Gasket is made of hard rubber.

(*iii*) Static seals are used in case of extreme speed.

(*iv*) Labyrinth seals cannot avoid rubbing contact.

(*v*) Non-contacting type of seals may be used where pressure difference is too much.

5

GENERALISED PROCEDURE OF INSTALLATION

Installation procedure of an engine or a machine involves a series of activities like –

(*a*) Location and layout;
(*b*) Positioning of machines;
(*c*) Foundation;
(*d*) Levelling and alignment;
(*e*) Grouting;
(*f*) Fitting of other parts, accessories, piping etc. ;
(*g*) Final levelling and test runs.

(*a*) Location and Layout

The location of an engine or machine depends on its purpose of installation, definite types and sizes and other special requirements, if any. As an instance, a centrifugal pump location should be as closer to the source of water as possible when suction lift is present, or for a compressor, location may be selected near the particular shop where maximum air will be consummed and at the same time a clear intake air will be available.

Moreover the location plan should permit to have required wide space all-around for continuing the erection work and should facilitate regular inspection, repairing etc.

The location should also be finalised in such a fashion that it will not disturb or obstruct the operation and maintenance work of other machines.

Once the location is finalised, the work of laying out the foundation plan is to be undertaken.

Laying out means marking of the foundation plan. It may be done with the help of chalk on a concrete floor and by a string with a number of pegs.

The general procedure is to indicate the outlines as per specification of the machine. The axis lines are to be drawn both longitudinally and crosswise to locate the centre of foundation.

Excavation of soil may be started only when the layout is completed as per the requirement.

(*b*) Positioning of machines

Positioning of the machine at the location is an important job, which deserves care, skill and an efficient teamwork.

An equipment may have the weight of a few tons. But it is to be loaded or unloaded, to be moved vertically or horizontally to bring it at the site and to place it on the foundation as well. Different types of lifting devices like pulley blocks, chain-hoists, overhead cranes etc. may be used as per the availability and requirement. When the load is light, rope pulley blocks are widely used and chain pulley blocks may be taken into consideration up to a load of one ton. When the load exceeds one ton, chain-hoists may be useful, but for a very heavy load, electric cranes are generally employed.

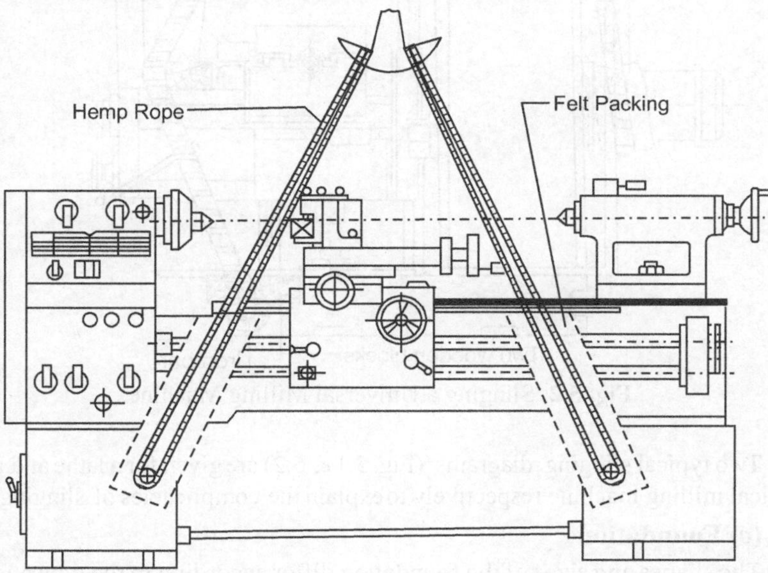

Fig. 5.1. Slinging a Centre Lathe.

However, the slinging should be done with much care to avoid any impact shock, to avoid formation of any scratches and breakages, etc.

For a machine of about 2 tons, hemp rope or Manila rope of diameter about 4 to 5 cm may be used instead of chains, wooden blocks and felts should be interposed between rope and the machine, to avoid damage to the finishing. Care must be taken to sling the machine upright with weight distributed evenly. Thus, the efficient fixing and passing of ropes or 'chains' through the machine structure deserves skill and experience.

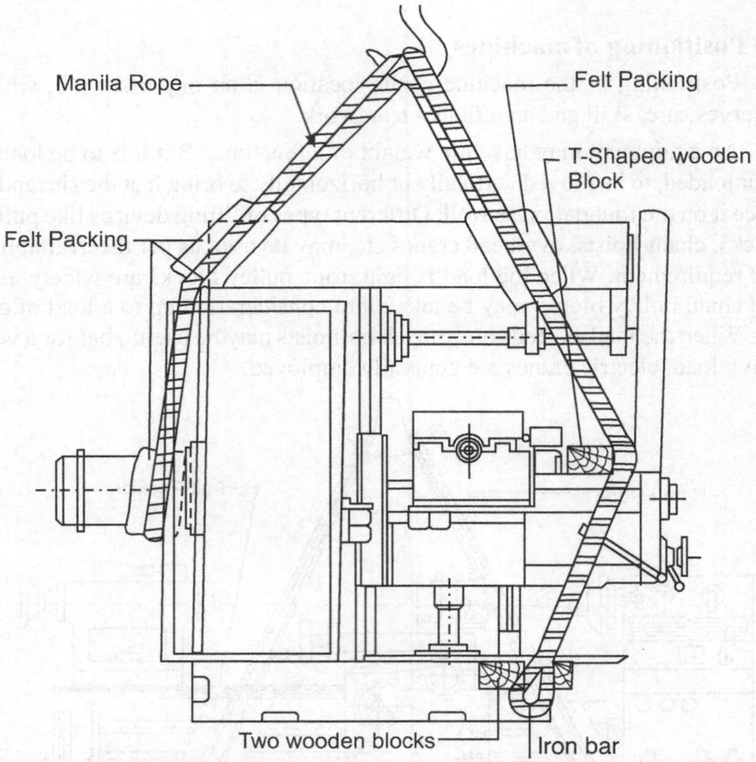

Fig. 5.2. Slinging a Universal Milling Machine.

Two typical slinging diagrams (Fig. 5.1 & 5.2) are given for a lathe and a vertical milling machine respectively to explain the complicacies of slinging.

(*c*) Foundation

The shapes and sizes of the foundation differ according to the type and size of the machines or engines. They are also dependent on the property of the subsoil and the dynamic loads of the machine during operation.

If the weight of the installed machine is not too much or if the dynamic loads are insignificant, the size of the foundation may be finalised on the basis of design considerations.

But when the dynamic loads predominate, the foundation should also serve the purpose to protect the machine from external vibration and to lower down the frequency of natural vibration by increasing its total mass.

Dynamic and Static Loads

For example, in metal-cutting machine tools, the static loads (which are equivalent to respective dynamic loads) can be calculated in the following manner:–

(*i*) For a rotary machine, the following empirical formula may be used to find out the equivalent static load (P_{static}) :

$$P_{static} = G/g\,.e.\,(\pi\, n/30)^2\,.K_{dynamic}$$

where G = Weight of rotating masses.

g = acceleration due to gravity.

e = eccentricity of rotating masses.

n = Speed of the work-piece.

and '$K_{dynamic}$' is the dynamic co-efficient generally taken as 1.5 to 2.

(*ii*) For a reciprocating machine, the general practice is to take the equivalent static force as five or six times the value of cutting force.

The dimensions of the foundation block may be determined, primarily finding out the weight of the foundation (W_F) with the following empirical formula:

$$W_F = K_F \cdot W_M$$

where, K_F is a factor, commonly taken as 2, 3 for the machine tools with dynamic load and W_M is the weight of the 'machine tool.' The height of the foundation

$$= \frac{\text{Weight of the foundation } (W_F)}{\text{Area of the base of foundation} \times \text{Specific weight of material}}$$

Of course, the height, so calculated should be sufficient enough to accommodate the foundation bolts etc.

The width and length of the foundation can be suitably calculated considering the foundation drawing, its height and weight.

The Indian standard recommendation is that, "As far as possible, the foundation should be so dimensioned that the resultant force due to the mass of the machine and the mass of the foundation passes through the centre of gravity of the basic contact area", especially for a rotary machine.

Other Design criteria of foundation

Besides the dimensions, as stated above, some other criteria are also considered for suitable foundation work.

(*i*) Ground Condition

The nature of soil is obviously a vital criterion.

For a hard soil or for a normal soil, the construction of a concrete bed does not entangle too many complicacies. It is rather most straightforward to consider the foundation plan supplied by the manufacturer.

But, for soft and loose soil, a large surface area with proper depth is needed for the foundation of a machine. The problems encountered due to vibration may force to construct a concrete raft (l) on piles (2), as shown in Fig. *5.3a.*

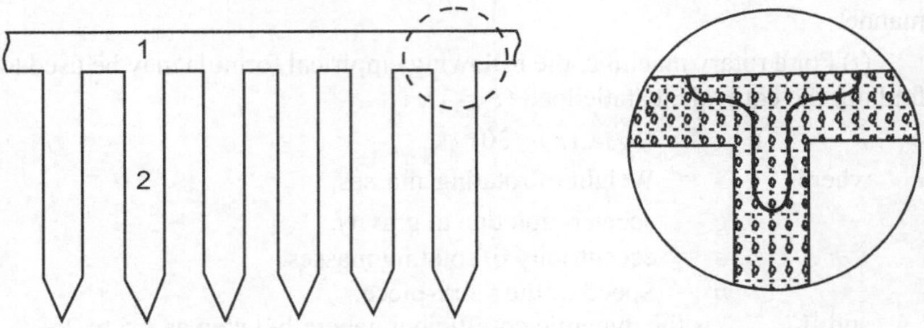

Fig. 5.3. (*a*) Raft on Piles, (*b*) Bondage of Raft with Pile.

The main structure of the machine or an engine is to be erected on the piles. The piles should have a stable and rigid bond with the raft as shown separately in the Fig. 5.3*b.* The bond may be a reinforced concrete having a steel-rod within it.

For water-logged soils also, mostly the pile driven raft construction is the means to avoid problems due to subsidence of soil.

(*ii*) Vibration Consideration

To avoid transmission of vibration to adjoining parts of buildings or other foundations, it is necessary to provide a suitable isolation between the equipment foundation and the adjoining structure. Usually a gap is maintained all around the foundation (like a trench), and is filled by sand to avoid such transmission of vibration. Any vibration isolating material, other than sand, such as rubber, lead sheet, felt etc. may also be used.

As a rule, the equipment foundation shall not be allowed to serve as support for other structures or for machineries not related to the particular equipment.

The impact type machines, like stamping press, drop and forge hammer, need special care during foundation. The depth of the foundation becomes very large to make the foundation heavy.

Two typical foundation blocks are shown in Fig. 5.4.

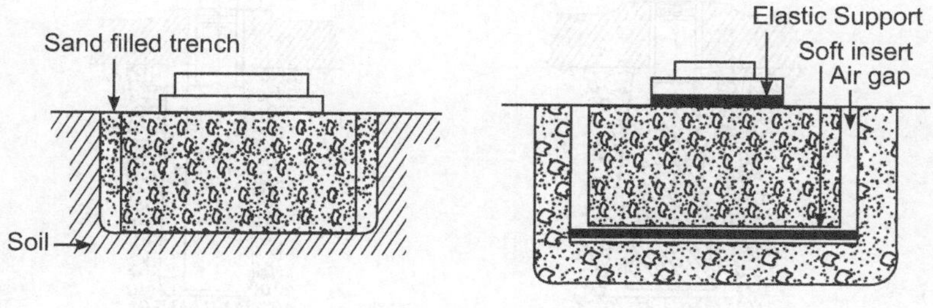

Fig. 5.4. (*a*) Sand-filled trench around foundation.
(*b*) Soft insert below foundation block.

Fig. 5.4*a* shows a sand-filled trench all-around the concrete foundation block. Such a foundation design reduces vibration transmission. Fig. 5.4*b* shows that the foundation block is placed on some soft inserts, below which the concrete (or reinforced concrete) base is there. The air gap is maintained all-around and at the same time there is an elastic support (to provide cushioning effect) in between the foundation block and the equipment. Such a design avoids vibration transmission and absorbs shock as well. The soft inserts may be cork or timber sleepers etc.

Foundation bolts

To install the machineries, foundation bolts will be specified and supplied by the manufacturers. Some of the foundation bolts become rigid on pouring concrete and some may be removable and adjustable bolts.

A few types of foundation bolts, very commonly used, are discussed below.

Eye Foundation bolt

They are generally made of mild steel or wrought-iron and can be readily forged according to the size. They are very much suitable to use for fastening the lids to tanks etc. (Fig. 5.5).

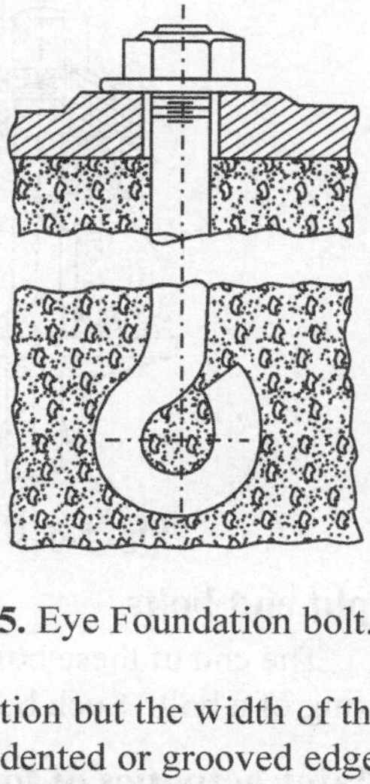

Fig. 5.5. Eye Foundation bolt.

Rag Bolt

Rag bolts (Fig. 5.6) are rectangular in cross-section but the width of the rectangle increases gradually. They are having an indented or grooved edge. After suspension of the bolts, the gap around them may be filled-up with molten lead. Such bolts cannot be removed easily after grounting.

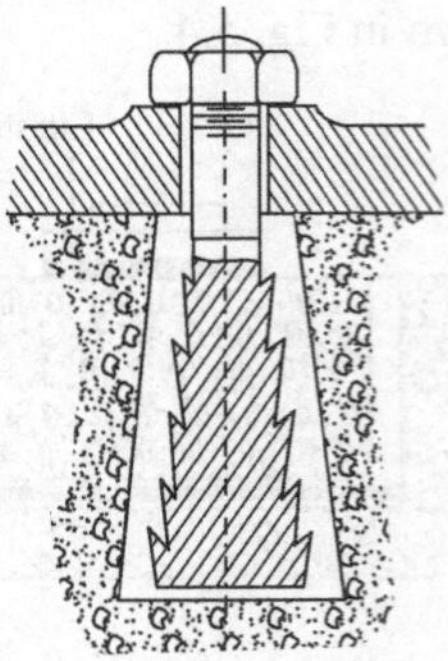

Fig. 5.6. Rag Bolt.

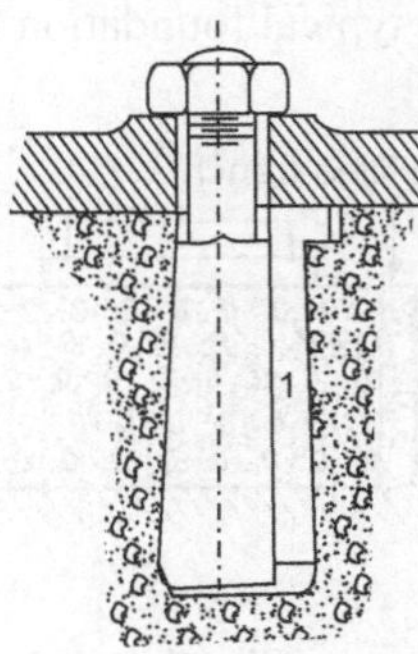

Fig. 5.7. Lewis Bolt.

Lewis Bolt

Such bolts can be removed easily, as a tapered key (l) is fitted through the side of the bolts to make it rigid. It may be noted that the key is fitted on the straight face of the bolt body. (Fig. 5.7)

Cotter Bolt

To install heavy-machines, 'cotter bolts' are commonly used (Fig. 5.8). A hand hole must be provided in the foundation block for the insertion of the cotter (1). The bolt possesses a rectangular slot to receive the cotter. A cast-iron washer (2) is also used as a bearing surface of the cotter.

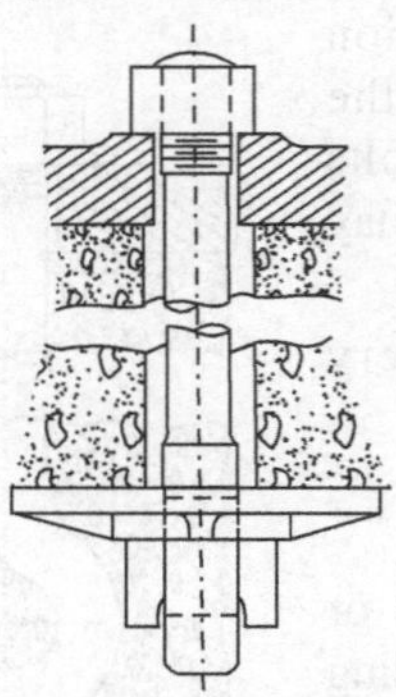

Fig. 5.8. Cotter Bolt.

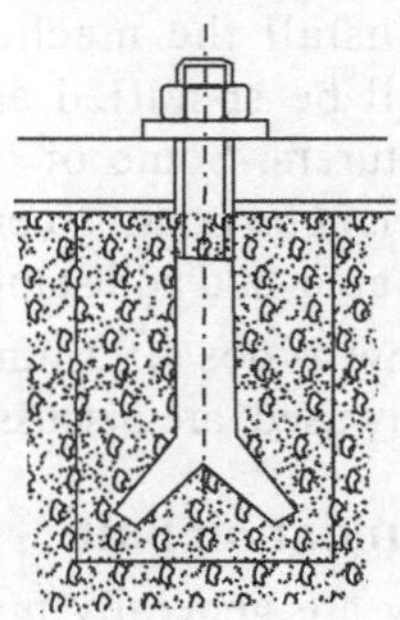

Fig. 5.9. Splitted End Bolts.

Split end bolts

The end of these bolts are made splitted. The bent split facilitates a rigid fixing of a bolt. Such bolts are mostly used for medium-size machine tools.

Major activities of foundation work

The foundation involves the activities like digging up the soil (excava tion), positioning of foundation bolts, concrete mixture pouring and setting.

Suppose the foundation plan of a 'centre lathe' is supplied by the manufacturer as shown in Fig. (5.10) with proper dimensioning.

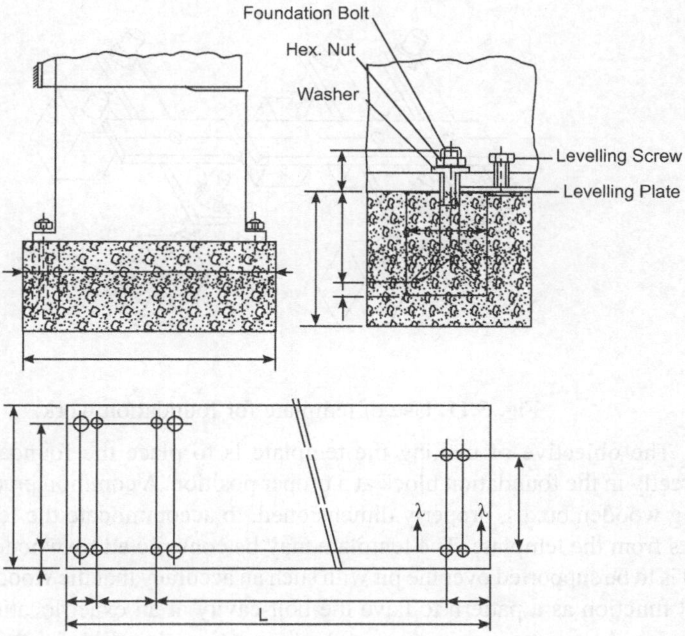

Fig. 5.10.

Excavation of the soil should be a bit larger at all sides than the planned foundation block. The tools required to have the pit may be listed like shovel, spade, crowbar, hand rammer etc.

Once the excavation is completed, a pad of rammed sand, broken stone or gravel is laid at the bottom to even-up the foundation footing. To have the final form of mould of the concrete foundation, smooth boards of wooden planks are erected on all vertical walls of the pit. After setting-up of the cement concrete these wall supports are to be taken out and the cavity thus formed, may be filled with sand etc.

Before pouring the concrete, provision for fixation of foundation bolts is to be made. A template, made during foundation work, as per the requirement, is generally used as shown in Fig. 5.11. with reference to the foundation plan shown in Fig. 5.10.

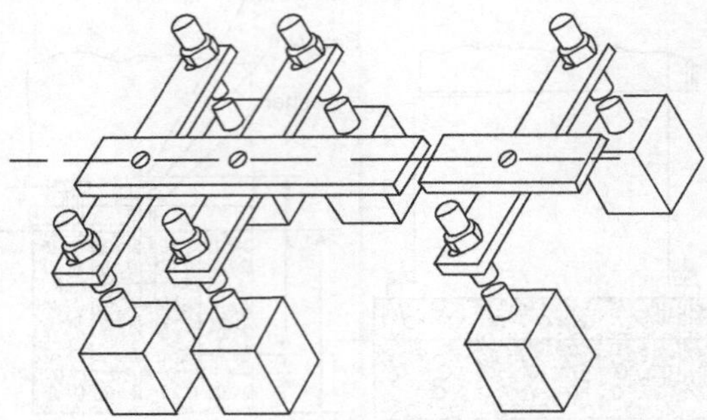

Fig. 5.11. Use of template for foundation work.

The objective of making the template is to place the foundation bolt 'directly in the foundation block at a proper position. A common practice is to hang wooden boxes, properly dimensioned, to accommodate the foundation bolts from the template. The template may be constructed by wooden planks and is to be supported over the pit with such an accuracy that the wooden boxes will function as a pattern to have the bolt-cavity at an exact location as per drawing. As per the requirement of the foundation plan (Fig. 5.10), six boxes are to be suspended from the template (Fig. 5.11). It is better to make the box sectional to facilitate its removal after concreting.

The concrete mixture at a definite proportion and quality will be made ready and is to be poured in such a way that there should not be any air-gap inside the block. For a heavy foundation block, mechanical vibrators may be used to remove the air pocket, if any within it. Before pouring, it is advisable to sprinkle water on all the surfaces of the pit to get them wetted. It may also be noted that the concrete is to be poured in horizontal layers without any interruption.

Sufficient time (depending on the shape and size of the block) should be allowed for the setting in of the block. Regular supervision on sprinkling water on the concrete surface is also essential at this period. The moistened surface will avoid to form any crack in the foundation block during the process of hardening.

The erection work is undertaken mostly when the foundation block has been set in and hardened.

The machine tool is placed on the foundation with the help of spacers or pads, levelling wedges, etc. Foundation plates are also supported similarly, for different engines or machines etc. At this time, a gap (minimum 50 to 70 mm) depending on the type of the machine is maintained between the top of the foundation block and the bottom of the machine or base plate.

The foundation bolts are positioned before positioning the machine and the exact location of the machine is guided by the insertion of the projected foundation-bolt ends through the holes, provided at the machine footings or base plate. The bolt-ends should remain sufficiently projected to accommodate the washers and nuts as well.

(*d*) Levelling and alignment

After having the machine on the foundation, the important job is to level and align it with other accessories. The levelling is performed with levelling wedges, shoes etc. as stated before.

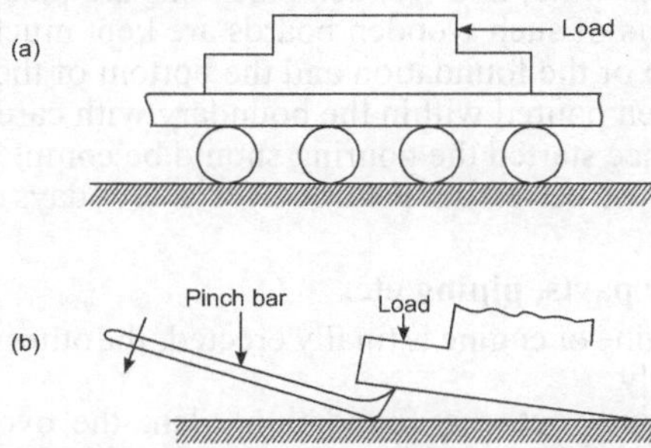

Fig. 5.12. (*a*) Rolling of Load, (*b*) Lifting of Load.

The horizontal and slight vertical movements of the heavy mass of machine is performed by pipes, rollers as shown in Fig. 5.12(*a*) or using pinch bars [Fig. 5.12(*b*)]. Straight edge, spirit level, dial indicator etc., are generally useful instruments to level the machine. The levelling is to be checked in the both, longitudinal and transverse direction.

The parallelity, perpendicularity of different faces, axes should be levelled and aligned, as are required. When levelling is completed, the foundation bolt cavity along with the bolt may be made concreted. Pouring of cement concrete is generally made through the gap provided at the top of the foundation, as shown in Fig. 5.13. This concreting will provide a bondage of the foundation bolts with the foundation block while performing grouting, discussed later. Whenever, the foundation block is continuously made for a connected accessory (say, the alternator with the turbine), the relative alignment should be established before concreting the foundation bolts.

(*e*) Grouting

Grouting is a procedure of connecting the machine with the foundation by a concrete mixture of plastic consistency or cement mortar. It is extensively

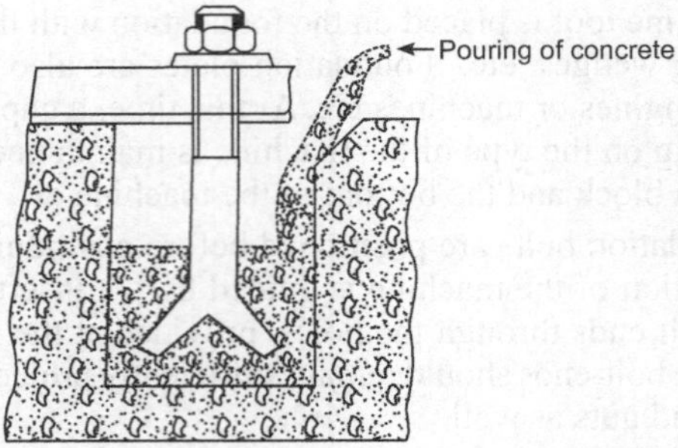

Fig. 5.13. Pouring concrete in Bolt cavity.

used in installing most of the machines. Generally, a quick setting cement is used to perform grouting. The top of the foundation block is made roughned, made moistened with water and wooden partitions are placed all-around the machine. The heights of such wooden boards are kept much higher than the gap between the top or the foundation and the bottom of the machine. Quick setting cement is then poured within the boundary with care to eliminate any air gap within it. Once started the pouring should be completed continuously and the machine must be left undisturbed for a few days after grouting to provide it time to set.

(*f*) Fitting of other parts, piping etc.

When the machine or engine is finally erected, the other accessories may be joined accordingly.

But, while laying out the foundation plan, the overall requirement should be kept in mind. The auxiliary structures *e.g.* in case of a heavy-duty diesel engine foundation, the structures for outer bearing pedestal, water pump blocks etc. should be planned at a time. This will minimise the problem of internal fittings.

(*g*) Final levelling and test runs

Accurate levelling can be carried out only when the grouting has set in after a few days. The machine is to be made cleaned and levelled then. Such levelling involves minor adjustments. Whenever provided, the levelling screws as shown in Fig. 5.10, may be operated to achieve the final level.

Everything should now be made ready to carry the test run. The style of testing will differ from machine to machine.

For instance, for metal-cutting machine tools, they are always sent out with a test chart.

'Test chart' may be defined as the specification of accuracies of major typical points like 'straightness of guideways', 'Trueness of spindles' etc. which should exist for the desired functioning of the respective machines.

The accuracy shown in the chart will be re-obtained only if the machine is correctly erected and levelled.

Two typical test charts, one for a 'general purpose lathe' and the other for a 'horizontal/universal milling machine' are given here. It is to be noted that both the machine tools are of definite specifications.

TEST CHART FOR GENERAL PURPOSE PARALLEL LATHES (LATHES WITH SWING OVER BED UP TO 800 MM)

Sl.No.	Figure (5.14–5.27)	Test Item	Measuring Instruments	Permissible Deviations
	2	3	4	5
1	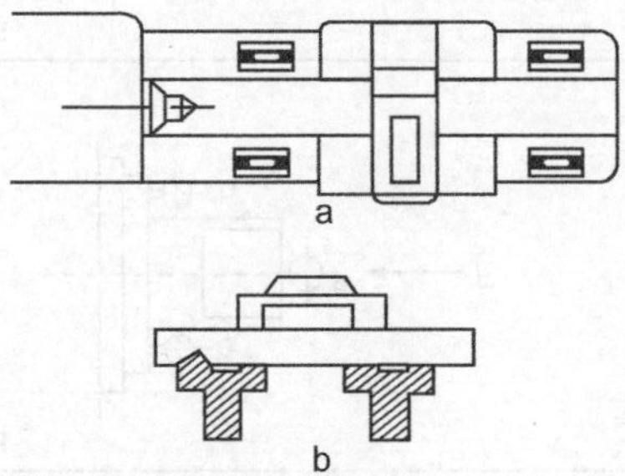	Straightness of carriage sideways: (*a*) In longitudinal direction: Straightness of sideways in vertical place	Precision level or any other optical or other instruments.	(*a*) DC < 500, 0.01 (Convex), 500 < DC < 1000, 0.02 (convex) Local tolerance 0.0075 for any length of 250 DC > 1000. For each supplementary length of 1000 fir DC beyond 1000, add the tolerance 0.01. Local tolerance 0.015 for any length of 500
		(*b*) In transverse direction	Precision level	(*b*) 0.04/1000
2	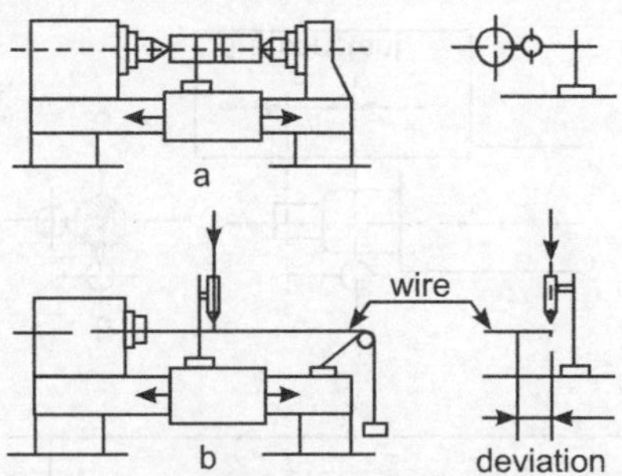	Straightness of carriage movement in a horizontal plane or possibly in a plane defined by the axis of the centres and the tool point.	(*a*) Dial gauge and test mandrel or straight edges with parallel faces between centres for DC < 1500. Length of mandrel to equal as close as possible the value of DC. (*b*) Taut wire and microscope or optical or any other instrument whatever be the value of DC.	DC < 500 0.015 500 < DC < 1000 0.02 DC > 1000 For each supplementary length of 1000, for DC beyond 1000, add 0.005 without exceeding the maximum deviation 0.03.

1	2	3	4	5
3	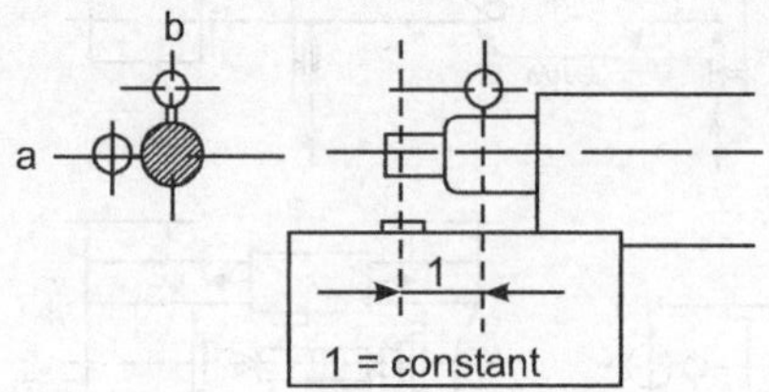	Parallelism of tailstock movement to the carriage movement: (*a*) in horizontal place (*b*) in vertical plane	Dial gauge	Both for (*a*) and (b) DC < 1500,0.03 with 0.02 for any length of 500, DC > 1500. 0.04 with 0.03 for any length of 500.
4	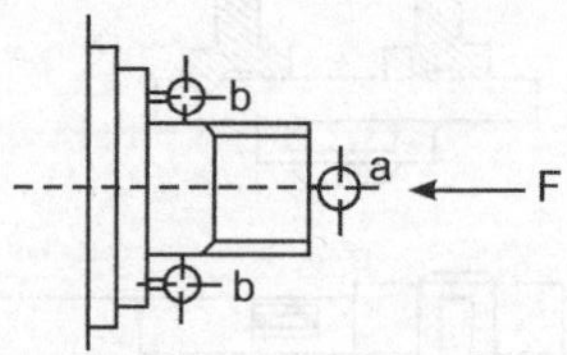	(*a*) Periodic axial slip (*b*) Camming of the face plate mounting surface	Dial guage F = Force = 0	(*a*) 0.01 (*b*) 0.02 including periodic axial slip
5	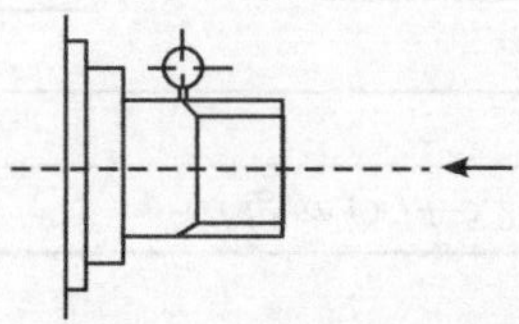	Renout of spindle nose centering sleeve or cone	Dial guage F = Force = 0	0.01

1	2	3	4	5
6	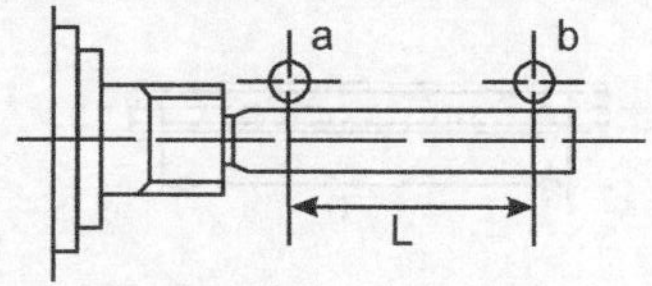	True running of the taper bore of the spindle: (*a*) near to the spindle nose (*b*) at a distance 1, where L = Da/2 or Max 300	Dial gauge and test mandrel	(*a*) 0.01 (*b*) 0.02 for 300
7	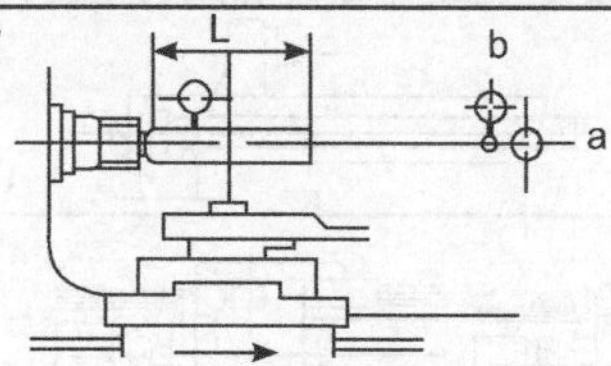	Parallelism of spindle axis to the carriage movement: (*a*) in horizontal plane (*b*) in vertical plane, where L = D.a/2 or Max 300	Dial gauge and test mandrel	(*a*) 0.015/300 free end of the mandrel inclinded towards the tool only. (*b*) 0.02/300 free end of the mandrel inclinded upwards only.
8	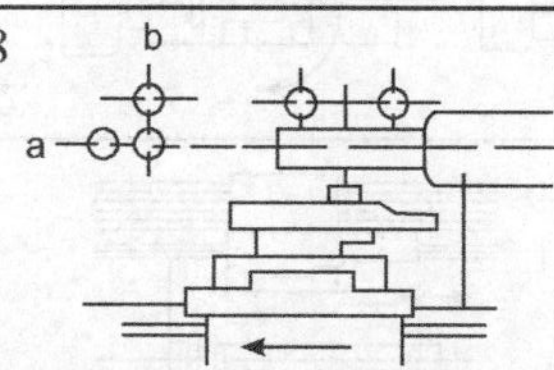	Parallelism of the external surface of a tailstock sleeve to the carriage movement (*a*) in horizontal plane (*b*) in vertical plane	Dial gauge	(*a*) 0.015/100 free end of the sleeve inclinded towards the tool only (*b*) 0.02/100 free end of the sleeve inclined upwards only.
9.	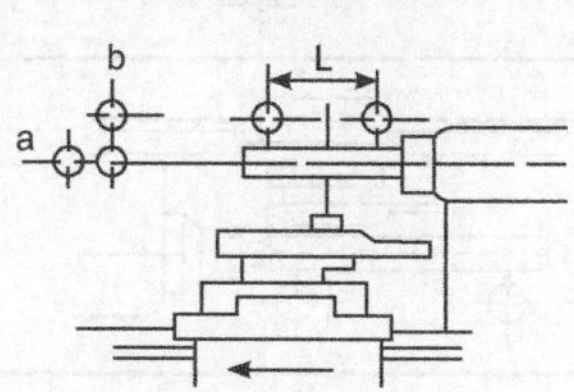	Parallelism of taper bore of tail stock sleeve to the carriage movement: (*a*) in horizontal plane (*b*) in vertical plane L = D.a/2 or Max 300	Dial gauge and test mandrel	(*a*) 0.03/300 free end of the mandrel inclined towards the tool only. (*b*) 0.03/300 free end of the mandrel inclined upwards only.

1	2	3	4	5
10	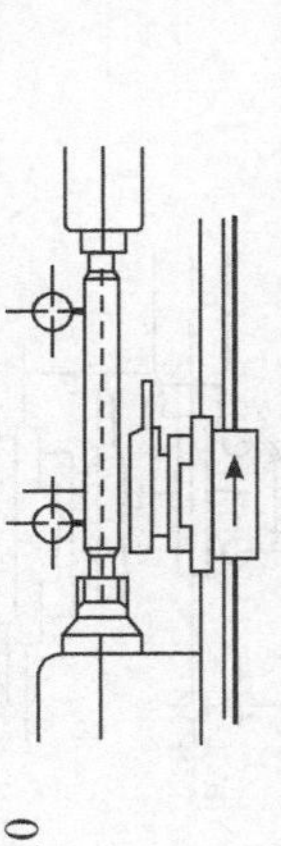	Difference in height between headstock and tailstock centres	Dial gauge test mandrel length of mandrel as close as possible to DC or a maximum of 1000	0.04 Tailstock centre higher than headstock centre.
11	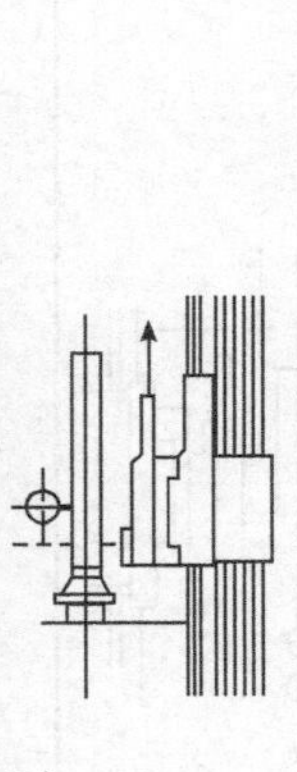	Parallelism of the longitudinal movement of the tool slide to the spindle axis.	Dial gauge and test mandrel	0.04/300 free end of the mandrel inclined upwards only.
12.	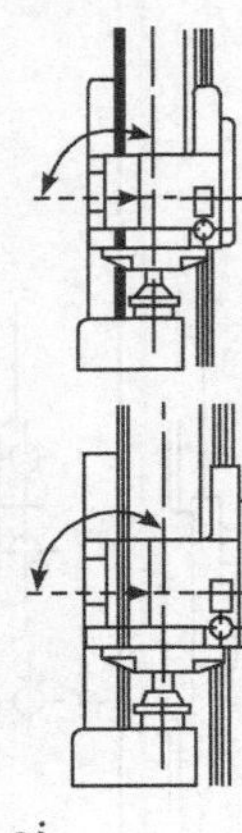	Squareness of the transverse movement of the cross slide to the spindle axis.	Dial gauge and flat ground disc or straight edge.	0.02/300 Direction of error ⊠ > 90°
13	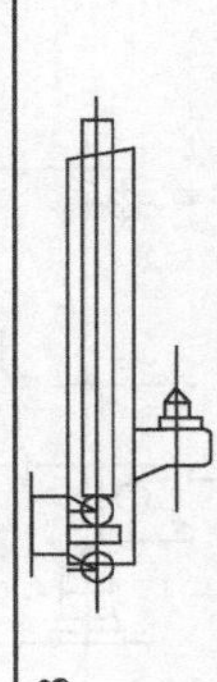	Axial slip	Dial gauge	0.015
14.	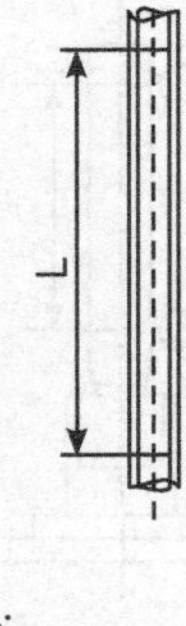	Accuracy of the pitch generated by the lead screw	Dial gauge and length bars	(*a*) 0.04 for any length of 300, for DC < 2000. For each supplementary length of 1000 for DC, beyond 2000. add 0.005 without exceeding maximum deviation 0.05 (*b*) 0.015 for any length of 50.

TEST CHART

FOR Horizontal/Universal Milling Machine

Sl.No.	Figure 5.28 to 5.42	Object	Permissible error
1	2	3	4
1	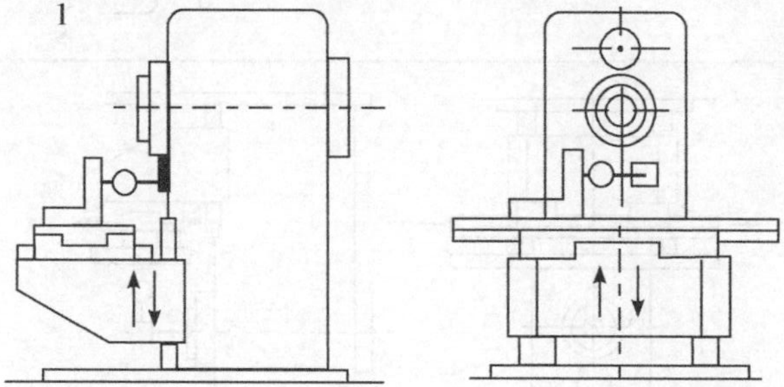	Squareness & straightness of the vertical movement of the knee to the table surface. (Knee not locked when taking measurements) (*a*) In the vertical plane of symmetry of the machine or transverse vertical plane. (*b*) In the plane perpendicular to the vertical plane of symmetry of the machine or traverse vertical plane.	(*a*) 0.02 for a measuring length of 300. (*b*) 0.02 for a measuring length of 300.
2	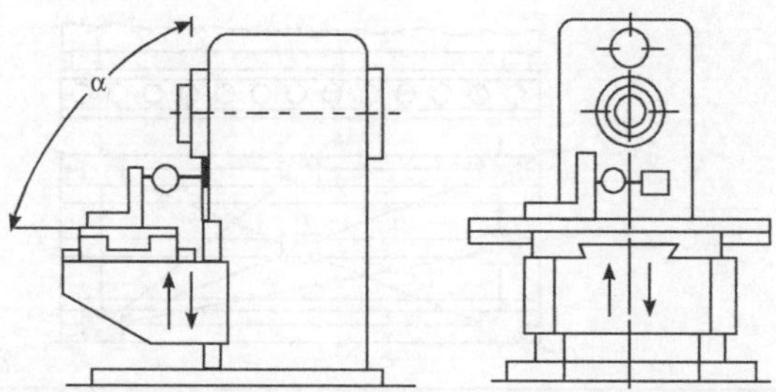	Squareness of the table surface to the column ways for knee (in three positions i.e. in the middle and near the extremities of the level), (Knee locked when taking measurements) (*a*) In the vertical plane of symmetry of the machine or transverse vertical plane. (*b*) In the plane perpendicular to the vertical plane of symmetry of the machine or transverse vertical plane.	(*a*) 0.02/300 with $\alpha \leq 90°$. (*b*) 0.02/300.

1	2	3	4
3	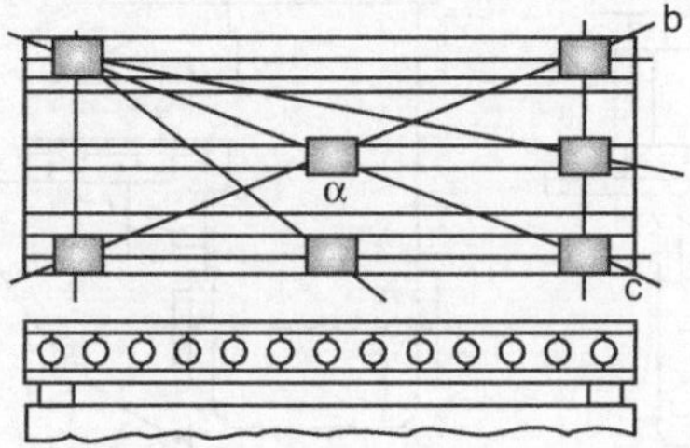	Flatness of table surface	0.04 up to 1000 Max. permissible = 0.05 Local tolerance 0.02 for any length of 300.
4	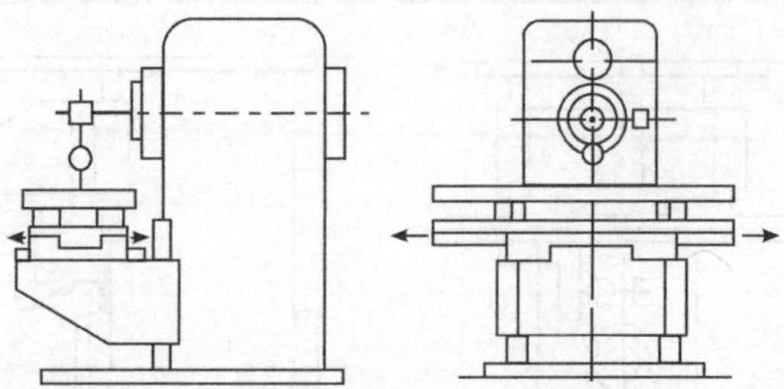	Parallelism of the table surface to its movement: (*a*) transversity (*b*) longitudinally	(*a*) 0.02 for any length of 300 (*b*) 0.02 for any length of 300 with a max. Permissible deviation of 0.05.
5	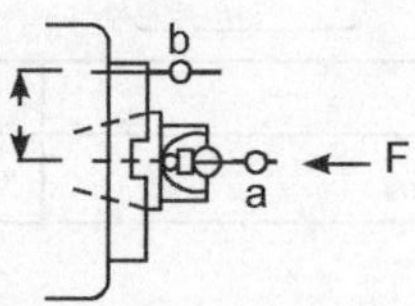	(*a*) Periodic axial slip (*b*) Axial runout (coming) of the face of the spindle nose.	(*a*) 0.01 (*b*) 0.02 including periodic axial slip.

1	2	3	4
6	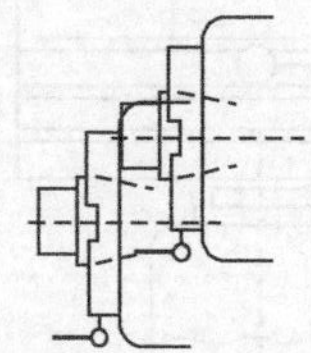	Run out of spindle-x nose centring surface.	0.01
7	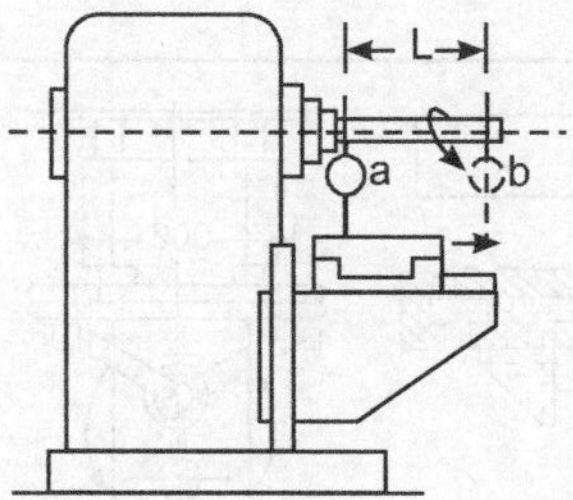	True running of the taper bore of the spindle: (*a*) Near to the spindle nose. (*b*) At a distance L when L = 300	(*a*) 0.01 (*b*) 0.02
8	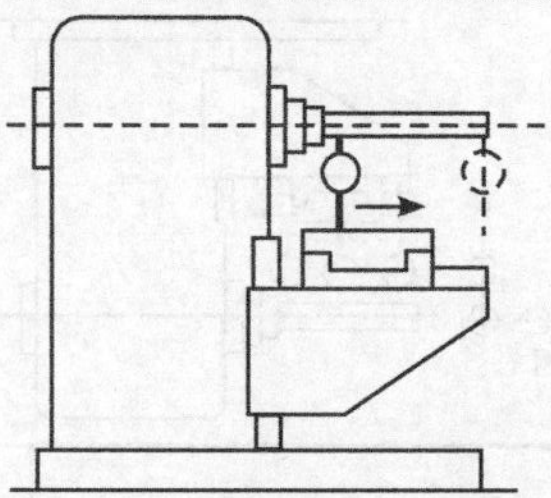	Parallelism of the spindle axis to the table surface.	0.025 for 300 (free end of the test mandrel inclinded downwards only).

1	2	3	4
9	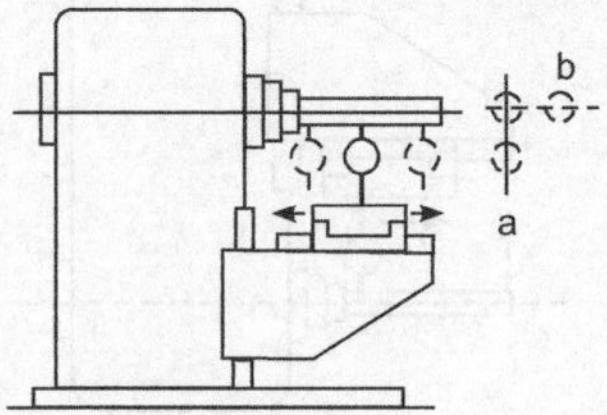	Parallelism of the spindle axis to the transverse movement of the table: (*a*) In the vertical plane (*b*) In the horizontal plane.	(*a*) 0.025 for 300 (free end of the test mandrel inclined downwards) (*b*) 0.025 for 300.
10	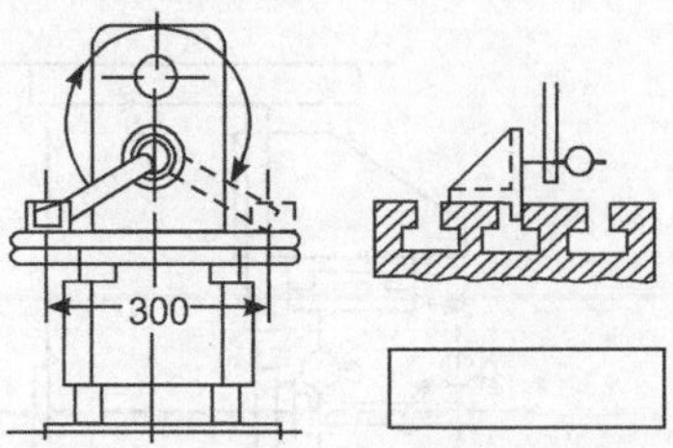	Squareness of the spindle axis to the median or reference T-Slot of the table.	0.02/300* *Distance between the two points touched.
11	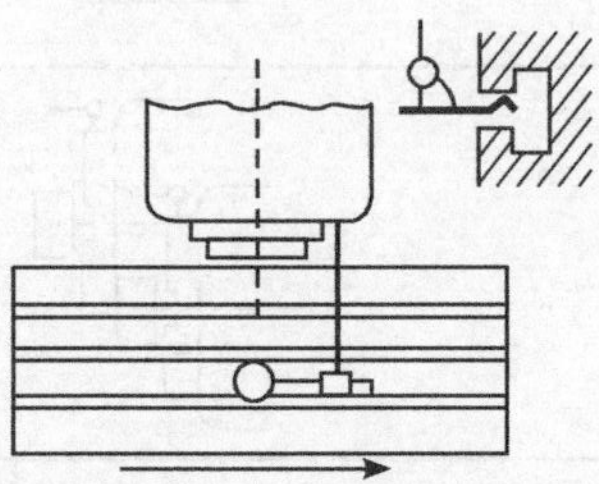	Parallelism of the median or reference T-Slot to the longitudinal movement of the table.	0.015 for any length of 300 with a max. permissible deviation of 0.04

1	2	3	4
12	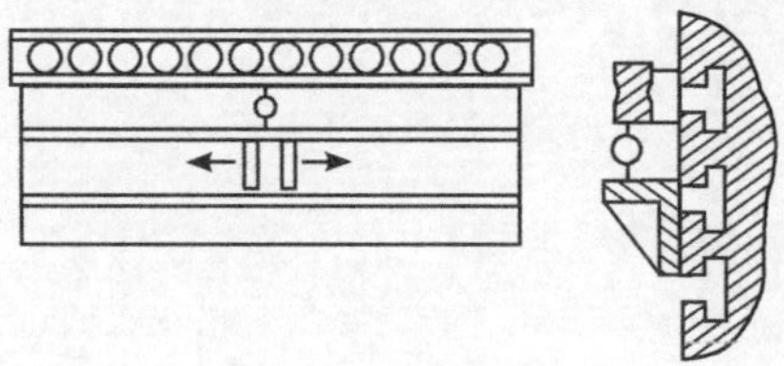	Straightness of the median of reference T-Slot of the table.	0.01 for any length of 500 with a max. permissible deviation of 0.03.
13	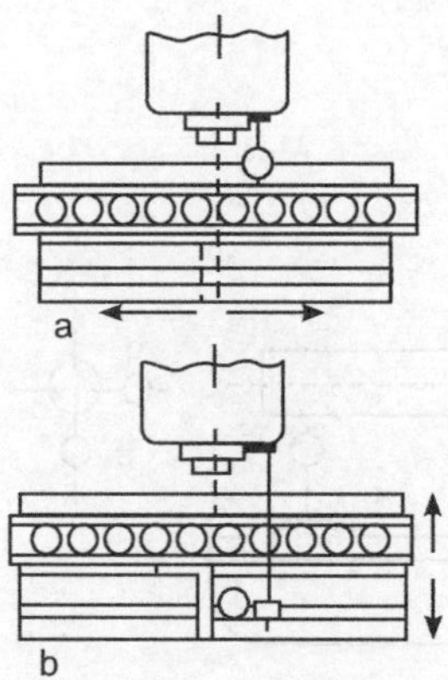	Squareness of the movement of the table transversely to its longitudinally movement.	0.02/300

1	2	3	4
14	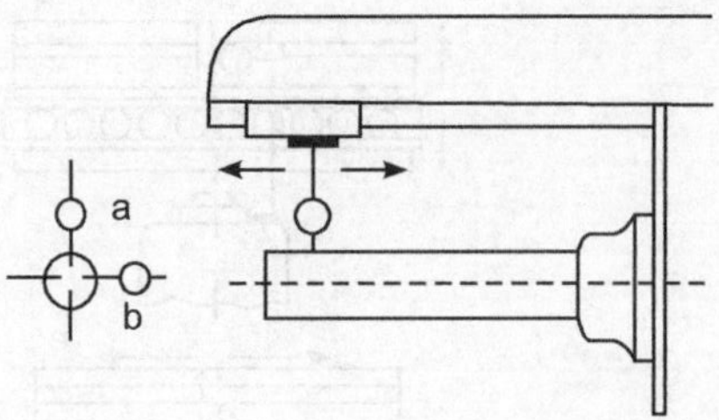 	Parallelism of arbor support guide on the over-arm to the spindle axis : (a) In the vertical plane (b) In the horizontal plane.	(a) 0.02 for 300 (overarm inclined downwards only) (b) 0.02 for 300

REVIEW QUESTIONS

1. State the operational steps to be followed in general, for installation of manufacturing machines in a workshop, along with equipments and materials generally needed for each step.
2. State the procedural steps for installation of an ordinary lathe. Sketch any three types of foundation bolts generally used for the purpose.
3. Explain with sketches the main features of a Technological Test Chart of a lathe.
4. Write short notes on:
   (*a*) Template for foundation work.
   (*b*) Levelling and alignment of a machine tool.
   (*c*) Procedure of grouting.
   (*d*) Vibration isolators.
   (*e*) Design criteria of foundation.
5. Select the appropriate word and rewrite the sentences :
   (*a*) Lewis bolts (can/cannot) be easily removed.
   (*b*) Cotter bolts are used for the installation of (heavy/light) machines.
   (*c*) The foundation-bolt-cavity may be made concrete (before/after) the levelling.
   (*d*) The top of a foundation block is made (roughened/smooth).
   (*e*) Piles are used for the installation in a (soft/hard) soil.
   (*f*) Straightness of flat surfaces are levelled with (Dial indicator/ Precision level).
6. Suggest possible means to avoid the following :
   (*a*) Subsidence of soil below the foundation block.
   (*b*) Transmission of vibration to the adjoining structure.
   (*c*) Wrong positioning of the foundation bolts.
   (*d*) Existence of an air gap in the grouting area.
   (*e*) Run out of spindle nose of a centre lathe during installation.

POWER TRANSMISSION DEVICES

Power is transmitted to a system through some devices. The reliability and efficiency of the device will ultimately maintain the continuity and requirements of a particular system. Proper mounting, alignment, maintenance of a power transmitting component are of major importance to establish a trouble-free situation.

However, the devices like keys, bearings, gears etc. are discussed in this chapter with respect to their general descriptions, mountings and maintenance criteria etc. A few repair techniques have also been suggested whenever it has been felt necessary.

KEYS

PINS:

Pin joints are frequently used for the purpose of fitting and fastening a lots of machine parts together. Pins used for this purpose may be of various types. The most common design is a cylindrical or a tapered pin.

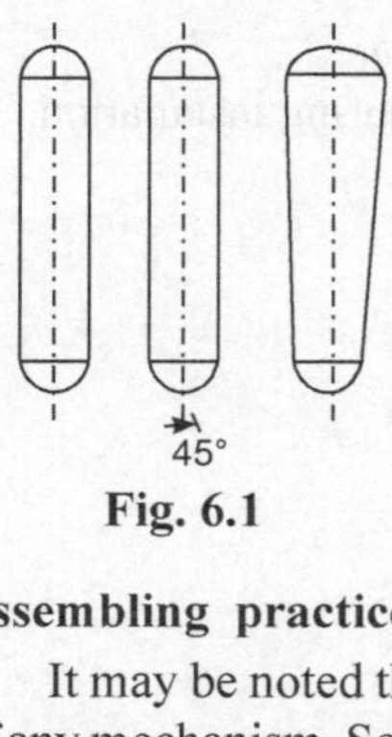

Fig. 6.1

The cylindrical pins may be having both ends rounded or otherwise it may be having one end rounded and the other end slightly conical. The taper of [Fig. 6.1] the tapered pin is mostly dependent on it's length and the purpose of it's use. A tapered pin is mostly favored for joining, as it offers a tight fit and it provides scope of reuse after repeated removal. But the cylindrical pins often get loose after a few reuse and it deserves to be replaced.

Assembling practice :

It may be noted that assembling of a pin is the last part of assembly work of any mechanism. So, any fault in it may create problem to the total assembly. Naturally utmost care is taken to perform the job.

The very first step is to assemble the parts to be joined and they may be clamped properly, maintaining the exact position of the parts on a drilling machine. The required hole is drilled in gradual steps. The reaming is then performed and the cylindrical pin may then be inserted into the cleaned hole with light hammer blows. It is better to get the pin slightly greased before insertion.

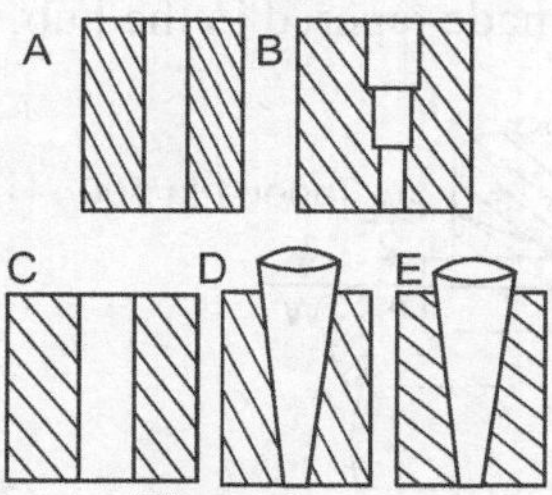

Fig. 6.2

For a tapered pin the tapered hole is drilled by enlarging the upper part (*b*) of a drilled hole (A) as shown in Fig. 6.2 and then the hole is reamed (*c*) by a reamer of proper size. The tapered pin is then assembled with slight pressure (D). The tapered pin is generally kept projected one third of it's length to enjoy the advantage to reuse it several times. (E).

Taking out pins :

If a pin is to be driven out from a joint, it is to be beaten up by a proper drift (A). The force is to be applied in the opposite side of the driven in end [Fig. 6.3]

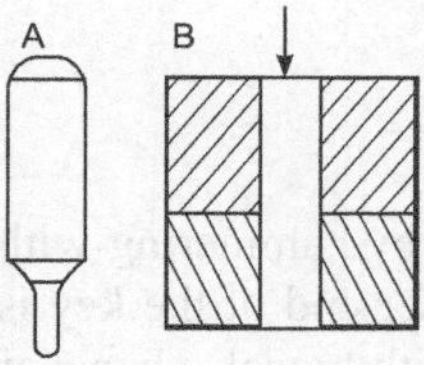

Fig. 6.3

It is easier to select the side to be beaten up for a tapered pin (*b*), but for a cylindrical pin, mild blows are applied to both ends of the pin to select the side, through which the pin can be easily taken out. 'Keys' connect two parts, *e.g.* a shaft with a pulley. No relative motion in between the parts are allowed to transmit power. They are generally subjected to shearing and torsional stresses. So keys are mostly made of steel.

A few common types of keys are discussed below :

Rectangular Taper Keys

Transmission of small forces are possible with such keys. It is tapered at the top to facilitate the sliding. The key (3) is placed in the flat keyway of the shaft (1) and the projected portion is connected with the hub (2) as shown in Fig. 6.4.

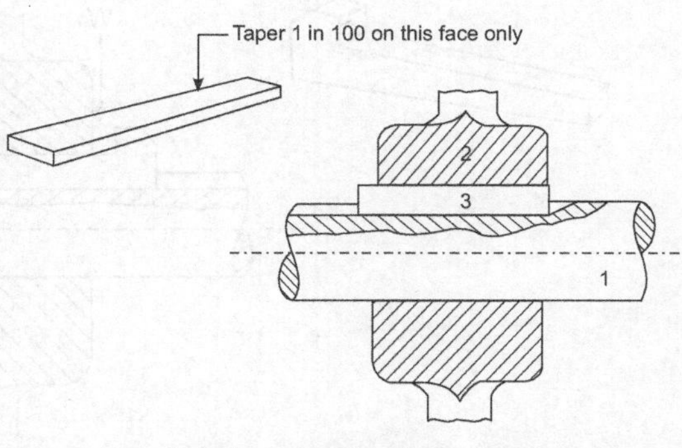

6.4. Rectangular taper key.

Sunk Taper Keys

To transmit high forces such keys may be used. Sunk taper keys may be of rectangular or square section and are used when neither of the sides of the key are accessible to the mating parts. It requires a tight fit. The keyway depth in the shaft is kept uniform whereas, it is made tapered in the hub.

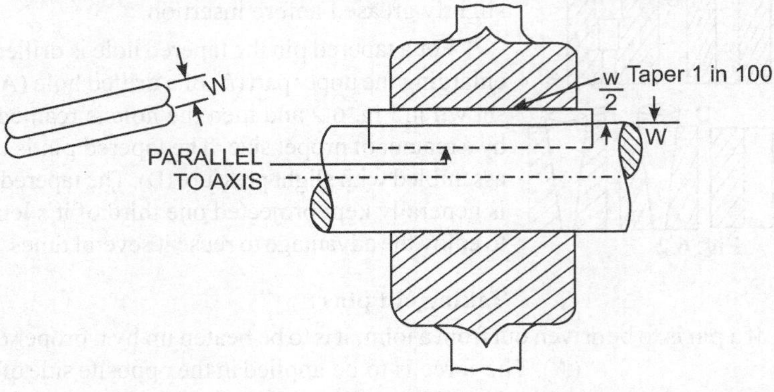

Fig. 6.5. Sunk taper key.

Gib-head keys

When a taper key is to be taken out, it is withdrawn by hammering with a drift from the smaller end of the key. But, when thinner end of the key is inaccessible, it is better to use gib-head keys for easy withdrawal whenever necessary, from the larger end, through which side, it was driven in. It can be taken out easily using a proper wedge at the position (*W*) shown in the Fig. 6.6. However, the projected gib, in such keys may create difficulties. A protective cap may be used over the gib wherever feasible.

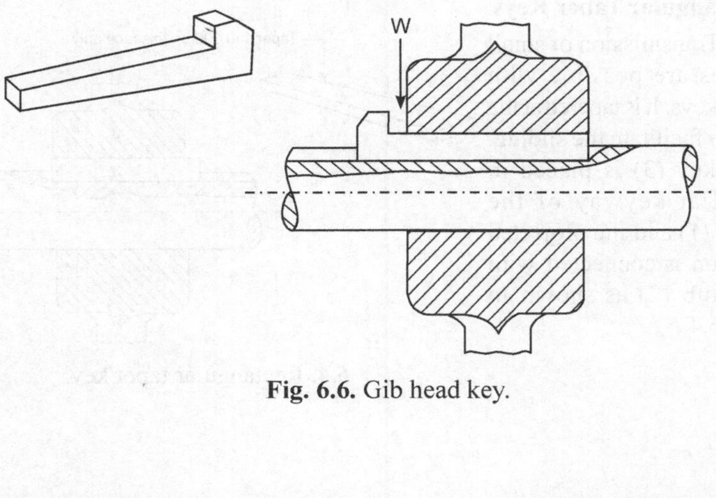

Fig. 6.6. Gib head key.

Feather Keys

Feather keys are also termed as plain parallel keys (Fig. 6.7). It is of uniform thickness and it allows sliding of axial movement of the mating parts while restricting the relative rotation. They are generally fitted with a firm push fit in the shaft and an easier push fit on the top and sides, to accommodate the hub.

They are sometimes fastened with the shaft or the hub with screw or setscrews, having countersunk heads. They may be of rectangular, or square or of dovetail cross-section.

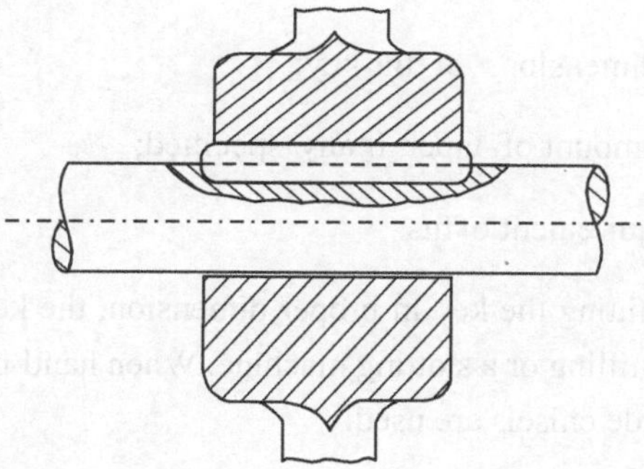

Fig. 6.7. Feather Key.

Woodruff keys

Woodruff keys are having the form of a segment of a circle (Fig. 6.8). They are capable of transmitting small and medium forces.

Of course, such keys facilitate the removal of pulleys from the shaft. They are never used as sliding keys. Slight rounding of the corners simplifies the withdrawal of keys from the shaft. It may be noted that shafts generally get weakned due to a greater depth of such keys.

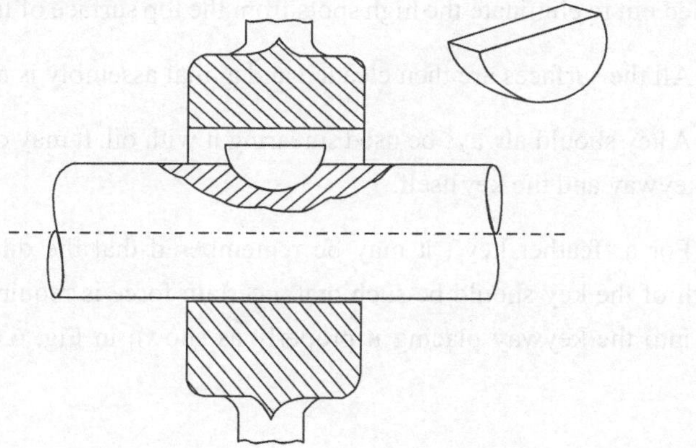

Fig. 6.8. Woodruff Keys.

Fitting of Keys

The basic features of matching different types of keys are almost same. The keys are generally machined oversize first. They are made to proper dimension by filing and by scraping.

The technical factors to be noticed carefully after the selection of keys are as follows:

(*a*) Exact dimensions of the key;

(*b*) Exact amount of taper, if any specified;

(*c*) The requirement of fits.

After machining the key in proper dimension, the key way is to be cut, desirably on a milling or a slotting machine. When hand-cutting is necessary, cross-cut and side chisels are used.

The sharp corner of the keys are removed. The inner edges of the keyways are also smoothed-off to avoid formation of any crack.

The key is then tried in the keyway with a coat of chalk or an ink. The high spots are identified and they are either filed or scrapped off. The test is repeated until a smooth surface is obtained.

Placing the key in the keyway, the hub is assembled, and similar tests are carried out to eliminate the high spots from the top surface of the keys.

All the surfaces are then cleaned and a final assembly is made.

A key should always be used smearing it with oil. It may otherwise score the keyway and the key itself.

For a 'feather key', it may be remembered that the dimension of the width of the key should be such that a certain force is required to press the key into the keyway placing it properly as shown in Fig. 6.9. For fitting a

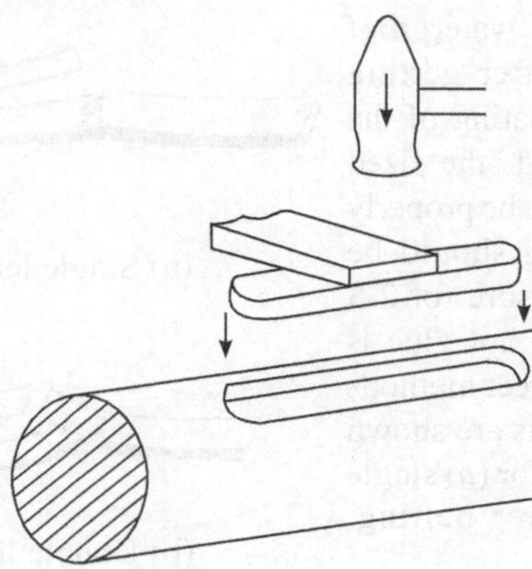

Fig. 6.9. Fitting of a feather key.

'woodruff key, it may be mentioned that the side faces should be machined first. The semicircular face may then be produced with the help of a radius gauge.

In the case of an assembly of sliding feather keys, controlling guide forks are used to maintain a certain position of the 'belt pulley' or any other rotating body' on the sliding feather keys.

BELT DRIVES

Belt and pulley drive is extensively used in industry for transmission of power. We will concentrate on both flat and V-belt drives.

Flat belts

The flat belts may be differentiated as follows with respect to the material of belts:

(*i*) Non waterproof oak leather.

(*ii*) Waterproof oak leather.

(*iii*) Mineral retained leather.

(*iv*) Combination of oak and mineral retained leather.

(*v*) Friction surface rubber belts and

(*vi*) Cotton belts.

Belt Joining

Belt Joining may be (*a*) Endless, (*b*) Laced or (*c*) Metallic.

Endless Method

Endless method is considered to be very much efficient. The leather belts offer the most favourable conditions for making them endless because it is of multiple joints. The ends are to be scrapped down to a thin edge. Primarily, they are to be sized with a waterproof Pyroxylin cement. After getting them dried, another coating of the cement may be applied the sized surface. Then they are to be properly a placed together and should be finally kept under pressure for 2-3 hours. Belt-joining by hot glue is also very popular. Correct methods of making endless joints are shown separately in Fig. 6.10 for (*a*) single and (*b*) double leather belting.

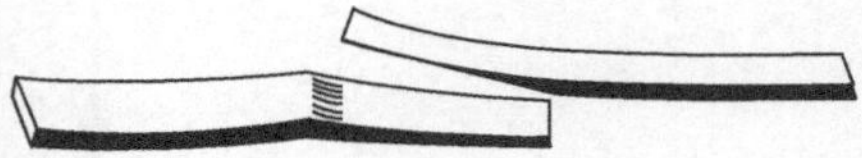

(a) Single leather belt joint.

(b) Double leather belt joint.

Fig. 6.10

Lace Joint

The laced method employed rawhide or wire. Rawhide lacing is applied manually in different design (one of which is only shown in Fig. 6.11) and the wire lacing may also be performed by machine.

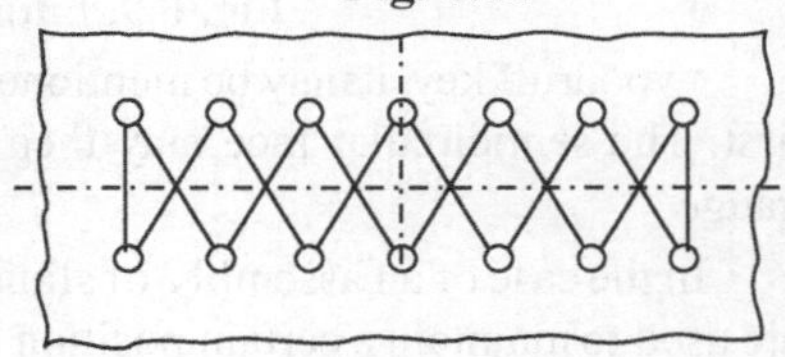

Fig. 6.11. Lacing of belt

Metallic Joints

Wire hook lacings are easy to apply by a portable or a stationary machine. They are pressed into the ends of belts so as to form a series of loops. Loops are meshed together and either rawhide or fibre pins are inserted. Hooks are generally of plain steel, stainless steel or monel metal.

Installation of Belts

Belts are to be installed naturally on pulleys which are mounted on the shaft. Installation procedure deserves perfectness of the total system. Moreover, the selection of belt, its width, its thickness are also very important. The shafts should be checked so as to be sure that they are running parallel and without any eccentricity. The squareness of the pulley may be checked by the help of a try square (1) as shown in Fig. 6.12 and then the alignment of the pulley may be checked. This can easily be done with the help of a string (2).

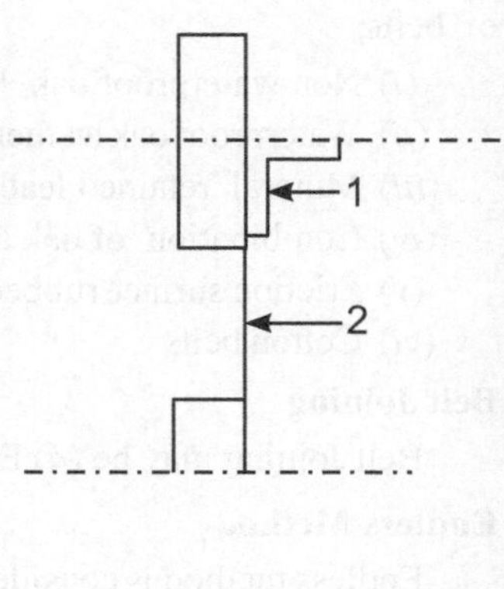

Fig. 6.12. Checking squarencess of pulley

Joining the belts, which has already been discussed before, is also a part of belt installation. The tension of the belt should be adjusted so as to transmit power without slippage. A too tight belt drive may spoil bearing.

Maintenance of belting

Systematic care and attention are absolutely necessary to maintain the efficiency of the operation. Some important points are as follows:

(*a*) Belting should be maintained in proper physical condition.

(*b*) Belting requires dressing or lubrication. The dressing should be of suitable character to accomplish the desirable results. The driving face of a belt should be maintained in a mastic state. This will enable the belt to run stickerandwilllengthen itsservice life.Amixtureofgoodtallow andcodliver oil or whale oil is used. A dressing compound with some curing ingredients may also be used. Mineral and vegetable oil should not be used. Rubber or a canvas belt do not require dressing.

(*c*) The belt surface should always be kept clean from grease or machine oil.

Selection of type of belts

To minimise the maintenance work the belting types may be selected during installation, paying positive weightage as per the following table:

Belting type	*Acid spillage*	*Acid fumes*	*Abrasive dusts*	*Steam and moisture*	*Temperature up to 140°F*	*Temperature up to 300°F*	*Oil and grease*
1	2	3	4	5	6	7	8
(*a*) Non-waterproof Oak leather	No	No	Yes	No	No	No	Yes
(*b*) Waterproof Oak Leather	No	No	Yes	Yes	Yes	No	Yes
(*c*) Mineral retanned leather	No	Yes	Yes	Yes	Yes	No	Yes
(*d*) Combination tannage leather	No	No	Yes	Yes	Yes	No	Yes
(*e*) Friction surface rubber	No	No	Yes	Yes	Yes	No	No
(*f*) Rubber covered friction duck	No	Yes	No	Yes	Yes	No	No
(*g*) Canvas	No	No	Yes	Yes	Yes	No	No

However, it may be noted that all the belts are suitable for normal atmospheric conditions.

V -Belt Drives

V-belt drives are widely used both for heavy and light duty operation.

Generally for heavy duty application, that is for industrial uses, when frequent heavy starting and stopping or some other shock loads are involved, multiple V-belt drives are installed. For a fractional horse power a single belt may also be used.

V-belts are of trapezoidal cross-sections and are made of chord and fabric impregnated with rubber. The chords may be of cotton or synthetic or of steel.

While selecting V-belts the following factors are taken into consideration:

(*a*) The H.P. required to be transmitted.

(*b*) Speed of smaller pulley.

(*c*) The difference between outside and pitch diameter of pulleys.

Besides the above factors, difference between the inside and outside length of the belt is also to be taken into consideration.

While selecting the belts, different service factors, suggested by the manufacturer, must be taken into account to ensure adequate capacity to offer long and continuous service.

Installation

(*i*) Select matched set of V-belts.

(*ii*) V-belts are never stretched or are forced into the grooves to avoid the rapture of an outside jacket or the inside chords or of both.

(*iii*) The prime mover is better to move towards the machine to be driven. The belts are placed in the pulley goroves and then the prime mover may be moved away to apply tension.

(*iv*) A V-belt may ride up to the top surface of the pulley groove. The incorrect installation procedures and a correct one are shown in

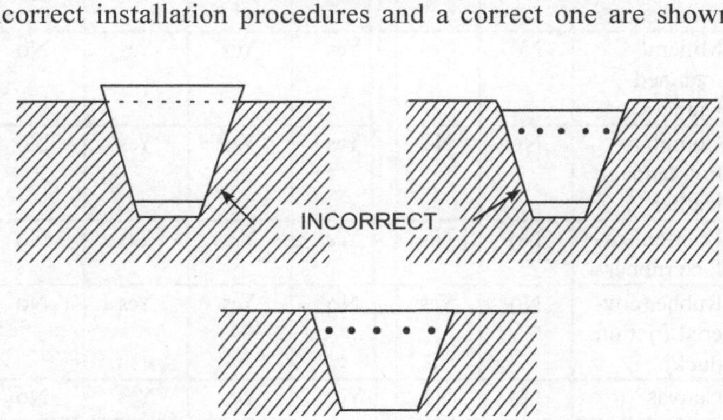

Fig. 6.13. Installation of V-belts

Fig. 6.13. Definite clearance should be provided between the base of the pulley groove and the base of the belt.

(*v*) If a idler is to be used to adjust belt tension, it should always be used on the inside of the belt.

(*vi*) The correct tensioning cannot be made standard. However the minimum tension, at which the belt does not slip with maximum operating load, is always preferable.

Maintenance

(*i*) Entry of any metal particles, sawdust etc. in the belt drive should be avoided to restrict the abrasion of belt. The belt drive may be made covered at the top and at the side for this purpose.

(*ii*) Excessive load which results in excessive stretch on the belt, should always be avoided.

(*iii*) V-belts should be preserved free from conditions, injurious to the rubber component.

(*iv*) Slippage if any in the drive, should be checked periodically to ensure the transmission of rated load.

(*v*) Belts of different bands should not be mixed. When replacing belts, total set of belts is to be replaced. New and used belts should not run together.

(*vi*) The belt wear results from its contact with the walls of the pulley groove. So, a periodic checking of grooves with templates is essential. The grooves should also be inspected for burrs, rough spots, dust, dirt etc.

CHAIN DRIVE

A positive transmission of power by the help of chains is very much common, specially for light and medium drives. Chain drive consists of endless chain links over sprockets, fitted on both the driving and driven shafts. The chain links are generally made of alloy steel. Mostly, two types of chains are used as standard chain drive.

(*a*) Roller chain drive; and (*b*) Silent chain drive.

Roller Chain Drives

Looking at the chains, it may be seen that the set of rollers are connected with the inner links (Fig. 6.14*a*). The Fig. (6.14*b*) describes the assembly. The link pin (4) is mounted tightly with the outer links (2), and the bushings (3) are fitted on the inner links (1) in such a way that the roller (5) can rotate round the pain.

While making assembly, two sets of rollers and bushings are placed in the inner link and it is connected with the other set of rollers by the help of the outer rings thorugh the pins.

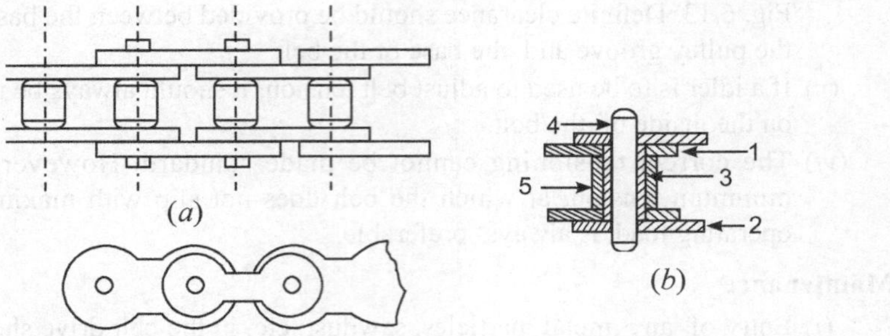

Fig. 6.14. (*a*) Rollers in a chain link (*b*) Roller assembly

To obtain a permissible operating speed, chain drives of shorter pitch of roller chains are preferred.

Silent Chain Drive

These are the inverted tooth-type links joined on pins one after another to form the endless chain (Fig. 6.15). Such a drive deserves to operate within casing having provision for lubrication.

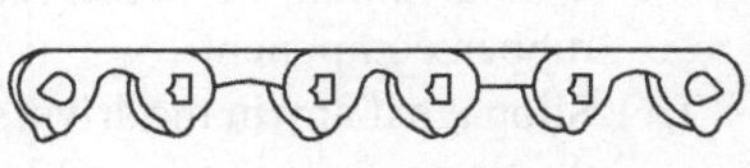

Fig. 6.15. Inverted tooth-type links

Erection

(*i*) The shafts on which the sprockets are rigidly mounted should be parallel and truly aligned.

(*ii*) If possible, it is better to mount sprockets close to be bearings.

(*iii*) Before starting the chain, connections are to be checked and to be properly lubricated.

Maintenance

(i) Scheduled lubrication should be employed to restrict wear of bushes and pins.

(*ii*) It is a good practice to check the adjustment and alignment about one month after the installation and then at a regular intervals of six months.

(*iii*) The chain wear may be inspected periodically, pulling it radially away from the larger sprocket.

(*iv*) Worn out sprocket may sometimes be used refitting it in a reverse manner, if the wear is not too high.

(*v*) New chains should never be installed over a worn out sprocket. The sprocket also should be replaced in such cases.

GEARS

The power transmission procedures by belts, ropes, chains etc. have the advantage to connect the shafts at long centre distances, but in case of such drives the drive is not positive. There is slip and creep. The gear drive on the other hand is a positive and smooth drive, but the shafts at comparatively short distances can only be connected by means of gears.

Common types of gears (Fig. 6.16), generally used in industrial application includes spur, helical, herringbone, bevel, worm gears etc. Spur gears are generally used to transmit power to a parallel shaft without end thrust and axial displacement at moderate speeds.

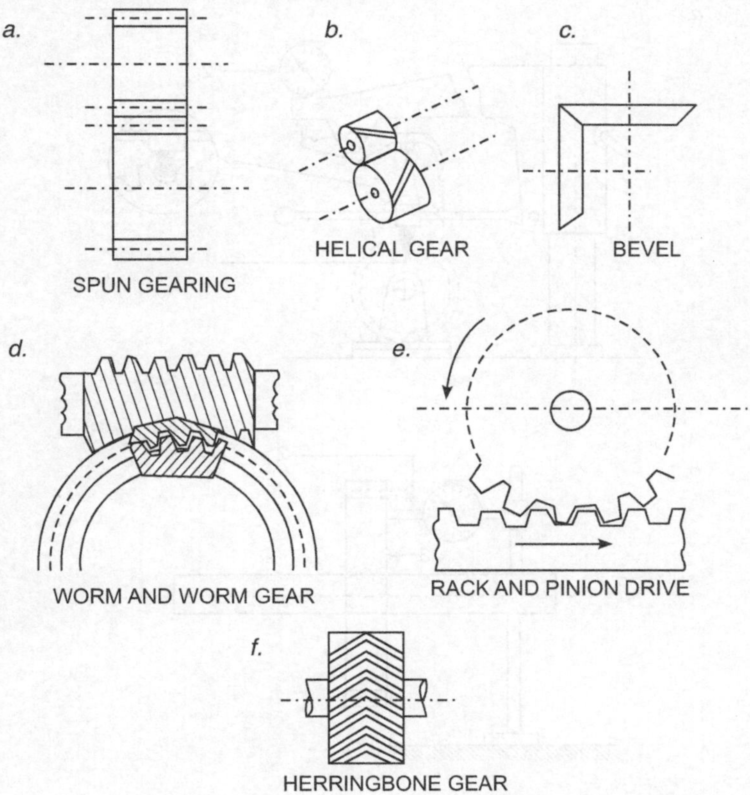

Fig. 6.16. Different types of gears

Helical gears allow to mesh several teeth at a time increasing the load carrying capacity and it reduces noise and vibration. It produces end thrust and that is why it is to be accommodated with thrust bearing.

Bevel gears are normally used to transmit power to a shaft at right angles.

Worm gears are very much popular due to its noise and vibration less compact service.

Gear Mounting

The fundamental steps are to (*i*) mount the gear on the shaft and then to (*ii*) fit the shaft into the bearing.

The gear is mounted on the shaft with a key, the type of which depends on the functioning of the assembly. The plain parallel key with a force-fit is generally used. Sliding feather keys are also used if the gears are to work at different positions.

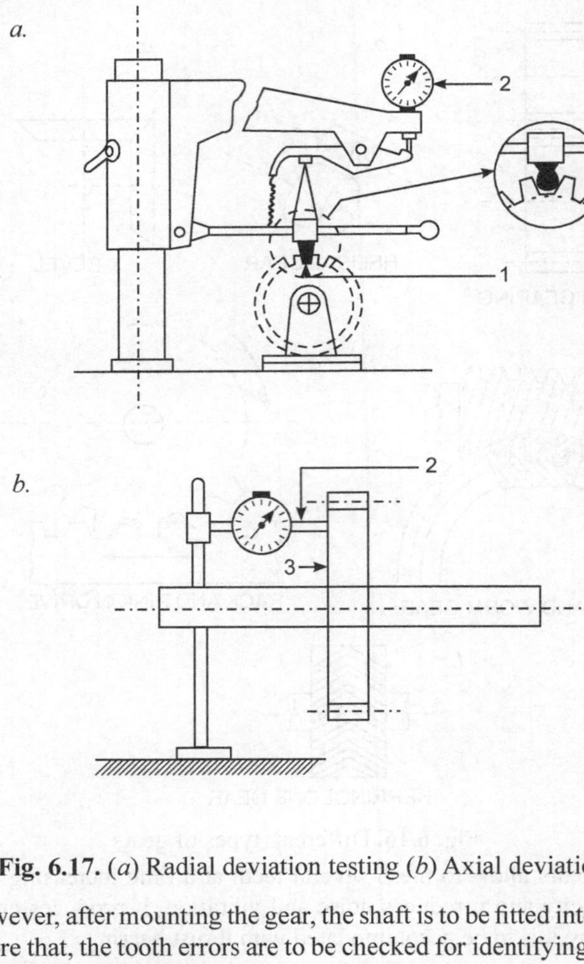

Fig. 6.17. (*a*) Radial deviation testing (*b*) Axial deviation testing

However, after mounting the gear, the shaft is to be fitted into the bearing. But before that, the tooth errors are to be checked for identifying the presence of radial or axial deviation of the gear teeth, if any.

The radial deviations may be tested by placing a measuring rod (1) in between gear teeth and positioning a dial indicator (2) on it (Fig. 6.17*a*). When the assembly will rotate, the degree of trueness of the circumference of the gear, that is, the radial errors may be identified, if the readings are compared placing the measuring rod one after the other tooth profile.

And the amount of trueness in the axial direction may be tested by placing the dial indicator (2) in the side face (3) of the gear, as shown in Fig. 6.17(*b*) and the reading is taken rotating the gear slowly.

Gear-tooth failure

The failure of a gear tooth may be of two types (*i*) Surface wear and (*ii*) Tooth breakage.

The surface deterioration, may be due to wear, plastic flow, scoring, surface fatigue, and the tooth breakage may be due to fatigue, heavy wear, overload or cracking.

Maintenance

The recommended lubricants and filter is to be used to avoid normal wearing. But, abrasive wear may be caused by fine particles passing through the contact surface of gear teeth. Scratching on the teeth is a mark of identification of abrasive wear. In such cases, oil should be drained stopping the unit. The gear teeth should be scrapped and cleaned, and care is to be taken to use dust-free lubricants. Plastic flow may occur under heavy load. 'Ridging' is a form of plastic flow, appears as diagonal lines. It is caused also due to heavy load and inadequate lubrication. 'Scoring' is the quick removal of metal from a tooth surface. It may be caused due to a rupture of the oil film. Extreme pressure lubricants may be used in this case. It starts from the tearing out of a small surface particle. Marks due to 'scoring' appear in the direction of sliding. 'Tooth surface fatigue' is also a type of failure caused due to repeatative surface stresses, beyond its endurance limit. Metals are removed from the surface and the cavities may be formed due to surface fatigue. 'Pitting' is a type of surface fatigue generally occurs in the initial stage of operation. Normally, gradual running sometimes polishes the surface reducing pitting.

Sometimes, grinding and polishing of tooth surface is to be done to avoid pitting.

Gear Inspection

Gear inspection is a part of the gear maintenance work for achieving satisfactory performance of a gear train.

The following major errors may be tested during annual servicing for gear trains, which undergo excessive loading.

Pitch error

This is an individual spacing error between successive teeth. This error is checked generally by a measuring head which consists of a fixed stopper and movable lever (Fig. 6.18).

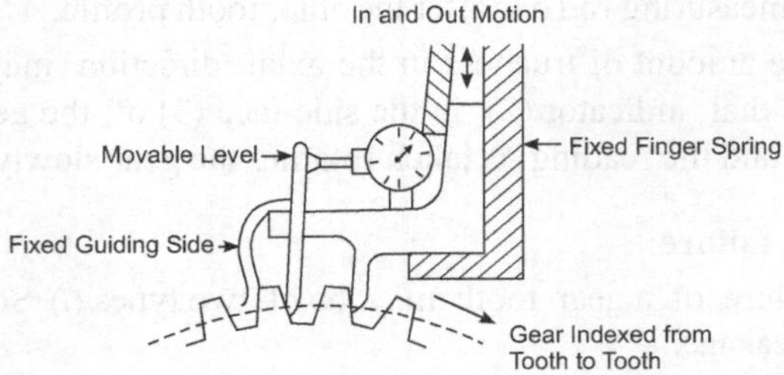

Fig. 6.18. Pitch error testing

Radial Runout or eccentricity error

Radial run out occurs due to the combined effects of teeth-irregularity, misalignment and eccentricity of teeth with reference to the axis of gear. This error can be measured by the process as described previously during gear mounting.

Tooth alignment error

This is the total deviation of the actual tooth trace from the theoretical or desired tooth trace on the reference cylinder. This is measured in transverse direction. The alignment error for a spur gear is measured (Fig. 6.19) by passing a stylus of a dial indicator on the tooth flank across the face-width of the gear, parallel to the gear axis and recording the readings.

The tooth thickness error, tooth profile error, should also be measured accordingly for a thorough inspection to take necessary measures for maintenance and repair.

Repair Method

(*i*) Gears running in one direction may be reversed 180° to take the help of a non-worn surface for better operation.

(*ii*) When a wear appears on both the sides of the tooth surface, it may be seen that a smaller meshing gear will wear more. This may be replaced by a new one. Sometimes, the depth of a tooth in the smaller gear is chosen slightly less than the normal and the outside diameter of the larger gear is reduced to match it.

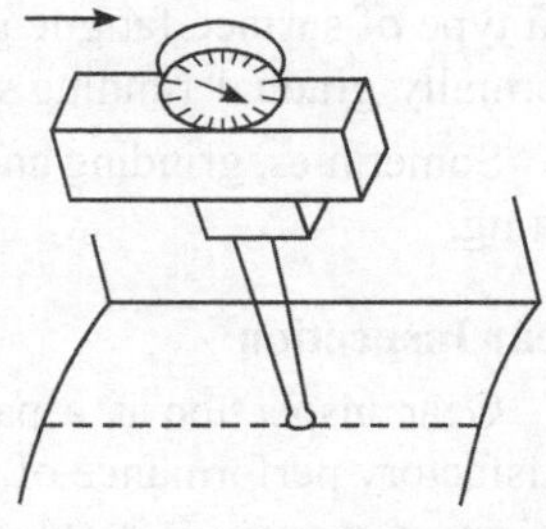

Fig. 6.19. Tooth alignment error testing.

(*iii*) If a particular tooth of a large gear is broken, it can be re-built by welding the deposit as described later in Chapter 8.

BEARINGS

Bearings are common machine components. They are the medium of contact with the rigid frame and the movable parts.

They may be of two types:

(*a*) Plain or sleeve bearings and

(*b*) Antifriction or rolling element bearings.

Plain bearings

Plain bearings may be used to support the rotating, reciprocating or oscillating shafts. According to the function, the plain bearings may be classified as

(*i*) Journal bearings;

(*ii*) Thrust bearings;

(*iii*) Guide bearings.

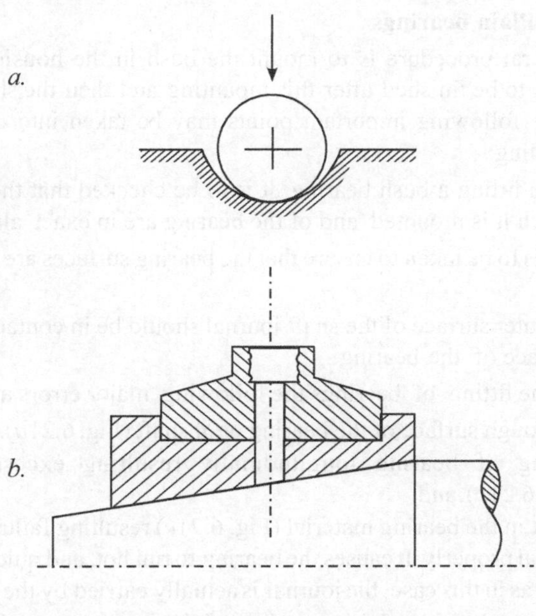

Fig. 6.20. (*a*) Journal bearing (*b*) Conical thrust bearing.

The mostly used journal bearing consists, in general, a shaft supported by a circular housing of somewhat softer materials (Fig. 6.20 *a*). The direction of load is at right angles to the shaft. The thrust bearings, say, a conical bearing

(Fig. 6.20*b*) may be used for a spindle which is having a load and also to experience a thrust. Whenever a shaft is placed in the vertical position it is termed as 'footstep bearings.'

Details of construction and designing criteria are outside the scope of this discussion. However, bearing construction materials are as follows:

Monometals: The general construction material primarily is a solid bar or tube of an aluminium or bronze alloy. They are generally used where loads are not too high.

Bimetals: Such bearings possess a steel back and a liner of babbitt, copper-lead or aluminium is bonded on it. They have outstanding embedability and surface action but with low fatigue strength. Babbitts are soft alloys of lead or tin with additives such as copper, antimony and arsenic.

Trimetals: For heavy-duty application it has been designed. A thin layer (about 0.001 inch) of a soft material is either electroplated or cast on to the layer of a bimetal.

The important points to be noted during assembly and the technique of mounting the plain bearings are discussed in procedural steps.

Assembly of Plain bearings

The general procedure is to mount the bush in the housing. The bore of the bush is to be finished after this mounting and then the shaft is placed within it. The following important points may be taken into consideration while assembling:

(*a*) While fitting a bush bearing, it is to be checked that the axes of the shaft on which it is mounted and of the bearing are in exact alignment.

(*b*) Care is to be taken to ensure that the bearing surfaces are not damaged in any way.

(*c*) The outer surface of the shaft journal should be in contact throughout the inner surface of the bearing.

During the fitting of bearings the following major errors are observed :

(*i*) The rough surface of the bearing or of shaft (Fig. 6.21*a*).

(*ii*) Tilting of bearing longitudinally resulting excessive friction (Fig. 6.21*b*), and

(*iii*) Upset in the bearing material (Fig. 6.21*c*) resulting failure to bear the journal properly. It causes the bearing to run hot, and quick wear takes place as in this case, the journal is actually carried by the upset points.

Proper assembly is to be ensured providing necessary radial clearance also. It is a common practice to maintain a radial clearance of 0.3 to 0.4 per cent of a journal diameter, if it is not otherwise specified. Axial clearance is also to be provided to avoid friction in the bearing shell.

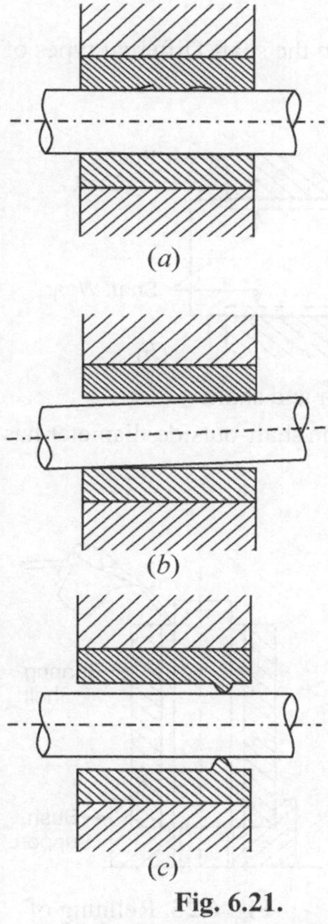

(*a*)

(*b*)

(*c*)

Fig. 6.21.

(*a*) Bearing on rough surface

(*b*) Tilting of bearing

(*c*) Upset in bearing material

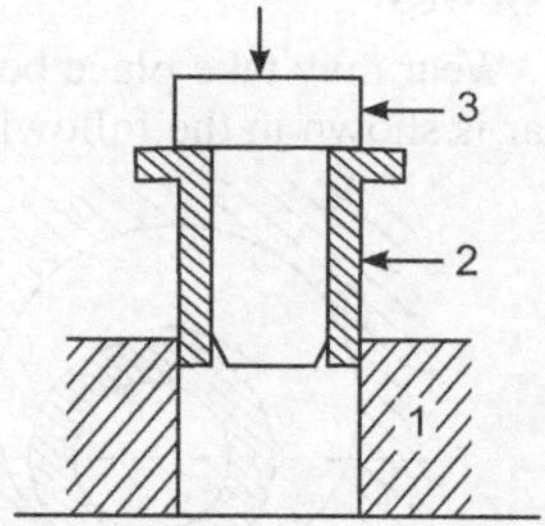

Fig. 6.22. Bearing bush fitting

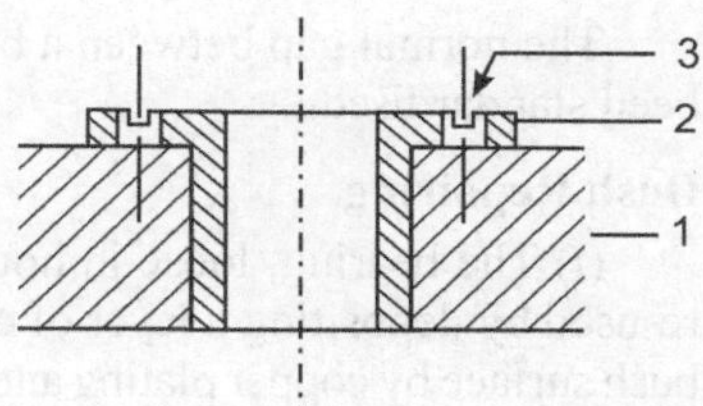

Fig. 6.23. Bearing bush fixing

Mounting Technique

The mounting work may be proceeded as below :

(*a*) Bearing bush is to be forced into bearing housing and to be secured against rotation and then the shaft is to be placed.

The housing (l) may be placed on a firm support and the bearing (2) may be inserted applying force through a definite drift (3) as shown in the Fig. 6.22.

(*b*) The rotation of bearing may be prevented by locking the collar of the bush (2) with the housing (1), by the help of fillistar head screws (3) as shown in Fig. 6.23. Tapered pins may also be used to secure the bearing bush against rotation.

(*c*) After the assembly, the oil holes may be drilled at a particular position.

(*d*) The bearing bushes may now be operated with a reamer to finish it to an accurate size of the shaft journal.

(*e*) The bearing and the shaft are then to be properly cleaned and the shaft may be fitted into the bearing bores.

Bush wear

Wear may take place both in the bush and in the shaft Different types of wear is shown in the following Figs. 6.24 *a* & *b*.

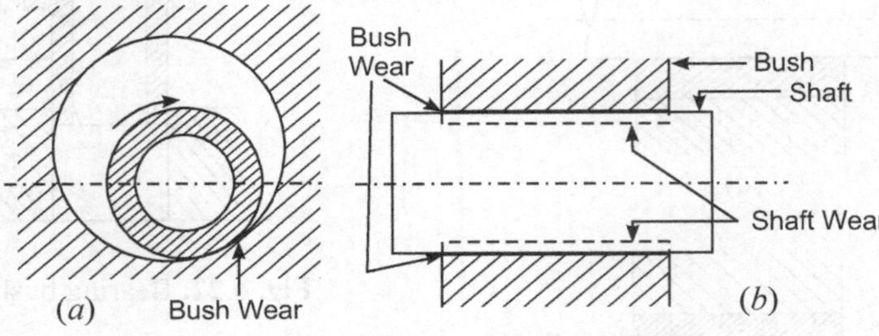

Fig. 6.24. Wearing in bush and shaft.

The normal gap between a bush-bore and the shaft-outside-diameter has been standardised.

Bush Repairing

(*i*) The bearing, loose in housing, can be re-used by depositing a layer of copper on the bush surface by copper plating and then turning the surface to the desired diameter.

(*ii*) Bearing lining can also be repaired by the process "relining." The bearing shell is placed on a support and a mandrel is put aligned inside. The bush is warmed up at a desirable temperature and a proper flux is used in the bore of the bush before pouring molten metal as shown in Fig. 6.25.

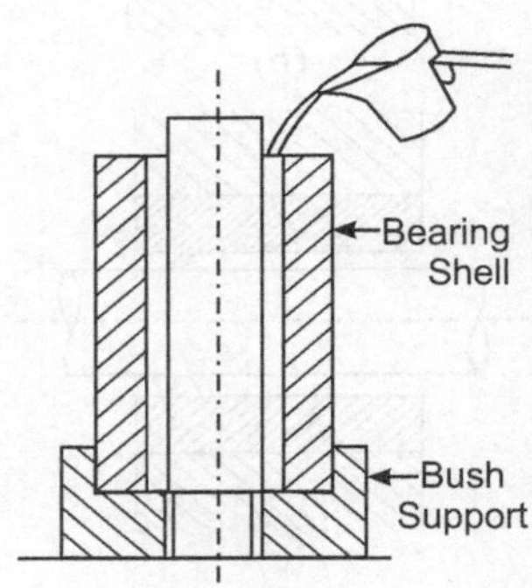

Fig. 6.25. Relining of bearing shell.

The lining is to be finally made concentric and at proper dimension by boring.

(*iii*) Worn bearings can also be repaired by sand moulding. The bush bore is turned or bored by 3 to 6 mm more than the nominal diameter. The sand mould is prepared with internal core and the molten babbit (Bimetal or alloys of Lead or Tin) is poured during moulding.

(*iv*) For splitted bush, the score marks or wearing is identified by the smearing technique and scraping is done properly. Minor wearing may be adjusted by this method, or otherwise re-babbiting of bearing halves may have to be done on the same way as described earlier.

Anti-friction Bearings

There is no doubt that the improvements of plain bearings have increased their efficiencies. Perhaps, it has reached in its limiting values. The very principle of dependence of plain bearings on sliding resistance of two solid surfaces, separating them bya film of oil, has been altered in case of antifriction bearings to attain further scope of improvements. The main motto is to reduce the friction and accordingly, the sliding friction has been substituted by a rolling friction in the antifriction bearings.

They are available in different varieties. They may be classified as :

Antifriction Bearing

Ball Bearing:

1. Deep groove ball bearing
2. Self aligning ball bearing
3. Angular contact ball bearing
4. Duplex bearing
5. Thrust ball bearing

Roller Bearing:

1. Cylindrical roller bearing
2. Spherical roller bearing
3. Taper roller bearing
4. Needle roller bearing
5. Cylindrical roller bearing
6. Needle roller thrust bearing
7. Spherical roller thrust bearing

Types and features

The special features of different types of bearings are explained in the following table

Ball Bearings

Types	*Figure (6.26)*	*Features*
(*a*) Deep groove Ball bearing	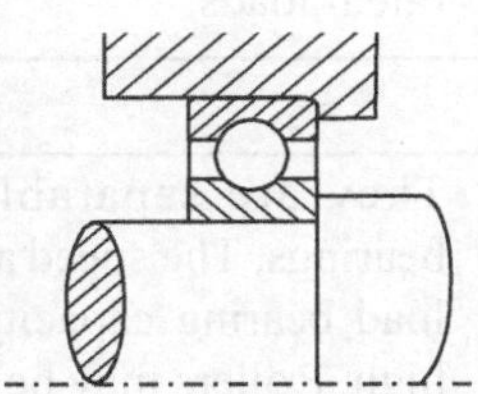	Most commonly used non-separable bearing. It takes both radial and axial loads. Double row deep groove ball bearing can carry higher radial loads in axial direction. However, the load carrying capacity is rather low. Suitable for high-speed operation and requires little maintenance.
(*b*) Self aligning ball bearing	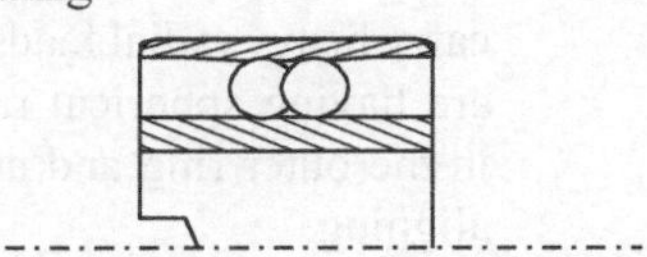	Double-row non separable bearing. They have spherical raceway in the outer ring and thus are adjustable to slight misalignment and shaft adjustment.

Types	*Figure (6.26)*	*Features*
(*c*) Angular contact ball bearing	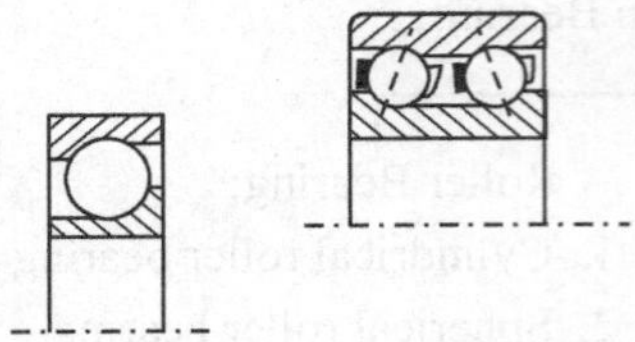	They can withstand both radial and thrust load. But, single row angular contact ball bearing can withstand thrust in one direction only, but double row bearings can absorb thrust loads on both the directions. They are suitable for high-speed operation.
(*d*) Duplex Bearing	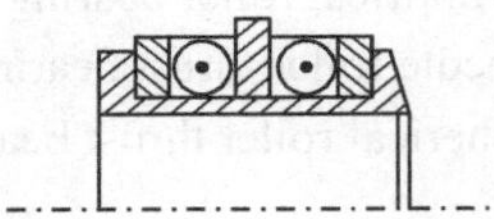	They are suitable for thrust in both directions. Two distinct tracks are provided in both inner and outer races to reduce the end play. The outer race is made in two parts to use maximum number of balls.
(*e*) Thrust Ball Bearing	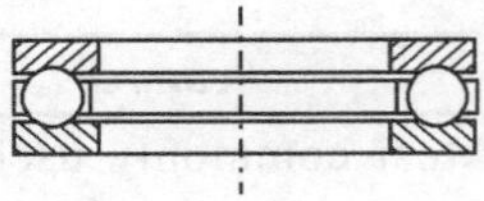	It withstands axial loads only. Single thrust ball bearing is used for loads acting in one direction only, whereas double thrust ball bearing may take axial loads from both the directions. It is not suitable for raidal loads.
Roller Element Bearings		
(*f*) Cylindrical Roller Bearing	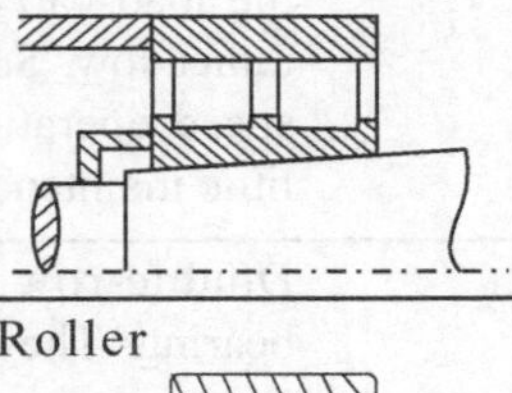	They are separable radial bearings. The speed and radial load bearing capacity is very high. Rollers may be arranged in single row or in double rows.
(*g*) Spherical Roller Bearing	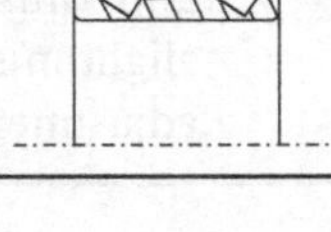	They are separable and can carry heavy radial loads. They are having spherical raceway in the outer ring and are self-aligning.

Types	*Figure (6.26)*	*Features*
(*h*) Taper Roller Bearing	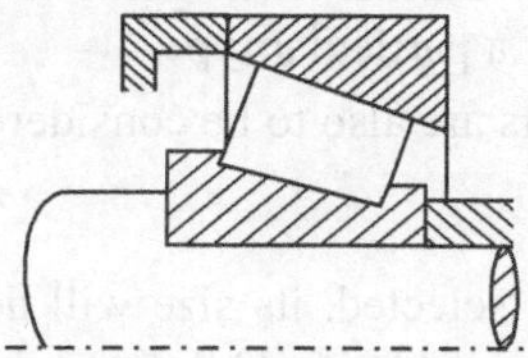	They are also separable bearings and used for combined radial and axial loads. Generally, for axial loads, it is adjusted by a second bearing and is mounted in opposition to each other.
(*i*) Needle Roller Bearing	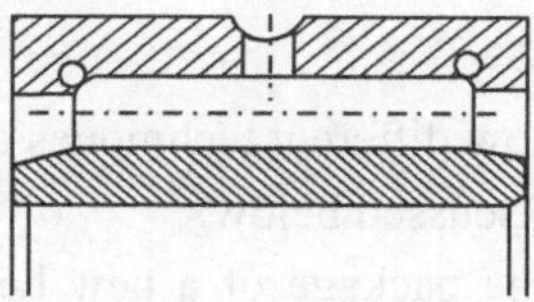	They are separable bearings and the inner ring can be mounted separately. They are used for radial loads only and are available both in single and double row.
(*j*) Spherical Roller Thrust Bearing	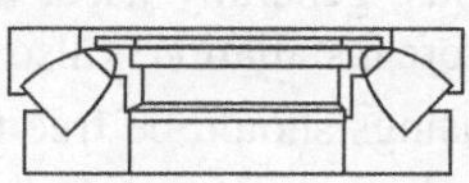	They are self-aligning and moderate speed bearings. They are used for heavy axial loads.

Selection of Bearing

The selection of a bearing for a particular purpose means, primarily to select the type and then the size of the bearing.

Selection of type

The type of bearings, that will be suitable for the situation may be selected considering the followoing characteristics :

(*a*) Load : The direction of load, that is whether it is radial or axial, is more important than the magnitude of the load. Of course, the magnitude will also be taken into account.

(*b*) Bearing location : The space available to accommodate the bearing is very important.

(*c*) Angular deflection : The resulting deflection is to match with the specified angular deflection of the bearing.

(*d*) Rigidity : The rigidity is a most decisive factor for machine-tool-spindle mounting.

(*e*) Axial displacement and mounting facilities : The axial displacement is to be checked before selection along with the mounting facilities available.

(*f*) Speed and accuracy : The required speed should be available from the selected bearing and accuracy requirement will guide to select the class of a particular type.

The initial and maintenance costs are also to be considered.

Bearing size selection

Once the type of the bearing is selected, its size will depend on some calculations with regard to the (*a*) dynamic load, (*b*) Rating life and (*c*) Service life with some available empirical relationships.

Antifriction Bearing Mounting

Different types of bearings follow different techniques of mounting.

However, a few of which are discussed below:

It is a general warning that, the package of a new bearing should be removed only during mounting. It should be kept wrapped since most of the modern rust-preventive compounds, generally used, react with petroleum-based lubricants. The cleaning before installation is also very much important.

The bearings, shafts and housings should be free from chips, burrs and dirt, etc.

(1) Small bearings may be mounted with the aid of a length of tubing. Direct blows should never be applied on the bearing and hammering should be made against the piece of pipe used and blow should be applied on the ring, which is going to be forced on the shaft or housing.

(2) Large bearings may be mounted by heating it upto 80° to 90°C in a oil bath. The recommended flash point of oil to be used for heating purpose should be below 250°C and the temperature of bearing should not exceed 120°C.

(3) When a roller bearing is mounted, the provision for thermal expansion or contraction of a shaft is maintained. The shaft is to be held and located axially. For a through shaft a specially designed nut is usually used (Fig. 6.27*a*) and when the bearing is mounted at the end of the shaft, a clamp plate is generally used (Fig. 6.27*b*).

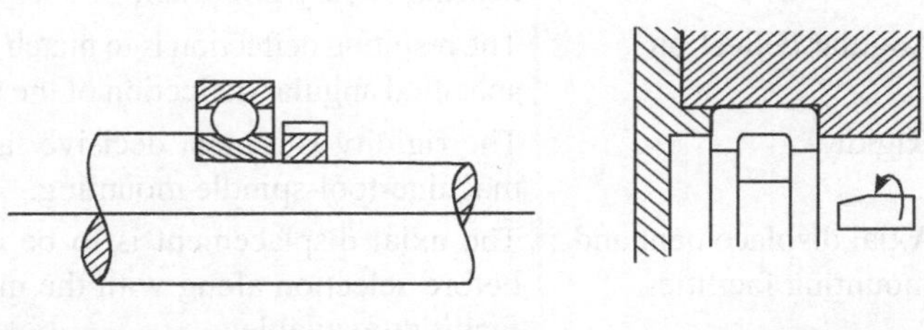

Fig. 6.27. (*a*) Bearing with a nut (*b*) Bearing with clamp plate.

(4) For a cylindrical roller bearing, the separable parts are mounted independently. The pushing of the loose ring into the roller set should not be done by force. The indentations on the raceway may damage the bearing. A guide sleeve may be used to ensure that the roller assembly is not placed at an angle.

(5) The thrust bearing races are always of different bore diameters. During mounting it should be noticed that the race with a larger bore should be in contact with a stationary body and the race with smaller bore should be mounted on the shaft collar (Fig. 6.28).

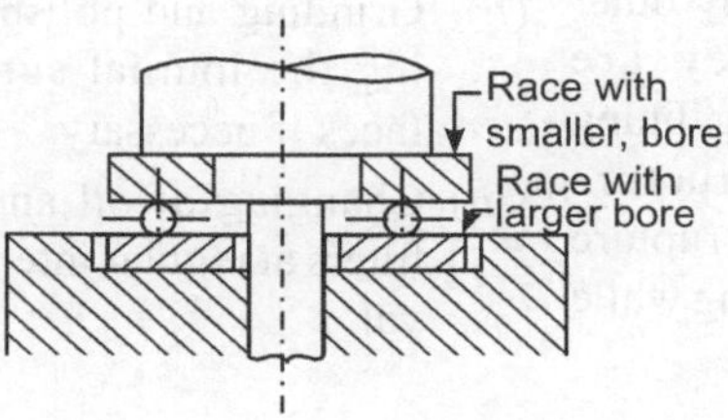

Fig. 6.28. Thrust bearing mounting.

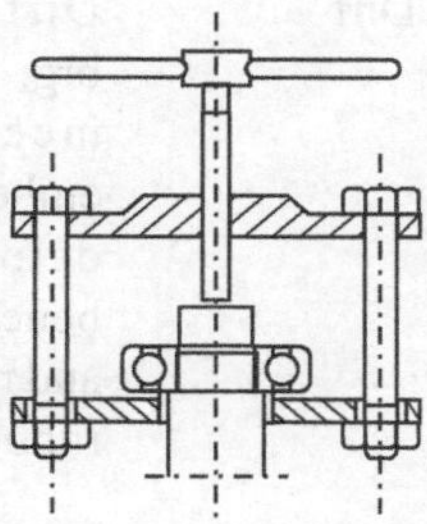

Fig. 6.29. Bearing Puller.

Dismounting of Bearing

It is to be remebered that the force for dismounting of bearing should always be applied on the ring whichever is of tighter fit. The pull applied should also be straight and square to save the bearing from any damage.

In most cases, a press or a puller (mechanical extractor) is used to remove bearings.

The bearing puller (Fig. 6.29) is having two hands through which pulling effort can be applied on the desired race.

Hydraulic presses are also available and is applied as shown in Fig. 6.30.

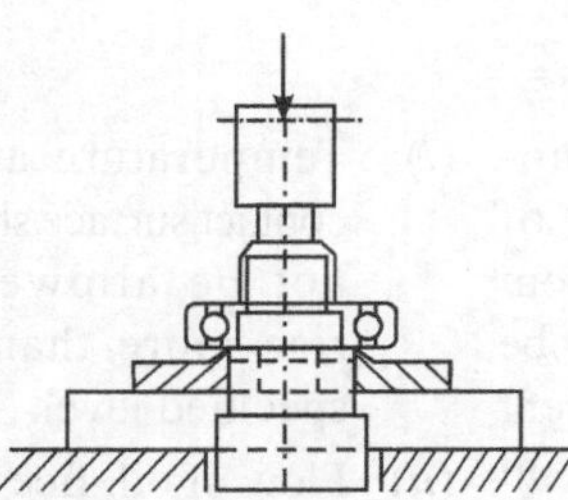

Fig. 6.30. Hydraulic Press.

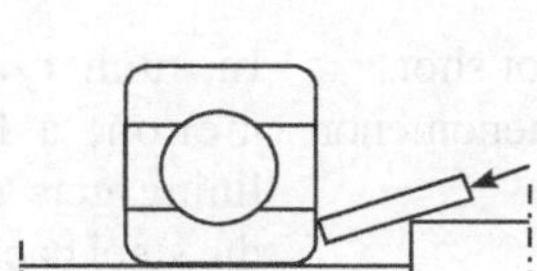

Fig. 6.31. Dismounting by hammering drift.

In case, when no such devices will be available, the dismounting of a bearing may be done with the help of a hammer and a short metal drift (Fig. 6.31). The blows should not be too strong and should be applied on the total circumference.

Major failure factors of bearings and corrective action:

Factors of failure	*Normal appearance and cause*	*Corrective action*
A. Dirt	Dirt is responsible for brg. failure than any other mechanism. They are embedded in the brg. lining deforming the structure beneath. It causes rupture and removal of lining while rubbing.	(*i*) Proper maintenance. (*ii*) Grinding and polishing the journal surfaces if necessary. (*iii*) Changing of oil and filters at regular interval.
B. Fatigue	When the running time exceeds the alloy's capability, the bearings fail due to fatigue. Fatigue cracks initiate in the bearing surface and propagate perpendicular to it. The causes may be due to load concentration for the presence of dirt, or poor shaft or bore geometry, misalignment etc. Such factor accelerates fast wearing.	(*i*) Proper lubrication. (*ii*) Maintenance of oil-film thickness at desired level
C. Hot shot phenomenon	In such type of phenomenon, a large area of lining gets removed from the steel back. This may be caused due to insufficient oil flow, dirt in the used oil or misalignment.	(*i*) Temperature at the contact surface should not be allowed to rise more than the specified level. (*ii*) Flow of oil, free from dirt should be ensured.

Factors of failure	*Normal appearance and cause*	*Corrective action*
D. Fretting	Fretting occurs due to load concentration. Fretting can be identified inspecting the crushing area, which causes relative movement between the bearing and the bore. This also appears to be highly polished area or pitted.	(*i*) Mating faces of the bearing assembly should be cleaned and made burr-free while installation. (*ii*) Attention to be given regarding load-carrying capacity.
E. Excessive Interference	Distoration of bearing bore causes overheating. Too tight fit and stragger at joint faces during assembly is its root cause. It is responsible for overheating of bearing.	Care should be taken during mounting specially regarding fit-selection.
F. Corrosion	It removes lead phase or creates organic acids by oxidation of lubricants in service.	(*i*) Oil suppliers may be consulted (*ii*) Investigation may be made for possible coolant leakage into oil.

COUPLINGS

Couplings

Rigid Coupling	Loose or disengaging	Flexible Coupling
(*i*) Box or muff coupling	(*i*) Claw coupling	(*i*) Bushed pin type.
(*ii*) Half lap coupling.	(*ii*) Conical friction coupling	(*ii*) Universal.
(*iii*) Split muff coupling		(*iii*) Oldham
(*iv*) Flange coupling.		

Couplings are used to connect shaft ends together for transmitting torques or rotary motion directly from one shaft to the other. The above referred transmission should occur without slip and at the same time should compensate for angular and/or parallel misalignment. A few types of couplings are shown below:

Muff coupling (Fig. 6.32)

It consists of a cylindrical muff and the two ends of the two shafts butt against each other. A long sunk taper key is driven through both the shafts.

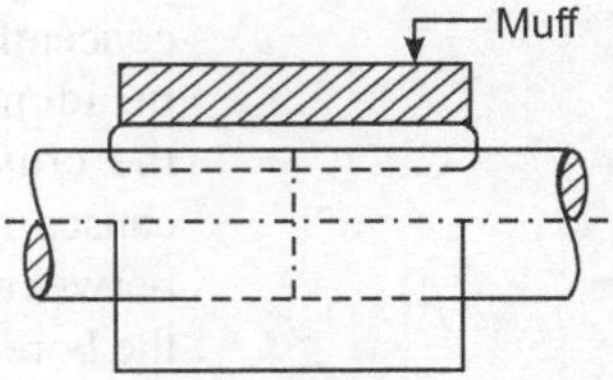

Fig. 6.32. Muff Coupling.

Half lap coupling

It is a muff coupling. Here the ends of the shafts are made to overlap each other (Fig. 6.33).

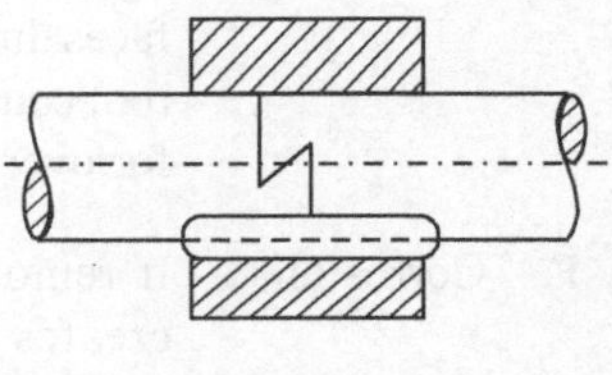

Fig. 6.33. Half lap coupling.

Split muff

The muff is made in two semicylindrical halves which are joined together by bolts and nuts.

Flanged Coupling

A standard extensively used coupling. It is having two cast-iron flanges, keyed to the ends of the two shafts and they are fastened together by 4/6 tight fitting bolts and nuts (Fig. 6.34).

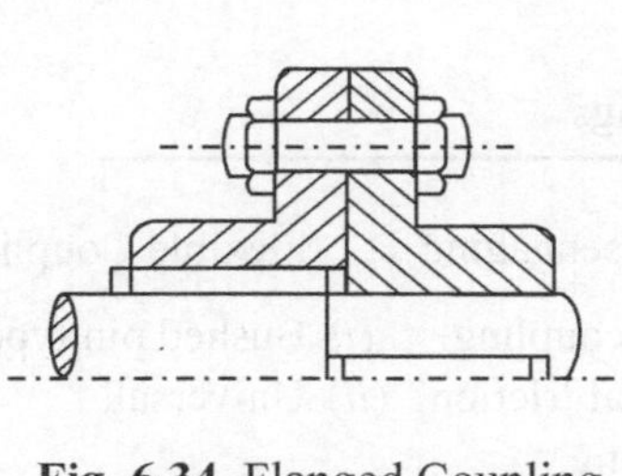

Fig. 6.34. Flanged Coupling.

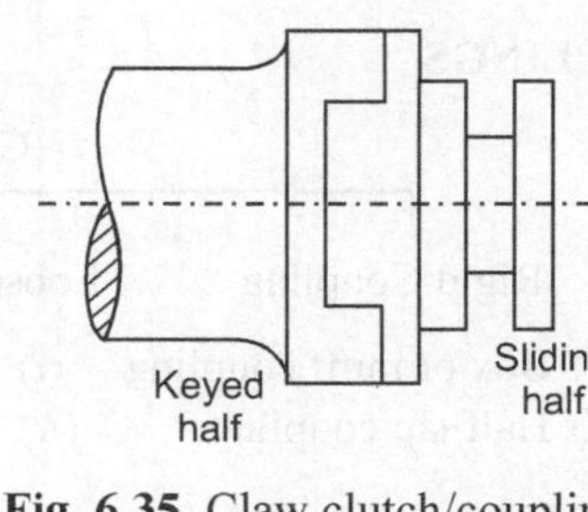

Fig. 6.35. Claw clutch/coupling.

Loose or disengaged type

Claw clutch/coupling (Fig. 6.35)

They are suitable for slow running shaft. One half of this coupling is keyed and another half slides over a feather. To make engagement the machinery must be stopped. For better life, the keyed half may be bushed. Adequate lubrication is necessary for proper functioning.

Flexible Coupling

Bushed Pin type

A flexible coupling permits alignment of relative rotation and variation in the alignment of shafts up to certain limits. The figure shows two flanges connected by a pin. The pin is rigidly fastened to one of the flanges and in an other flange it is connected through a rubber bush (Fig. 6.36). Such types of coupling is generally used for directly connecting an electric motor to a machine.

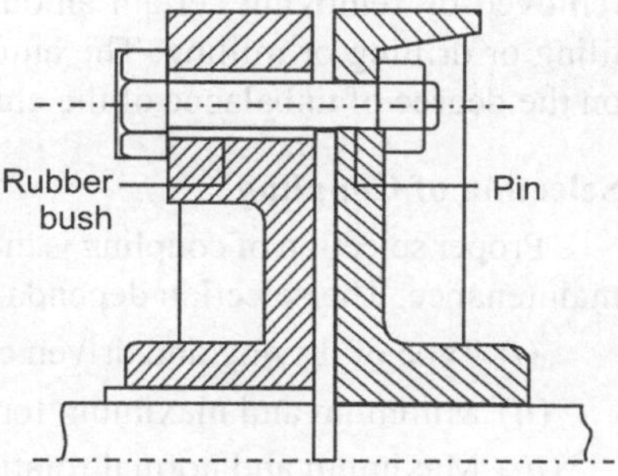

Fig. 6.36 Bushed Pin coupling

Coupling Installation

In most of the cases the couplings are dismantled before installation and the arrangements of the components are to be noted carefully to assemble it again. They are to be cleaned properly to make them free from dirt, burrs etc. and the keys are also to be checked for proper fit. While fitting hubs in the shaft, they may be heated in water, oil or in a furnace. Spot heating must be avoided.

The equipment should be brought into its approximate operating position and then the couplings are assembled.

Finally the alignment and concentricity are to be checked. Then it requires final balancing.

The concentricity may be checked (Fig. 6:37) with the help of a dial indicator (4) which should be placed on a marking plate (3). The two shafts (1) and the coupling (2) assembly is to be held against two centres and is to be rotated. The eccentric rotation will be immediately indicated in the marking gauge.

The balancing is generally done rotating the same assembly by holding the shafts by two bearings. If the centre of gravity is not exactly placed in the axis of rotation, the assembly is said to be unbalanced. In such a case the heavy portion of the clutch should be in the lowest position when the assembly will come to rest. The lowest part of the clutch may be marked and may be rotated repeatedly. If the marked area always appears in the lowest position,

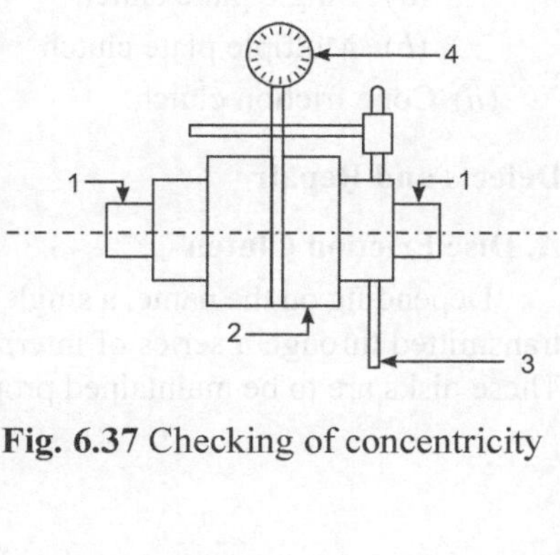

Fig. 6.37 Checking of concentricity

whenever the system comes to rest, the unbalance is detected and may be removed by removing certain amount of material from the marked part by filing or drilling or milling. The amount of material to be removed depends on the degree of unbalance of the clutch or coupling.

Selection of Coupling

Proper selection of coupling is the first-step to enjoy the benefit of efficient maintenance. The selection depends on various factors,

(*i*) Type of driving and driven equipment;

(*ii*) Minimum and maximum torque;

(*iii*) Maximum and normal rotating speed;

(*iv*) The distance between the shaft ends;

(*v*) Ambient condition;

(*vi*) Shaft size etc.

Coupling failures

(*i*) Failure of coupling may occur due to several causes. The concentricity and the squareness of the mating parts play a major role for good functioning of coupling. The alignment is always to be maintained.

(*ii*) Besides defective materials during construction, the provided tolerance may also be the cause of failure.

(*iii*) Coupling failures may also be caused due to improper lubrication.

(*iv*) Failure may be due to improper selection of coupling.

CLUTCHES

A driving shaft may be connected or disconnected with a driven shaft, whenever required in the running condition, through a device, called 'clutch.'

They may be of different types. The mostly used clutches are as follows:

(*i*) Disc friction clutch

  (*a*) Single plate clutch

  (*b*) Multiple plate clutch

(*ii*) Cone friction clutch.

Defects and Repair

A. Disc Friction Clutch

Depending on the name, a single plate may be there or the torque may be transmitted through a series of internal (2) and external (l) disks (Fig. 6.38). These disks are to be maintained properly.

(*i*) The plates may be having excessive wear creating problems in adjustment. New disks are to be added for adjustment.

(*ii*) If distortion of disks are too much, they are to be replaced.

(*iii*) For minor distortion grinding of plate may be done without disturbing the hardened surface layer (within 0.24 mm). Deposition of weld and subsequent filing may also be done.

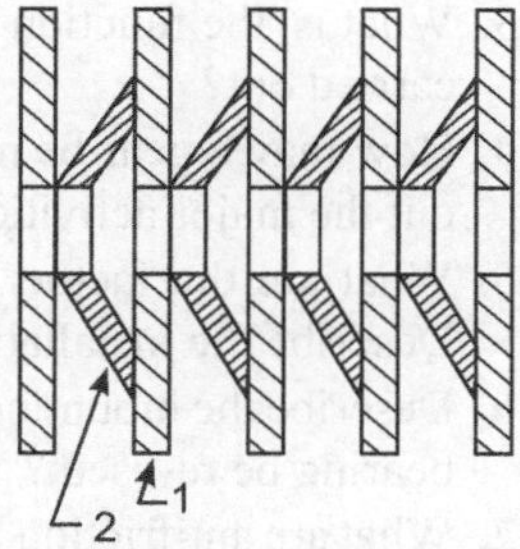

Fig. 6.38. Multiple plate clutch

(*iv*) If disengagemept of a clutch becomes a problem, the grade of lubricating oil may be checked, as a thick oil may become sticky and produces such defects.

B. Cone Friction Clutch

(*i*) The maximum wearing takes place in the tapered surface and ultimately the faces may touch each other. Primarily, by rubbing the taper surface with emery cloth and cleaning it, the faces may be machined to provide desired clearance. Sometimes, compensating rings may be pressed or adhered with the tapered surface and the mating surfaces may be scrapped with respect to each other.

(*ii*) When the size of the coupling is small, it is advisable to change one or both the cones by a new one. Fig. 6.39. Cone friction clutch.

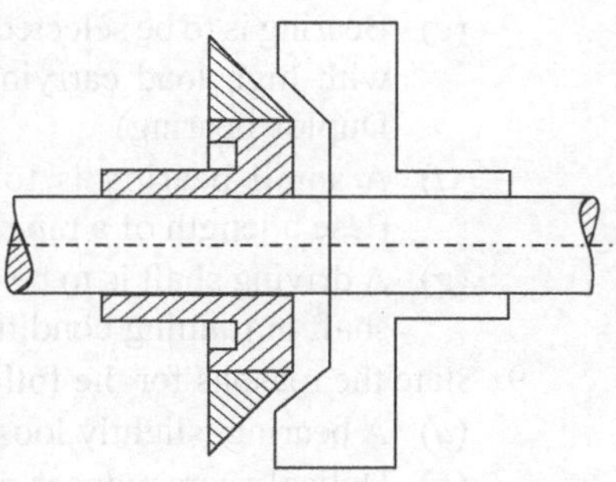

Fig. 6.39. Cone friction clutch

The mounting techniques of clutches along with other testings are the same as those of couplings as described before.

REVIEW QUESTIONS

1. What is fundamental difference betwen 'couplings' and 'clutches?' Show by neat sketch the defective surfaces preventing effective engagement in case of a 'cone clutch.' State how the defective surfaces may be repaired.
2. Name the various motion transmission devices from one shaft to another for different relative positions of their axes. Give neat sketch for each of the devices.

3. What is' the function of a key? How normally the fittings of keys are carried out?
4. How may a gear be mounted on a specific position of a shaft? Point out the major activities regarding the maintenance of a gear.
5. What are the factors to be considered for the selection of V-belts? Describe the installation and maintenance procedure of V-belts.
6. Describe the mounting techniques of a plain bearing. How maya worn bearing be re-used ?
7. What are antifriction bearings? How are they selected? State the steps of mounting and dismounting of such bearings.
8. Select the appropriate device in the following cases:
   (*a*) A small force is to be transmitted (sunk taper key, woodruff (key).
   (*b*) Heavy loads are to be transmitted with frequent starting and stopping. (Flat belt/V-belts).
   (*c*) Stippage is to be avoided during transmission of power (Belt drive/Chain drive).
   (*d*) An oscillating shaft is to be supported. (Plain bearing/ Antifriction bearing).
   (*e*) Bearing is to be selected to withstand thrust in both the directions with high load carrying capacity. (Deep groove ball bearing/ Duplex bearing)
   (*f*) A small bearing is to be mounted with suitable technique. (Use a length of a pipe to apply force/Use a hot oil bath)
   (*g*) A driving shaft is to be connected or disconnected with a driven shaft in running condition. (Coupling/Clutch).
9. State the reasons for the following:
   (*a*) A bearing, slightly loose in housing should not be thrown out.
   (*b*) Helical gears reduces noise and vibration.
   (*c*) Repeated dismantling of threaded joints are always avoided.
   (*e*) The unbalance of a coupling may be easily identified by rotating it.
   (*f*) The driving face of a belt should be maintained in a mastic state.

7

MAINTENANCE AND REPAIR OF GUIDE SURFACES

Introduction

The guideways of Machine Tools play the key role to achieve the accuracy of a job. They are properly checked while manufacturing and they deserve periodic checking and proper maintenance while in use as well.

The major components, mechanisms and useful devices are being carried and moved longitudinally or crosswise by the guide ways. Whenever the components slide over the guideways, a frictional force is developed resulting a shearing effect on the surface layer and consequently this effect may change the geometrical properties at the surfaces of contact.

Practically, it may be observed that the slider slides over a particular zone of the total length of guideways. Mostly the non-uniformity of wear is caused by this phenomenon. The guideways get concaved and the slider becomes convex in the middle.

Before going into the details of maintenance of guideways it may be remembered firstly that they are mostly of four types.

Flat	Fig. 7.a.
V-shaped	Fig. 7.b.
Inverted 'V' shaped	Fig. 7.c.
Dovetail	Fig. 7.d.

Secondly, the mechanical wear seen on both guideways and slider may be mostly due to:

(*i*) Mis-lignment of mating parts.

(*ii*) Lack of proper lubrication.

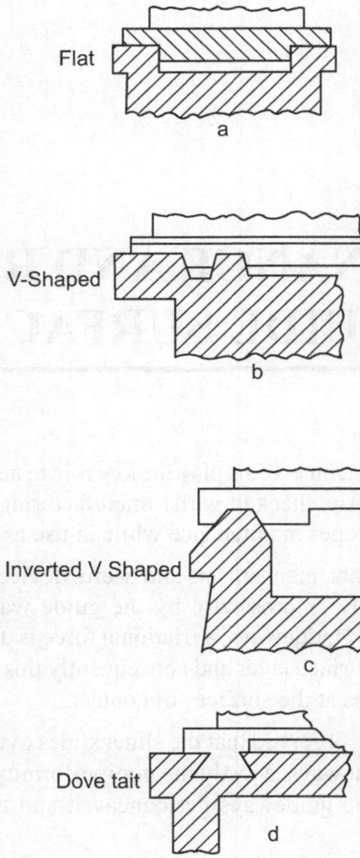

Fig. 7.1. Types of Guideways.

(*iii*) Excessive load, developed on contact surfaces.

(*iv*) Presence of foreign particles in lubricant.

(*v*) Accumulation of dust on guideways.

(*vi*) Absence of proper inspection and maintenance.

The term maintenance implies two major activities 'regular inspection' and 'measuring the developed wear' and finally, the repair of guideways is to be taken into consideration if the amount of wear goes beyond permissible limit.

Methods of measuring the extent of wear

The general practical methods of identification and measuring the extent of wear is as follows:

(*i*) Spirit level;
(*ii*) Straight edge;
(*iii*) Smearing technique;
(*iv*) Hydrostatic procedures;

(*i*) Spirit level

For the short and narrow guides, the most common method is to use spirit level lengthwise and the insertion of a feeler gauge to keep it at proper level indicates the amount of wear. To measure the wear in this process, the spirit level should be placed at regular intervals of 500 mm.

(*ii*) Straight edge

A straight edge can also be used for short guides. The straight edge is placed along the guide surface along with a spirit level on the straight edge and the gap may be measured by a feeler gauge. As the Fig. 7.2(*a*) shows, there is a guide surface (1) on which the straight edge (2) is placed and the feeler gauge or slip gauge (3) measures the wear.

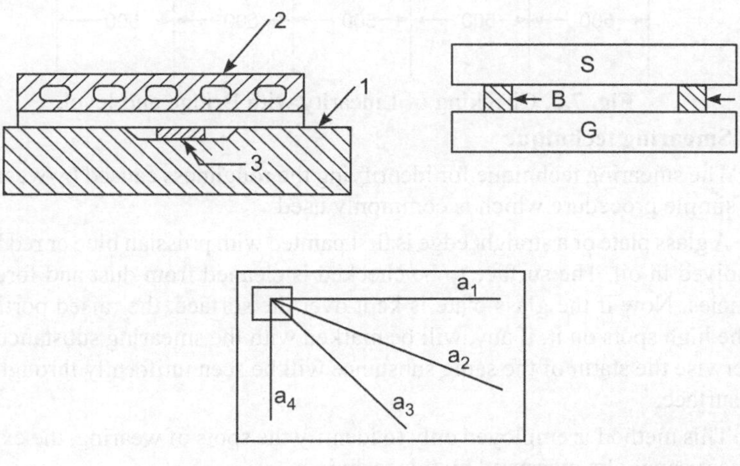

Fig. 7.2. Extent of Wear Measuring.

The straight edges can also be used on two equal thickness blocks (B). The gap between the guide surface (G) and the straight edge (S) is measured at several places. The difference of gap at different places will indicate the amount of wear. The same experiment is to be repeated at different directions a_1, a_2, a_3, a_4 and so on over the guide surface (Fig. 7.2*c*).

For long guides too, measurement of wear may be done with the help of a straight edge and spirit level. The straight edge may also be replaced by a suitable bridge as shown in the figure. The spirit level (1) is placed on the bridge (2) as shown in Fig. 7.3. Before conducting the experiment, the bed may be marked by chalk at an interval of 500 mm. The spirit level reading is noted repeating the experiment in the same manner for every 500 mm interval.

Say, firstly the level reading is R_1. For the second 500 mm, it is R_2, and next it is R_3 and so on. If the least count of the spirit level is 'L' mm/metre, the level difference for each 500 mm will be $\frac{R_1 \times L}{2}, \frac{R_2 \times L}{2}, \frac{R_3 \times L}{2}$ and so on.

The maximum wear can be identified plotting the above readings on a graph paper.

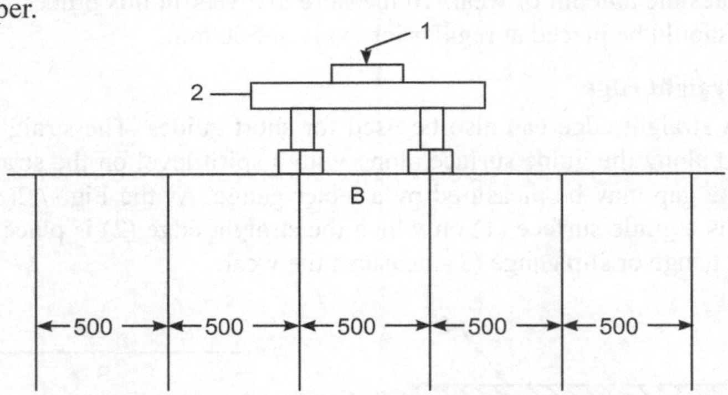

Fig. 7.3. Checking of Linearity with Bridge and level.

(*iii*) Smearing technique

The smearing technique for identifying the roughness caused by wearing is a simple procedure which is commonly used.

A glass plate or a straight edge is first painted with prussian blue or red lead dissolved in oil. The surface to be checked is cleaned from dust and foreign particles. Now if the glass plate is kept over the surface, the raised portions or the high spots on it, if any, will be marked with the smearing substance, or otherwise the statin of the same substance will be seen uniformly throughout the surface.

This method is employed only to identify the spots of wearing, the extent of wear cannot be measured by this technique.

(*iv*) Hydrostatic Procedure

The hydrostatic procedures are the most accurate methods to measure the amount of wear. Generally, these techniques are used for long guides, beds or plates. One technique may be to have two identical measuring glasses (G), filled with kerosene oil and connected with two flexible tubes (T), say polythene tubes. Both the glasses are fitted with a micrometre at the top (Fig. 7.4).

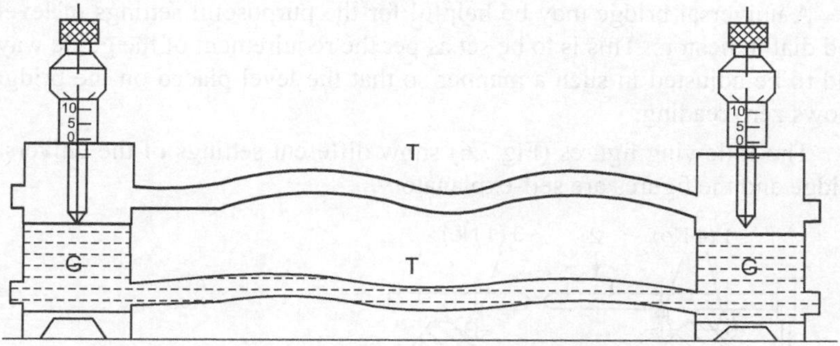

Fig. 7.4. Hydrostatic checking of wear.

To measure the wear, one glass is kept fixed at the middle of the guide surface and the other glass is kept first adjacent to it. The micrometre is adjusted so that the micrometre spindle ends just touch the oil.

Then one of the measuring glasses are moved all-around the fixed glass. The micrometre is adjusted in a similar fashion and the readings are noted. This will ultimately measure the amount of wear and the maximum wear can also be identified and measured. For a long bed the same procedure can be repeated throughout the surface. It may be noted that measuring glasses can be placed only on flat surfaces but the same method can also be employed on prism guides using a suitable bridge. The same procedure can also be followed with a measuring block (B) as shown in the Fig. 7.5. Measuring block is placed on the guide surface and micrometre readings can be taken at short intervals along the entire guide length.

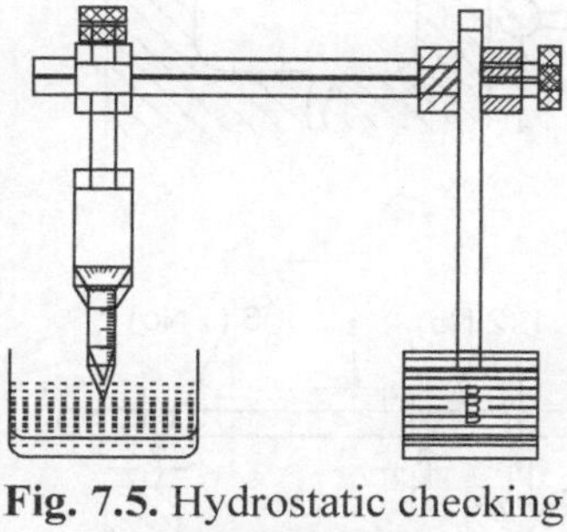

Fig. 7.5. Hydrostatic checking with Measuring Block.

Once the measuring of wear is over, the suitable repair method is to be selected to get it repaired. The depth of wear on guide surface may provide guidelines to select the proper method of repairing. An accepted guideline is given later to restore the worn guide ways.

Setting of Universal Bridge

Both the dial-indicator and level are efficient instruments to check the guide surfaces. But, it is practically difficult to take direct readings, placing along different profiles.

A universal bridge may be helpful for the purposeful settings of levels and dial indicators. This is to be set as per the requirement of the guide ways and to be adjusted in such a manner so that the level placed on the bridge, shows zero reading.

The following figures (Fig.7.6) show different settings of the universal bridge and the figures are self-explanatory.

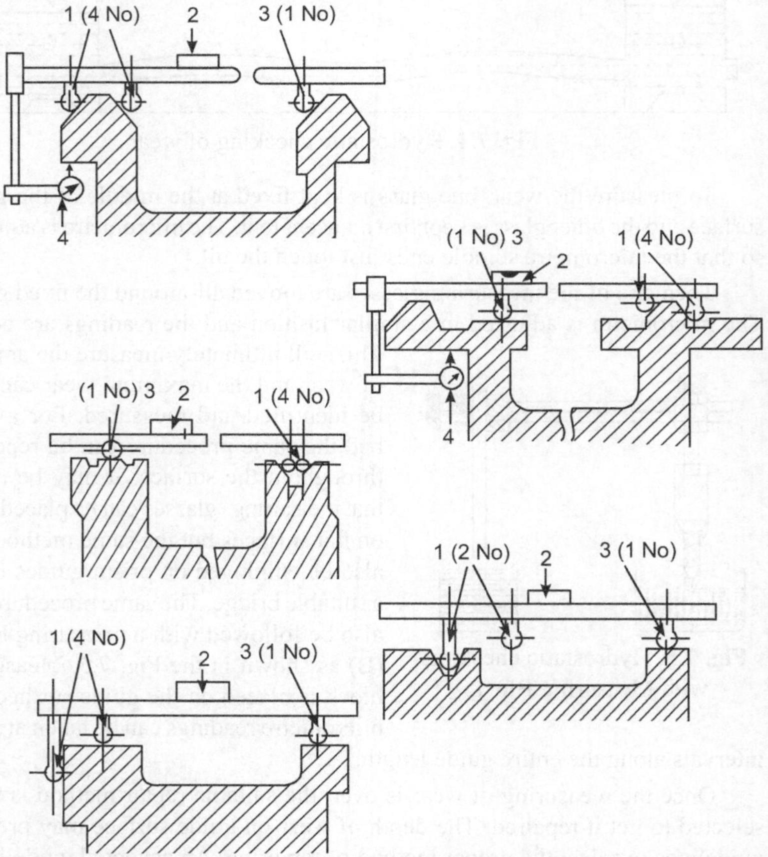

Fig. 7.6. Setting of Universal Bridge on (*a*) Prism Guides, (*b*) Prism and Flat, (*c*), (*d*) Vee and Flat and Prism Guides (1 & 3 – Bridge support, 2 –Level, 4–Dial Indicator).

Checking the guideways

1. Straightness of guideways

The straightness or the non-linearity can be checked by placing the precision level (Fig. 7.7) along the guide length. This can also be checked by a dial indicator placing it on a universal bridge. The bridge is to be moved along the guideways and the dial indicator readings are to be noted at certain intervals.

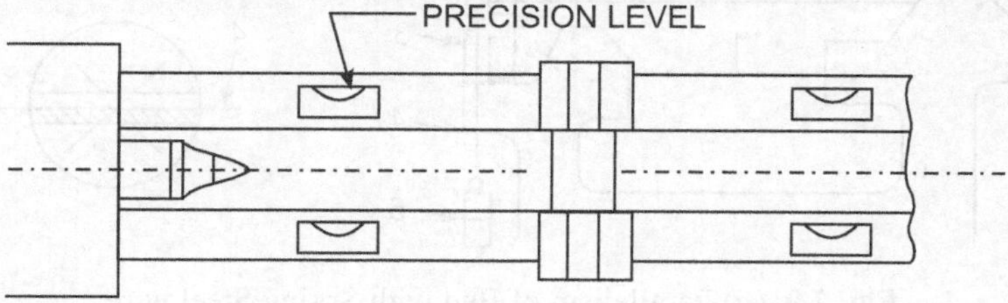

Fig. 7.7. Checking of Nonlinearity of guideways.

2. Spiral twist

Spiral twist is the measure of non-linearity across the guideways. The bridge (1) is set accordingly across the bed (Fig. 7.8) and the precision level (2) is placed in the middle of the bridge. The bridge is moved along the bed and the level readings are compared.

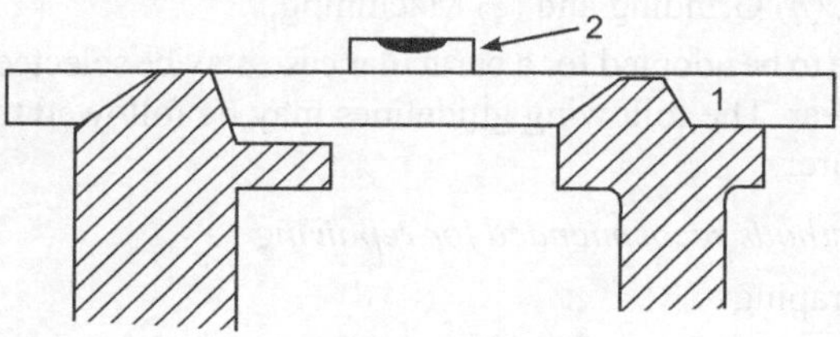

Fig. 7.8. Checking of Spiral Twist of Guideways.

3. Parallelism of guideways

Parallelism of two surfaces is required to be measured both in the (*a*) Vertical and (*b*) Horizontal planes.

(*a*) **Along the Vertical plane:** This can be checked in the same way as the non-linearity is measured.

It is checked usually by a dial indicator.

(*b*) **Along the horizontal plane:** The parallelism of the guide surfaces along the length can effectively be checked by the help of a spring steel wire (Fig. 7.9).

A bridge (1) is placed along the guide. At one end of the bridge an optical microscope (2) is mounted. The wire (4) is stretched over two brackets (3) fixed perfectly vertical at the front and rear edges of the bed. A dead load (5) is applied at one end of the wire to keep it stretched.

The bridge is moved along the guide surface and the microscope readings (Fig. 7.9*b*) are noted to determine the level in the horizontal plane. The microscope is to be set turn by turn at both ends.

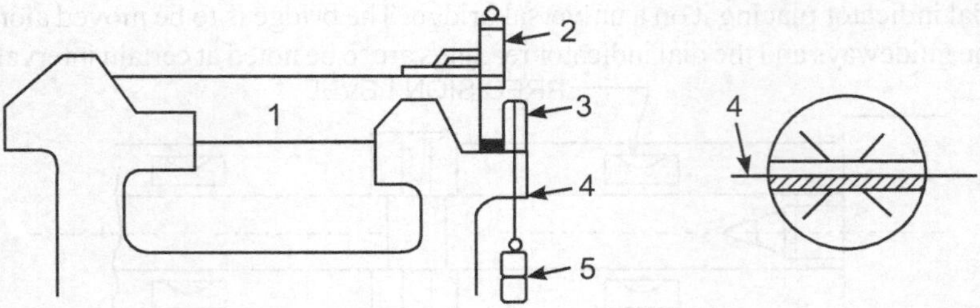

Fig. 7.9. (*a*) Parallelism of Bed with Spring Steel wire
(*b*) Position of Wire in Microscope Reading.

Methods of Repairing in General

The common methods of repairing of Guide Surfaces may be of three types, such as (*a*) Scraping, (*b*) Grinding and (*c*) Machining.

However, the procedure to be adopted for a particular case may be selected considering the extent of wear. The following guidelines may be followed to select the repairing procedure:

Extent of wear	*Methods recommended for repairing*
Up to 0.2 mm	Scraping
From 0.2 to 0.3 mm	Grinding
Over 0.3 mm	Machining (Planning and subsequent scraping or machining and grinding)

The general procedures of repairing of guide surfaces are discussed below. Different guide surfaces of a lathe have only been taken into consideration in this chapter to explain repairing methods along with a column of a horizontal milling machine.

Scraping

Scraping is a laborious procedure and deserves skill and care to finish a job. An effective cutting-angle may be approximately 20° as shown in Fig. 7.10. The point '1' shows the application of thrust from the right hand and '2' indicates the direction of application of thrust from left hand.

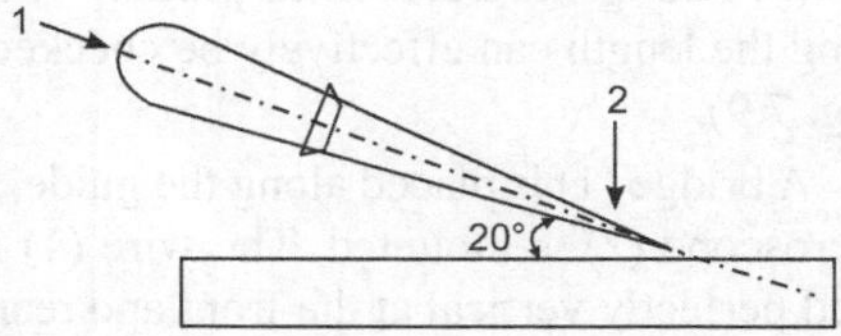

Fig. 7.10. Cutting Parameters of Scraper.

To scrap a guide surface, a straight edge of higher accuracy say 16-20 points/sq. inch may be used to obtain an accuracy of 10-12 points/sq. inch. Guide surface is to be cleaned off from oil or dirts. With the help of smearing technique, that is applying Prussian blue beneath the straight edge, the raised points may be identified and are scraped until the desired accuracy is achieved. Then the oil packets may be made throughout the bed surface by the scraper in the form of chess board as shown in Fig. 7.11.

Fig. 7.11. Formation of Oil pockets on bed.

Grinding

Grinding is needed when the extent of wear is in the range of 0.2 to 0.3 mm as stated above.

If the bed length is a shorter one, the bed may be ground placing, fixing and levelling it on a planer. An attachment is to be made to mount the grinding wheel on the tool-head of the planer. The attachment is nothing but a suitable electric motor (1) capable of working in the vertical plane and placed on a base plate (2). The grinding wheel (3) is mounted on the motor shaft. For rough grinding the wheel is to be mounted at an angle of 3° - 4° (Fig. 7.12*a*) and the finishing is performed keeping the grinding wheel perfectly vertical (Fig. 7.12*b*).

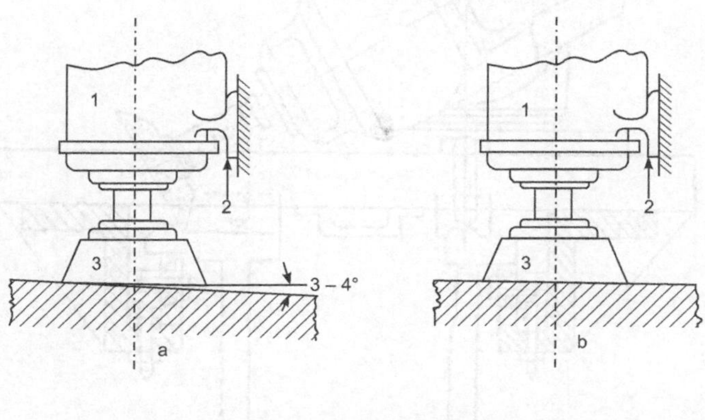

Fig. 7.12. Position of Grinding Wheel
(*a*) For Rough Grinding; (*b*) For Finishing

The feed is to be kept constant for both the forward and backward stroke.

But when the length of the bed is too long, say above 2.5 metres, it is ground with portable grinders. A separate attachment (Fig. 7.13) is needed again. Such an attachment consists of a base plate (1), which is screwed to two fixture plates (2), designed specially in conformity to the shape of the guide

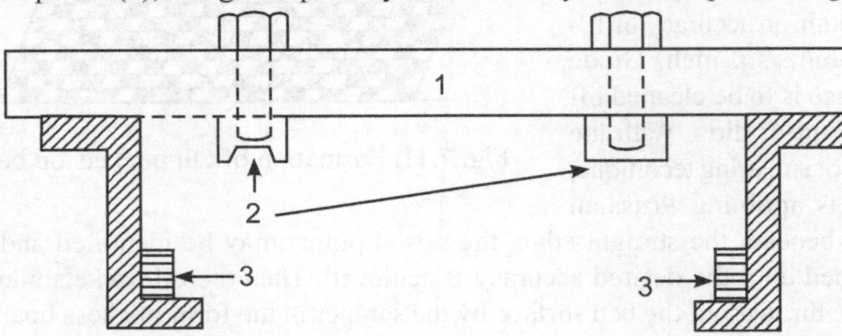

Fig. 7.13. Attachment to accommodate Grinder to grind long guideways.

paths on which it will slide. Four rollers (3) are fixed at the bottom of the base plate to slide it along the guide surface. Used ball bearings of proper size may be used as rollers. Finally, a portable grinder is to be mounted on the base plate.

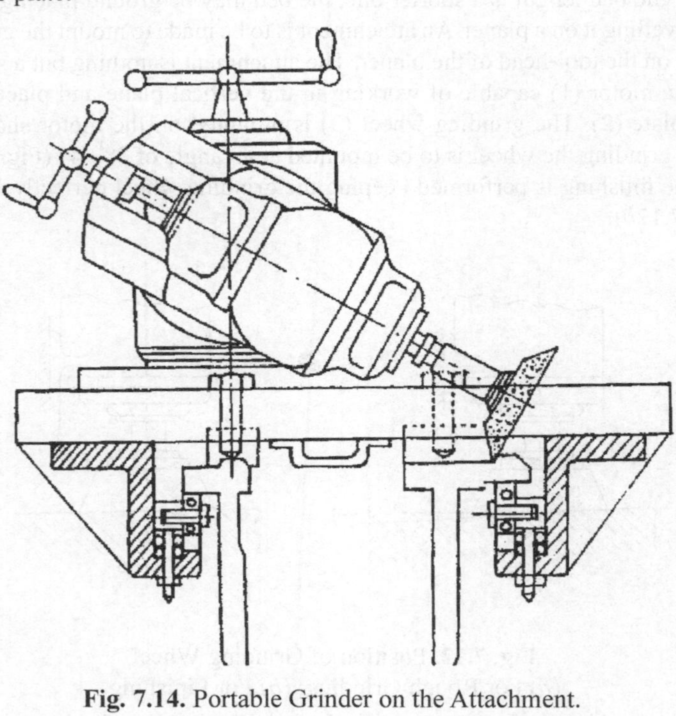

Fig. 7.14. Portable Grinder on the Attachment.

The grinder may be given feed manually by pushing the base plate or by pulling it by a wire which may be wound around a drum. A typical position of the grinder is shown in Fig. 7.14.

Position of grinding wheels to grind different profiles are shown in Fig. 7.15.

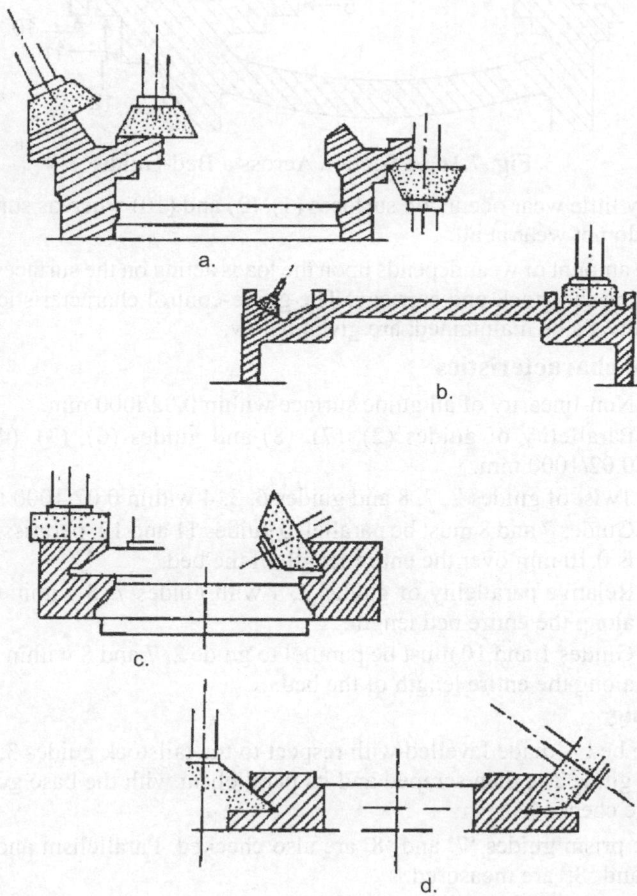

Fig. 7.15. Grinding of different profiles.

Lathe-bed-Guide – Repair

Application of the afore-mentioned procedures are discussed here with respect to a Lathe-bed-Guide and with control characteristics in detail.

The surfaces 3, 4 and 6 (Fig. 7.16) carrying the tailstock is subjected to less wear. But, considerable wear takes place on surface 7 and 8 which support front guideway of the carriage.

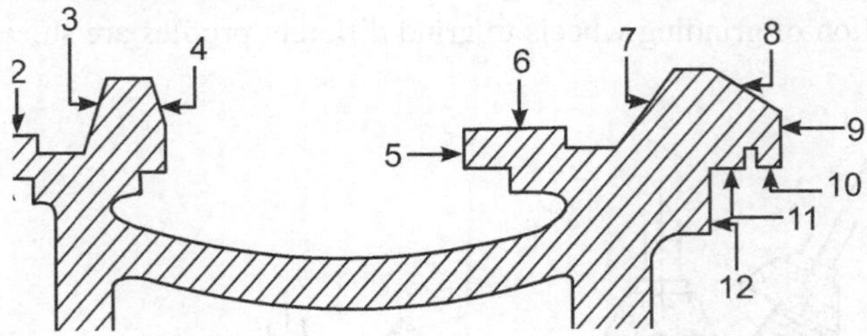

Fig. 7.16. A Section Across a Bed-Guide

Very little wear occurs on surfaces (1), (2) and (10) whereas surfaces (5) and (9) do not wear at all.

The amount of wear depends upon the loads acting on the surfaces through the moving tail stock and carriage. The guide-control characteristics, which should always be maintained, are given below:

Control characteristics

1. Non-linearity of all guide surface within 0.02/l000 mm.
2. Parallelity of guides (2), (7), (8) and guides (6), (3), (4) within 0.02/1000 mm.
3. Twist of guides 2, 7, 8 and guides 6, 3, 4 within 0.02/1000 mm
4. Guides 7 and 8 must be parallel to guides 11 and 12; permissible error is 0.10 mm over the entire length of the bed.
5. Relative parallelity of guides 3, 4 with guides 7, 8 within 0.03 mm along the entire bed length.
6. Guides 1 and 10 must be parallel to guide 2, 7 and 8 within 0.03 mm along the entire length of the bed.

Scraping

The bed is made levelled with respect to the tailstock guides 3, 4 and 6. The flat guide 2 is then scraped and its parallelism with the base guides 3, 4 and 6 are checked.

The prism guides '7' and '8' are also checked. Parallelism and twist of '2', '7' and '8,' are measured.

Parallelity of guides '7' and '8' with the base guides '3', '4' and '6' are also to be established. Oil pockets, as described earlier may then be formed.

Grinding

Grinding is a very rational method today. The bed may be mounted and levelled on the planer and the grinding is performed as discussed earlier with the help of a grinding attachment.

No scraping is generally necessary after grinding operation.

Machining

When the wear depth exceeds 0.3 mm, the bed is to be machined and then scraped.

A template is made on the unused portion of the main guides 2, 7 and 8 at an extreme end. A counter template which is required during machining is then prepared. Installing the bed on the planer it is levelled. The extent of wear is measured by a dial indicator on the surfaces 2, 7 and 8. Taking maximum wear depth plus an allowance of say 0.05 mm and bolting the counter template on the approaching end of the bed the machining operation is performed.

For machining guides 7 and 8, the depth of cut will be (cosine of the surface angle) cos a-times the previous depth of cut. The surfaces 3, 4 and 6 are also to be cleaned and machined. Then scraping is done as explained before.

Carriage Guide Repairing

A carriage guide is shown in the figure 7.17. The carriage guide paths 'cgf' is shown and the flat surface of the rear side (R) with which carriage rests on bed is also shown here.

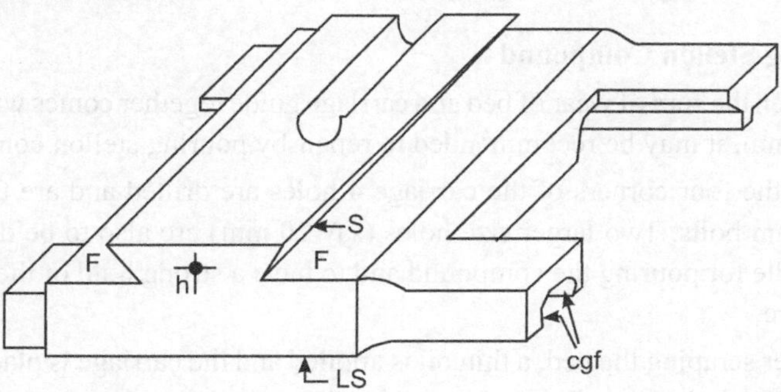

Fig. 7.17. A Carriage Guide.

The surface maintenance and repair for carriage guide includes the flat surfaces 'F', slant surfaces 'S', guideways 'cgf' and the flat portion of the rear end (R).

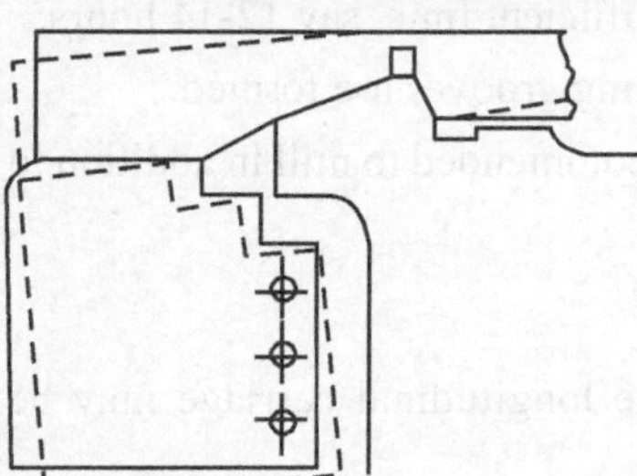

Fig. 7.18. Inclination of Apron Due to Wear of Guideway.

It is to be noticed that the surfaces 'F' and 'S' wear out more in the middle and as a result, their parallelism with the hole (*h*)-axis gets disturbed. The wear of guide paths 'egf' and rear-end flat surface (R) result an inclination of the apron towards the bed which is fixed on the lower surface (LS) as shown in Fig. 7.18.

Now for longitudinal guides, the pathways 'cgf' and the rear end flat surface (R) is scraped to an accuracy 8-10 points/sq. inch. The rear-end flat surface is generally scraped slightly more to keep the carriage slightly up on the apron side to accommodate the fast wearing of carriage guide paths 'cgf.'

For transverse guide, one of the flat surfaces (F) is scraped against a straight edge up to an accuracy of 10-12 points/sq. inch. At the same time parallelity with the other flat surface (F) is maintained within 0.05 mm over the entire carriage length and the parallelity of the flat surface with the axis of the lead-screw hole (h) is checked in the vertical planes.

The slant surfaces (S) are scraped one by one against a 55° angular straight edge. The parallelity between the slant surfaces is checked with a universal bridge. The parallelity of each slant surface with the lead screw hole (*h*)-axis is also checked.

All parallelity is maintained within 0.05 mm over the entire length.

Finer repairing of Carriage guides

(*a*) Using Stellon Compound

When the sum of wear of bed and carriage guide together comes within 0.1 to 0.25 mm, it may be recommended to repair by pouring stellon compound.

On the four corners of the carriage 4 holes are drilled and are threaded for 12 mm bolts. Two larger size holes (say 20 mm) are also to be drilled at the middle for pouring the compound and to have a strong hold of the poured substance.

After scraping the bed, a thin oil is applied and the carriage is placed on a film of the oil. It is levelled and installed. The co-axiality with feed shaft and leadscrew is also established. Wax is used all-around the carriage to prevent leakage of compound when it is poured. The compound is filled till it fills the higher size hole and is allowed to set for a sufficient time, say 12-14 hours.

Scraping is finally done and the lubricating grooves are formed.

For long guides two or three slots are recommended to mill in addition to the pouring holes of higher diameters.

(*b*) Machining

When total wear exceeds 0.25 mm the longitudinal carriage may be machined installing it on a planer or shaper.

Surfaces 1, 2 and 3 on the Fig. 7.19 are the carriage main guides. Apron is fixed on the face 6. The compound slide guides have been indicated by 5, 6 and 7 while 8 and 9 are the cross traversed feed screws.

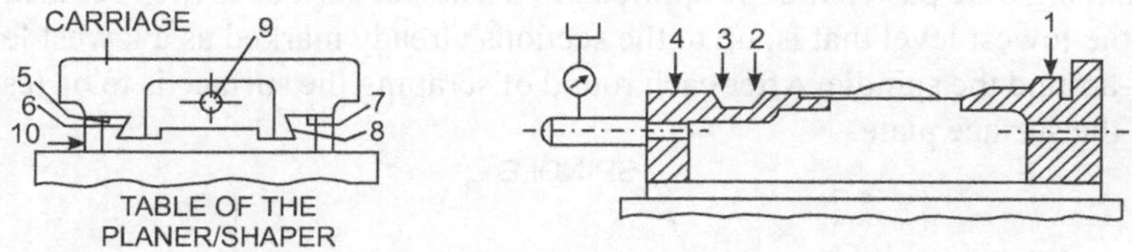

Fig. 7.19. Carroage Guide.

It may be noted that the carriage is placed on equal thickness pieces (10) as shown in Fig. 7.19.

Using Hylam packings

Minor wear (say about .0.25 mm) may also be rectified by applying hylam packings. Such packings may be pasted with araldite compound when the thickness of packing is within 4 mm. But, for packings of higher thickness a few countersunk brass screws may be installed in addition to araldite pasting.

Compound slide guides

It is advisable to restore a compound slide guide along with the transverse carriage guide. After getting the wedge surface of the transverse carriage guide repaired, the smearing technique may be applied to identify the raised points on the compound slide dovetail guides (Fig. 7.20) and scraping is done accordingly. Generally, an accuracy of 8-10 points/sq. inch is maintained.

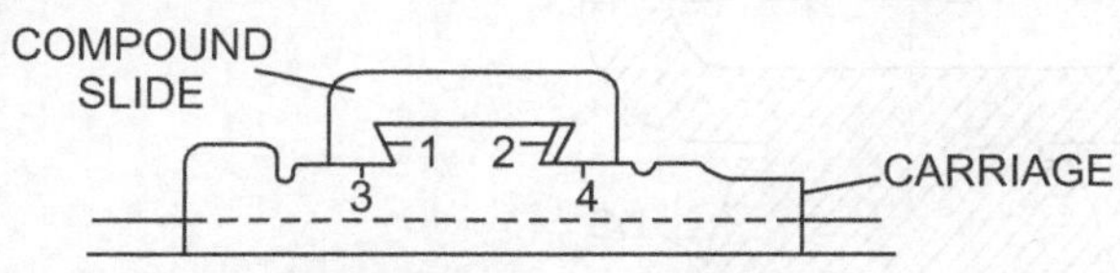

Fig. 7.20. Compound Slide Guide.

Repair of Column guides of Horizontal Milling Machine

The column guide of horizontal milling machine contains Flat Surface (FS) and Dovetail Guides (DS) as shown in Fig. 7.21(*a*). Before proceeding to the repair of flat guides, extent of scraping should be finalised. This is done by installing a dial indicator on the spindle and the reading is taken at a maximum radius as possible.

If the spindle is rotated one complete turn, the lowest level can be identified

(Fig.7.21 *b*). All-around the spindle, at an interval of 90°, three more sections are scraped, which are made equal to the level of the lowest level section. Now the bed is removed and Flat Surfaces (FS) are placed on a surface plate, which is having blue paste, already applied on it. The flat surface is then scraped up to the lowest level that is, up to the sections already marked as a lowest level all-around the spindle. After each round of scraping the surface is to be tested on the surface plate.

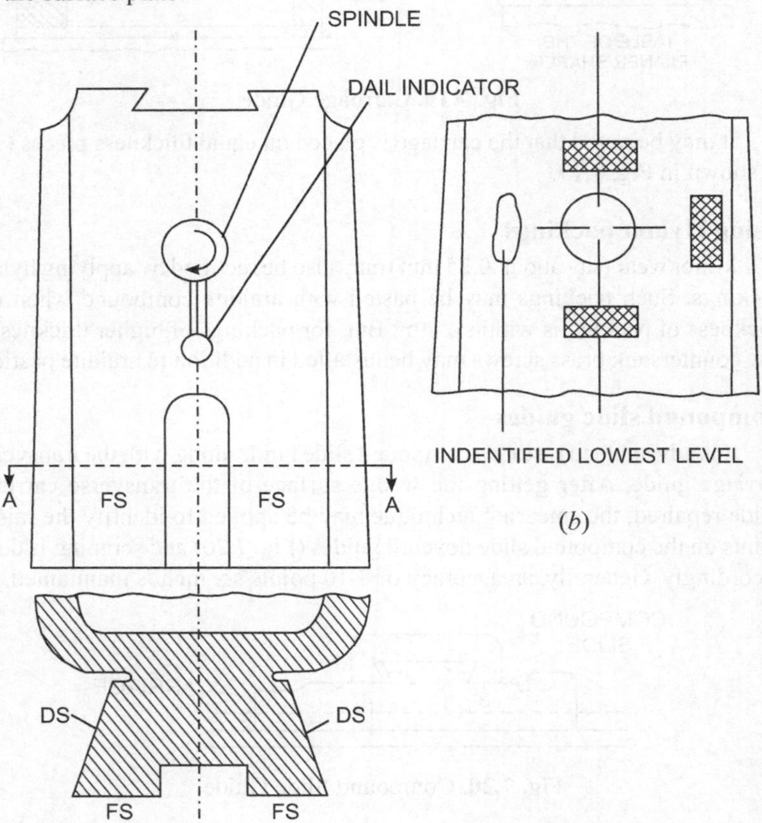

Fig. 7.21. Column guide of horizontal machine.

When blue paste marking will be found on the four identified sections (Fig. 7.21 *b*) along with the overall surface of the bed, the flat surface may be taken as completely repaired. An accuracy of 10-12 points/sq. inch is generally maintained in this regard.

The dovetail surfaces are then to be scraped with the help of 55 degree angular straight edge. The parallelity in between the two dovetail faces may be checked installing dial indicators on a suitably designed universal bridge.

The linearity and parallelity of guides are generally kept within 0.02/1000 mm and the perpendicularity of the flat surface with respect to the spindle is maintained within 0.02/300 mm.

REVIEW QUESTIONS

1. What are the causes of mechanical wear of guideways and slides? Explain the methods, generally used in measuring the extent of wear of the guide surfaces.
2. What are the common procedures to recover worn surfaces? Describe in detail.
3. Describe with neat sketches the specific features of maintenance of guide surfaces of machine tools. What check-up should be done in repairing a lathe saddle?
4. How are the wear on column guides of horizontal milling machine identified? How can it be repaired?
5. With the help of a line diagram, show the region on a lathe bed, where much wear may take place. Describe procedures to restore worn lathe bed with special reference to the control characteristics.
6. Fill up the gaps in the following sentences:
   (*a*) To take readings of dial indicator on different profiles of guideway.............................. is mostly used during measuring extent of wear.
   (*b*) The non-linearity across the guideways is called.......................
   (*c*) If the extent of wear is more than 0.3 mm........................ is done.
   (*d*) The grinding is recommended generally when the extent of wear is from.............................. to.............................. mm.
   (*e*) The permissible limit of non-linearity of all guide surfaces is within.............................. per.............................. mm.
7. Rewrite the following sentences with the appropriate word.
   (*a*) Scraping is done when the extent of wear is (below/above) 0.2 mm.
   (*b*) The hydrostatic procedure is employed to measure wear on (long/short) guide surfaces.
   (*c*) Scraping is not necessary generally after (grinding/machining).
   (*d*) Smearing technique is used to (measure/identify) wear on guideways.
   (*e*) Hylam packing may be pasted to restore worn guide surface in case of (major/minor) wear.

8

MACHINE PARTS: REPAIR, RESTORATION AND RECTIFICATION

Introduction

Maintenance of any machine tool, engine or other devices involves the repairing or replacement of any part of a component or the component as a whole.

The general procedures for various types of repairing works may be filing, scraping, machining, grinding, welding, surface restoring and so on.

It is just impossible to specify any repair technique in a generalised way. However, if the faults detected during the inspection or testing is of severe nature and do not allow to re-use the component after repairing, the replacing of the component will be essential. Other than that required repairing work is always undertaken to re-use the component.

But whenever filing or machining will be done, the removal of metal will certainly make it undersize, however small it may be. So, sometimes, the situation deserves to deposit metals by welding or by some other suitable methods primarily and then the surface may be regenerated by proper machining.

We will discuss repairing of a few components or parts. But before going into the definite repair methods the metal deposition procedure should be discussed. They may be:

(*i*) Depositing weld materials

(*ii*) Spraying procedures etc.

Deposition of Weld Material

The gas-metal arc welding procedure is an efficient way to deposit materials as a repairing measure. The very principle of the gas-metal arc welding is to utilise the heat of an arc formed between the work and the electrode, being continuously fed. The heat generated from the arc melts the work surface and the end of the electrode, resulting deposition of the molten electrode material on the work surface through the arc. The same principle and procedure is followed to rebuild a work surface providing an overlay with required properties.

Such processes are very much common to salvage and restore the original condition of a worn-out shaft.

This procedure improves service performance and extends the life of the surface.

For a circumferential weld deposit the process should be closely followed.

The most important criteria are to select the particular electrode of a definite composition for a definite use and the speed of welding along with the offset distance to be maintained accurately to perform welding.

However, skill and experience will simplify all these aspects. Machining is to be provided finally to achieve the exact dimension.

Splines, keyways etc. can be easily repaired by this process.

Metal Surfacing or Metalizing

Metalization is a thermal spraying process. Two consumable electrode wires are fed through nozzles. An outline of the procedure is shown in Fig. 8.1.

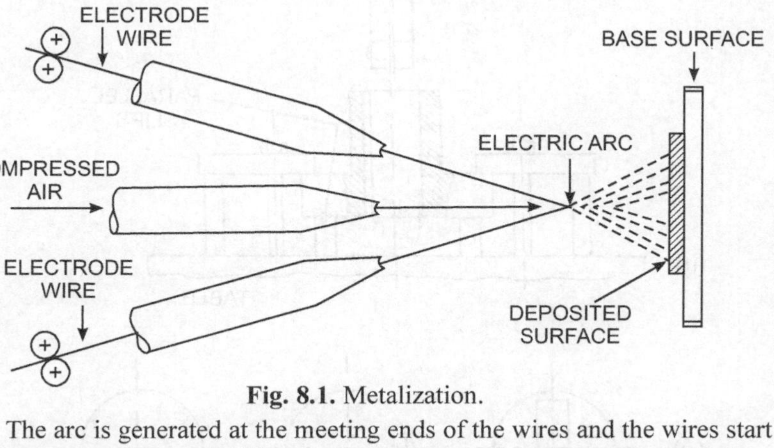

Fig. 8.1. Metalization.

The arc is generated at the meeting ends of the wires and the wires start melting due to the produced heat. A compressed air jet is located just in line with the junction of wires. The molten metal is forced to adhere on the base surface by the compressed air forming a lining on it. The thickness of deposit may be raised as per the requirement.

Such metalised coatings adhere to steels, aluminium alloys, monel metal, nickel etc. Machining can also be performed on the resulting surface.

The power source for producing the arc may be a direct current constant voltage welding machine in case of an electric arc welding or may be an oxy-acetylene or Oxy-hydrogen flame.

It may be noted that before metal spray the surface is to be made free from all dirts or oils etc., and is to be made roughened. Rough turning or knurling are usually done for shafts etc.

Repairing of Keyways

When a key is worn out, it is better to replace it. But when the keyway is also to be restored the following methods may be followed. The given example is to restore the keyway of a gear wheel.

(*a*) Weld material may be poured in the worn-out region of the keyway.

(*b*) The keyway is then to be marked off with the help of a vee-block and scriber. Centre punch may be used to punch the corners of the marked outline of the keyway.

(*c*) The job is set on a slotter table on parallel slips. The space below will act as tool-clearance (Fig. 8.2*a*).

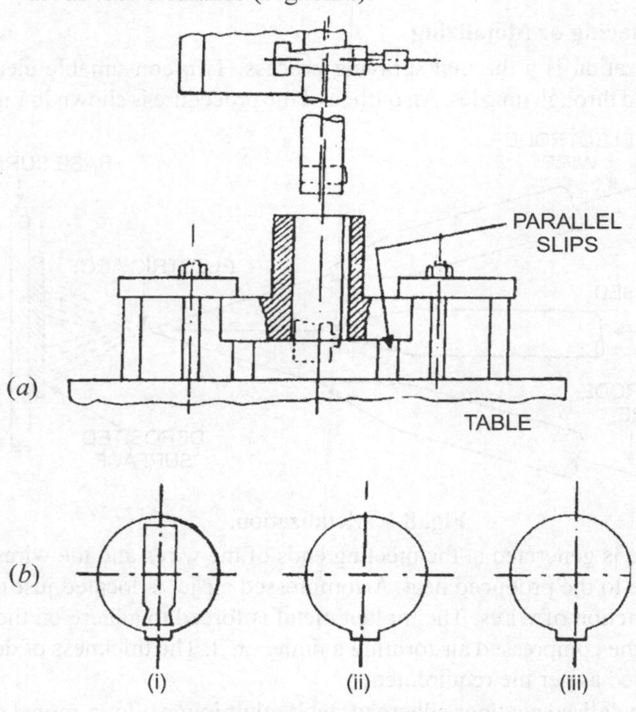

Fig. 8.2. Repairing of keyways.

(*d*) Set the job vertically aligning the axis of the stroke

(*e*) The tool is lined exactly on the marked off keyway

(*f*) The keyway is cut upto the centre punched outline

The tool setting is 'perfect for the Fig. 8.2.b (*i*), whereas a result of wrong tool setting is shown in Fig. 8.2.b (*ii*) and (*iii*).

However, if the situation allows, without going into the weld deposition, an enlarged keyway may be made and a step key may be used as a temporary measure.

Gear tooth rebuilding

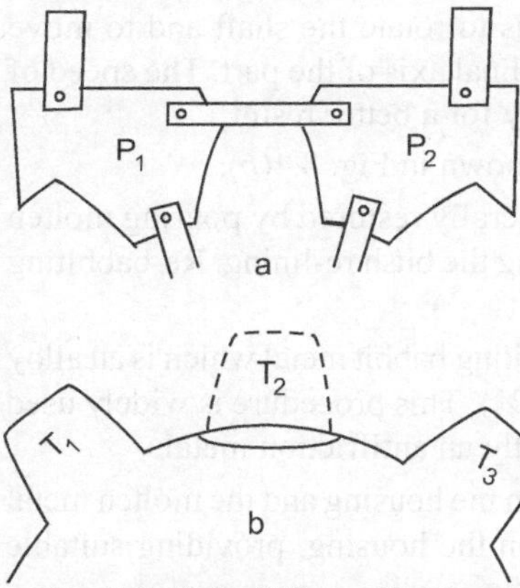

Fig. 8.3. Rebuilding of broken gear tooth.

If a tooth of a gear goes broken, that can be regenerated by welding deposits. Suppose, out of the teeth T_1, T_2 and T_3 (Fig. 8.3*b*) the tooth T_2 has gone broken. To deposit the metal in a particular form a template may be designed as shown in Fig. 8.3*a*. Generally, the templates are designed in two pieces (P_1 and P_2). For setting the templates in the exact place, different connecting plates may be arranged. The Fig. 8.3a shows, fixing plate among the templates P_1 and P_2, and fixing plates to fix them with the rim of the gear and with some other external support, if necessary. The gear is to be held rigidly, before placing the templates on it.

Before laying the weld material, the base surface is to be made roughened and properly cleaned and the weld material may be deposited through the gap in between the two templates.

The gear is to be allowed to cool down slowly and necessary finishing operations may be done on the re-built tooth. There are limitations of such re-building of teeth.

It can only be made for larger gears and it should be done where no impact load is present.

Bearing Assembly Repairing

Wearing mostly occurs in the inner race of the bearing or on the definite portion of the shaft. However, the result is an improper surface contact, which generates noise and produces heat while in operation.

If the extent of wear is too small, a very temporary arrangement may be done providing knurling on the shaft surface in an extraordinary situation for a small period of time. But, usually a metal deposition by welding (Fig. 8.4*a*) or by the process

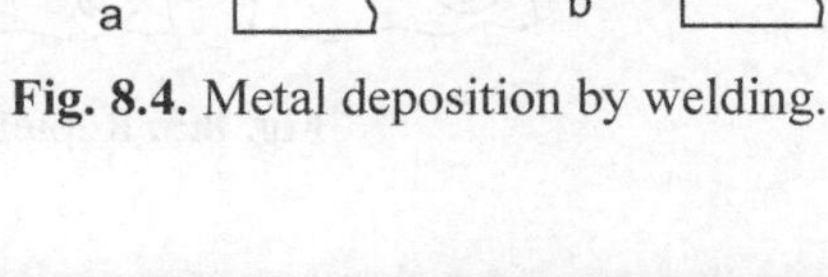

Fig. 8.4. Metal deposition by welding.

metalisation is provided on the shaft. It is possible to form a layer of upto 5 to 6 mm thickness.

The surface is to be made cleaned and to be made roughened. Better holding of deposited metal is obtained by performing knurling on the shaft beforehand. Generally, alloy steel is deposited by metalisation.

Care is to be taken during welding deposition to deposit materials on alternate sides continuously to heat the shaft shank uniformly and after the cooling of the shaft, the stress developed due to a rise of temperature is to be relieved by annealing. The usual practice is to rotate the shaft and to move the welding gun with respect to the longitudinal axis of the part. The speed of weld deposition is to be selected judiciously for a better result.

The metal deposited finished shaft is shown in Fig. 8.4(*b*).

Of course, the bearing assembly is generally restored by pouring molten metal as described in the chapter 6, regarding the bush re-lining. Re-babbiting is a popular term in this regard.

Re-babbitting is the procedure of depositing babbit metal which is an alloy of copper (4%), tin (88%) and antimony (8%). This procedure is widely used to restore the bearings as a babbit is basically an antifriction metal.

The procedure is to use a mandrel within the housing and the molten metal is poured through the drilled hole, made on the housing, providing suitable sealing in the sides to avoid leakages of metal.

Before re-babbiting, the old babbit is to be scraped out for necessary cleaning.

The mandrel may be painted with a mixture of graphite and gasoline for an easy removal.

Repairing of cracks on C.I. Body

The cracks developed on different surfaces due to various reasons may be-repaired by several means, as befitting. But, the cracks, especially in C.I. parts are not recommended for welding. Rivetting with headless copper screws is generally followed. Drilling of holes and placing the rivets are to follow a definite sequence, as has been illustrated in the Fig. 8.5(*a*), (*b*) and (*c*).

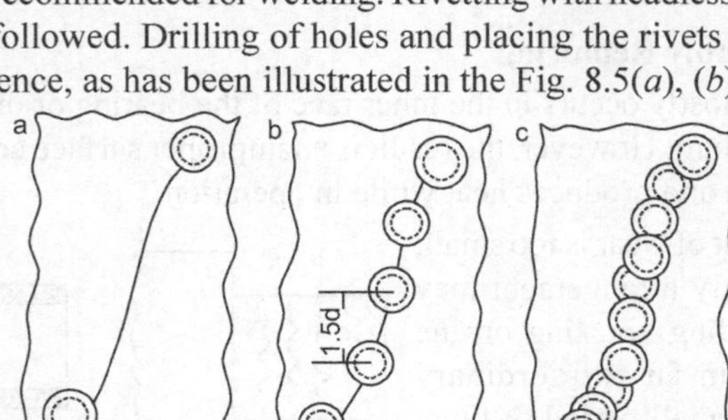

Fig. 8.5. Repairing of crack.

First of all, the hole according to the bolt diameter (*d*) is drilled at two extreme ends. The depth of the drill may be about 1.5*d*. The intermediate holes are drilled at the distance of 1.5d following the line of the crack. The copper screws are fixed and the projected ends are hammered to form a rivet head. The next step is to 'drill a hole and fixing rivets' in between the rivets already fixed (Fig. 8.5*c*). The first step, as described above, will arrest the probability of crack expansion and the third step will provide more strength of riveting to re-use the part.

Hot Clamping

Hot clamping procedure is efficiently applied to repair cracks. The clamping is usually done by fixing metal plates with bolts and nuts in hot condition on the cracked surface. When they cool down, their shrinkage will tightly close the crack together.

Two examples are illustrated with figures ahead, following the principle explained.

(*i*) Machine bed

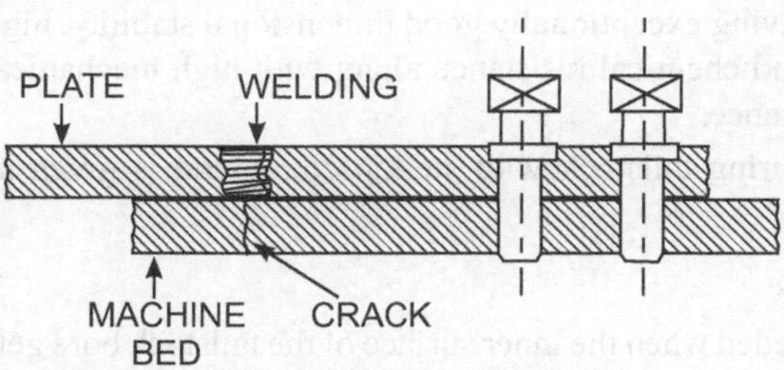

Fig. 8.6. Hot clamping on machine bed.

When a crack is seen in C.I. machine bed, it can be repaired by hot clamping. The two plates are placed as shown in the figure 8.6 and are bolted. These are special bolts and are tightened till the shank portion shears off.

The welding is done between the end faces. When it will be allowed to cool down, the shrinkage of the plates will bring the crack closed up

(*ii*) Anvil Crack

Due to the force of hammering, anvil cracks may occur. It is to be repaired in such a way that the anvil can withstand shock loads. The thick plates (may be 40 to 50 mm) are placed as shown in the Fig. 8.7. Bolts are usually heated separately and are tightened with nuts. Local heating is also permissible. The nut bases may be welded for rigid holding. While the cooling will take place, the bolts will contract and will make the cracks closed-up.

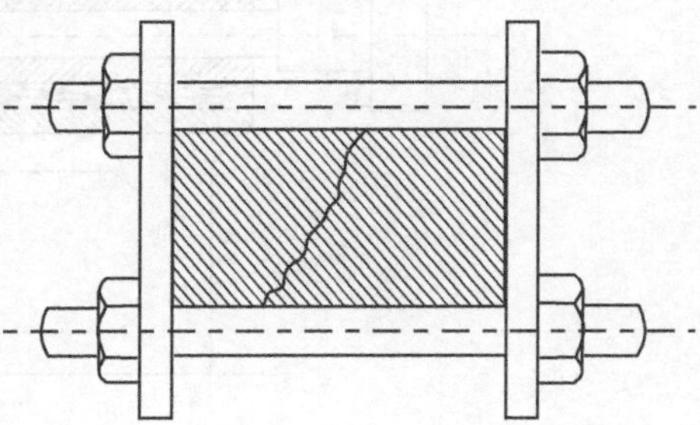

Fig. 8.7. Anvil crack repairing.

Repairing by Araldite or Stellon compound

Araldite: Araldite mixture is popularly used for a temporary measure to repair cracks on surfaces. Without using a rivet (as discussed previously), the drilled-hole at the two ends of the crack may be filled with sufficient araldite mixture. Glass tape is generally laid on the araldite. Again, araldite is poured on it and a second layer of glass tape is made. For each layer of glass tape it should properly be pressed and a few repeated layers of araldite will produce satisfactory repairing.

Cracks on pipes may also be repaired with such techniques.

Stellon: Stellon compounds are most efficiently used to-day to form a finished surface and to repair cracks etc.

Stellon compound consists of methyl methacrylate as the main component along with a small amount of other acrylates and crosslinking agent (initiator), such as tertiary butyl peroxide or dicumyl peroxide. A minute quantity of MEHO is also present. On heating, the mixture readily polymerises to form linear polymers initially, which then form a three dimensional solid network (cross linked) polymer having exceptionally good dimensional stability, high impact resistance, heat and chemical resistance along with high mechanical strength and shatter resistance.

An example of repairing tailstock with stellon compound is given as follows:

Repairing of Tailstock

Repairing may be needed when the inner surface of the tailstock bore gets worn or sometimes when the repairing of guide surface causes misalignment, lowering down the height of the axis of the tail stock. The simplest re-using technique of the same tailstock is to drill a larger size hole and to use a new suitable spindle of larger diameter.

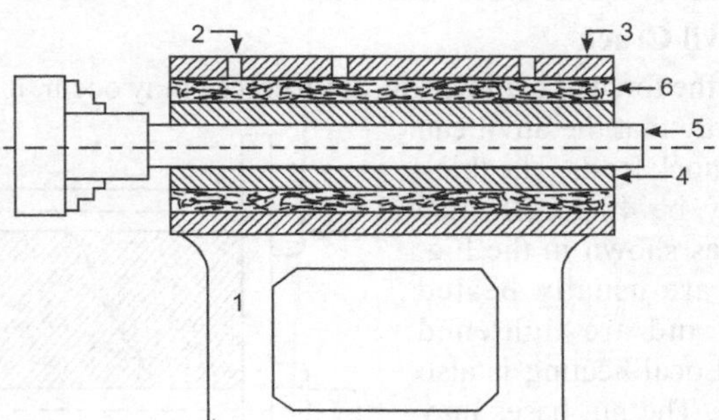

Fig. 8.8.

However, it is rather better to rebuild the inner surface of the bore. This can be done by depositing stellon compound in the inner surface.

The process (Fig. 8.8) is to take the tail stock (1) in between the headstock and the saddle. Re-boring is done through the existing hole of the tailstock to make it sufficiently large, considering the thickness of the stellon to be deposited.

A few small holes (2) are drilled at the top of the housing (3) to pour the stellon compound. The spindle (4) is placed inside the bore in alignment with the lathe axis. Generally, a tapered mandrel (5) is used to hold the tailstock spindle and the exact alignment may be achieved by holding the mandrel by a chuck properly.

The two sides of the tailstock bore may be temporarily sealed to avoid any leakage. The stellon compound is then poured through one of the small drilled holes continuously, until and unless the gap between the inner surface of the bore and the spindle is completely filled-up with the compound. The leakage of the compounds through other small holes at the top will ascertain it.

The compound is then to be allowed to become dry for a sufficient period of time, say about 12 to 15 hours.

The deposited surface (6) need not be machined any further and the tail stock may be recommended to re-use.

RECTIFICATION OF MACHINE BREAKDOWN

While in operation the machine tools will face a lot of problems. A definite trouble-shooting guide may be provided to rectify the breakdown by the shop floor people.

This is undoubtedly a major part of regular maintenance work. However, a guide line on probable problems are generally provided by the manufacturer of the machine tools.

Rectification guides for a centre lathe and for a horizontal universal milling machine is given as follows.

Of course, the following tables are for a particular size and type of each machine tool.

BREAKDOWN RECTIFICATION GUIDE FOR A CENTRE LATHE

	Apparent Problem	*Probable Causes*	*Suggested Remedies*
1.	HEAD STOCK		
	(*a*) Oil jet not visible when the motor is started.	(*i*) Pump is not working or pipe to oil sight glass might have been blocked.	(*i*) Remove top cover and clean pump and piping.
		(*ii*) No oil in the pump.	(*ii*) Check-up oil level and replenish.

	Apparent Problem	*Probable Causes*	*Suggested Remedies*
	(*b*) Noise in Head stock	Gearas damaged due to shifting while running.	Replace the damaged gears. Shifting the gears should be done only after the main spindle stops.
	(*c*) Spindle bearings, heating up excessively.	Lack of lubrication.	Check-up for proper lubrication and set the oil points properly.
	(*d*) Vibration and noise in Main spindle.	(*i*) Bearings running loose.	(*i*) Adjust the belt tension.
		(*ii*) Bearings damaged	(*ii*) Replace bearings.
	(*e*) Main spindle stops under load.	V-belts slack.	Adjust bearings.
	(*f*) Pully heating up excessively.	Belt tension more.	Adjust the belt tension.
	(*g*) Speed changing lever not enagaging properly.	Gears not sliding due to damaged shafts.	Check-up for damaged shafts and replace, if necessary.
2.	FEED BOX		
	(*a*) Gears damaged.	(*i*) Shifting gears while running.	(*i*) Replace damaged gears. Do not shift gears when in motion.
		(*ii*) Lack of lubrication.	(*ii*) Check-up oil level in the sight glass and replenish.
	(*b*) Feed change gears jammed.	Forcible gears shifting when mating gears are out of mesh.	Replace the broken parts. Use the feed selector levers gently.
3.	Lead Screw/Feed rod not revolving.	Shear pin failed due to overloading.	Replace the Shear pin. Check-up the load or any obstruction to saddle movement.
4.	Vibration in machine.	(*i*) Incorrect levelling.	(*i*) Check-up and level as per test chart.
		(*ii*) Incorrect balancing of job when mounted on Face Plate.	(*ii*) Balance it properly.
5.	Chattering during parting off operation.	(*i*) Too much clearance between keeper plate and bed guideways.	(*i*) Adjust the clearance to the optimum.
		(*ii*) Top slide gib is loose.	(*ii*) Adjust the gib properly.
		(*iii*) Saddle gib is loose.	(*iii*) Adjust the gib properly.

	Apparent Problem	*Probable Causes*	*Suggested Remedies*
		(*iv*) Saddle clamping is not effective.	(*iv*) Set right clamping.
		(*v*) Tool overhang too much.	(*v*) Reduce the tool overhang.
		(*vi*) Wrong cutting angle of tool or blunt tool.	(*vi*) Replace tool.
		(*vii*) Main spindle being faulty.	(*vii*) Adjust the bearings.
		(*viii*) Work piece holding not rigid.	(*viii*) Check-up and tighten the chuck jaws.
		(*ix*) Vibration in the machine.	(*ix*) Reduce vibration as described in item 4.
		(*x*) Incorrect speed and feed.	(*x*) Select proper speed and feed.
		(*xi*) Improper tool height.	(*xi*) Set the correct tool height.
6.	(*a*) Taper or run out in jobs held in chuck.	Misalignment in head stock and bearing adjustment not proper.	Adjust bearing and alogn head stock properly.
	(*b*) Taper in jobs between centres.	Tailstock misalignment.	Re-align Tailstock as per test chart.
7.	(*a*) Main drive motor does not start.	(*i*) Failure of incoming supply	(*i*) Check-up incoming voltage at terminals.
		(*ii*) Electrical troubles like, failure of control voltage, fuses blown out, or contractors not functioning.	(*ii*) Check-up contactors, fuses etc.
	(*b*) Main Motor noisy.	(*i*) Fan of the motor may be loose.	(*i*) Tighten fan.
		(*ii*) Motor bearing failure.	(*ii*) Replace bearings.
	(*c*) Burning of motor.	Failue of overload protection.	Check overload relay and its setting.
	(*d*) Main motor starts immediately when isolating switch is put on.	Permanent contact established in Push Buttons or Contactors.	Check the spring action of all the Push Buttons and Contactors.
	(*e*) Main motor starts but coolant pump does not start.	Failure of supply of motor.	Check the voltage at Rey Rolle plug, check for blown up fuse and also check the Isolator.

BREAKDOWN AND REMEDIES FOR A HORIZONTAL UNIVERSAL MILLING MACHINE

Apparent Problem	*Probable Causes*	*Suggested Remedies*
COLUMN AND MAIN DRIVE		
(*a*) Execessive heating of main spindle.	Improper adjustment of axial and radial plays of the spindle bearings.	Adjust the plays of the spindle taper roller bearings.
	Less or excessive quantity of lubrication.	Check the pipe connection from pump to the bearing housing. Proper greasing of V.M. Head spindle bearings for 'V' machine.
(*b*) Lubrication Pump noisy or does not pump sufficient quantity of oil.	Pump plunger loose in the bore after a few years running.	Replace worn-out parts if necessary.
(*c*) Too much instantaneous noise near main motor at starting.	Flexible coupling rubber bushes might have failed after a few years.	Replace the rubber bushes of the coupling.
(*d*) Main Spindle does not stop quickly after switching off the main motor.	Brake not functioning	Check all the relevant electrical connections and tighten them.
	Excessive 'Armature play' of the electromagnetic brake.	Adjust the 'Armature Play' to the recommended value.
KNEE		
(*a*) Knee movement jerky.	Knee gib screws may be too loose or too tight. Improper lubircation.	Adjust the screws for easy movement. Ensure correct lubrication.
(*b*) Vertical feed screw heats up or ceases when the knee is moved up or down.	Insufficient lubrication.	Check for proper lubrication.
(*c*) Power feed movements occur when the crank handle is in vertical or corss cranking dial.	Interlocking limit switches in knee not fixed or connected properly.	Check the connections.
(*d*) Vertical power drive engaging lever does not stay in engaged lower position.	The spring loaded locating bush does not have sufficient force.	Adjust the spring force.

Apparent Problem	*Probable Causes*	*Suggested Remedies*
(*e*) Loud noise in knee at starting the vertical power drive or rapid.	Improper backlash setting of feed box to knee drive spur gear pair.	Adjust the play with two screws at the top of feed box after taking out the taper pin. Lock with larger dia taper pin.
FEED BOX		
(*a*) Lubrication pump noisy or does not pump sufficient quantity of oil.	Pump plunger loose in the bore after a few years running.	Replace worn out parts, if necessary.
(*b*) Too much instantaneous noise near feed motor at starting.	Flexible coupling rubber bushes might have failed after a few years.	Replace the rubber bushes of the coupling.
(*c*) Feed/Rapid electromagnetic clutch not engaging to give feed/rapid.	Faulty electrical connections.	Check all the relevant connection of clutches, contactor, limit switches and push buttons etc.
(*d*) The feed drive is not being braked properly when switched off.	The brake clutch in the feed box, which should energise for braking action is not functioning properly.	
SADDLE		
(*a*) Saddle hand movement very tight.	Tool less a clearance in the gib.	Adjust the gib.
OR		
Jerky movement in power feed.	Cross feed screw driving gear keyway is blocked by dirt or small chips.	Clean the parts and assemble back.
	Improper lubrication.	Provide proper lubrication. Check the lubrication pump in knee and all pipe connections.
TABLE AND OVERARM		
(*a*) Chatter when heavy cut is given.	Excessive axial and radial play of main spindle.	Adjust axial and raidal play.

Apparent Problem	*Probable Causes*	*Suggested Remedies*
	Gibs of table, saddle of knee may be loose.	Adjust the gibs properly.
	Arbor support bushes running loose in the bronze bearings of the overarm bracket.	Adjust ring nut on bronze bearing.
	Clamping of knee, saddle or table not effective.	(*i*) Replace damaged gears. Do not shift gears when in motion.
	Improper foundation and bed levelling.	Follow the instructions on foundation, level the machine as per chart and tighten all grouting bolts.
	Execessive backlash in longitudinal screw feed drive.	Replace worn-out nut.
(*b*) Overarm support bracket too tight on the guideways.	Nuts my be tight.	Loosen the nuts.
GENERAL		
(*a*) Lubrication distribution not effective.	Dirt in the pump.	Clean the pump and filter.
(*b*) Coolant not flowing.	Coolant motor running in opposite direction.	Interchange any two phases in the electrical connection of coolant motor.
	Dirt in the pump	Clean the pump.
	Dirt clogged in the mesh.	Clean the mesh.
(*c*) Electrical motors do not start.	Low voltage supply.	Check-up the volate supply.
	Blown out fuse.	Replace fuse.
	Loose contact in circuit	Tighten all screws and electrical contancts.
	Defective contact points in the contactor.	Clean or replace contactors, if necessary.
(*d*) Indicating light on electrical cabinet does not light up in spite of isolating Switch being in 'ON' position.	Loose connections.	Check and tighten the electrical connections
	Light bulb may be fused	Replace the bulb.
(*e*) Excessive noise of electric motor and excessive drive vibrations.	Faulty motor.	Replace the motor.

REVIEW QUESTIONS

1. Explain with sketches the repairing techniques of cracks on cast iron body.
2. Describe with sketch how worn out tailstock spindle and hole can be restored.
3. Suggest means to restore machine components like keys and journal hearings. How may a broken tooth of a large gear be restored?
4. Write short notes on:
   (*a*) Metal Spraying
   (*b*) Weld deposition
   (*c*) Repairing of cracks by Araldite and Stellon compound.
5. Identify probable causes and suggest remedies for the following:
   (*a*) Main spindle of a lathe stops under load.
   (*b*) Noise and Vibration in the main spindle of a lathe.
   (*c*) Excessive heat is generated on main spindle of a horizontal Milling machine.
   (*d*) Electrical motor of a machine tool is not being started.
   (*e*) Vibration is occurring in a centre lathe.
   (*f*) During metal spraying molten metal is not being adhered on the surface.

9

LUBRICATION

Introduction

Metal surfaces are mostly irregular in nature and when two surfaces are kept in contact with each other, they touch only at the tips of their asperities as discussed previously (chapter 3). The real area of contact is much smaller than the apparent. At this stage, a strong adhesion is produced due to the attraction of atoms of one surface with those of the other. While sliding, this adhesion force is to be overcome by shearing the junction of contacts. This junction shearing is actually the basic cause of friction in clean metals. The wear and tear of substance is also caused due to this phenomenon.

Lubrication is a procedure to separate the surfaces with a film of lubricant to minimise friction and to restrict wear and tear as well. The substances used for this purpose are called 'lubricants.'

Functions of Lubrication

From the above discussion, the major function of lubrication may be summarised as:

(*a*) To minimise friction in the mating surfaces;

(*b*) To minimise wear and tear of the surfaces;

(*c*) To carry out the heat generated due to friction, leaving the assembly cool as far as practicable; and

(*d*) To lengthen the service-life of the components by performing the above functions.

It may be noted that the lubrication can produce the desired result, depending on the:

(*a*) Physical, chemical and mechanical properties of the selected lubricants, and

(*b*) The lubricating methods.

Properties of lubricants

The major important properties of lubricants are:

Viscosity

It is the indication of the degree of fluidity of lubricating oil. It only focuses how the liquid/lubricating oil will flow under the temperature at which it is to function. Viscosity enables the lubricant to withstand load as because it is a resistance forming and retaining property. The rate of change of viscosity with temperature is called *viscosity index*. Low range of variation of viscosity with temperature identifies good viscosity-index.

Flash point and fire point

Flash point is the temperature at which the lubricants produce sufficient vapour at the application of a flame and the *fire point* is to some extent, higher temperature at which the generated vapour will ignite and the oil will burn steadily.

They are actually the properties of the lubricants to withstand temperature of the environment where it is applied.

Pour point

This is the property to know the behaviour of lubricants at low temperatures. The pour point can be specified by the temperature at which the lubricant ceases to flow. It is of major importance in the refrigeration grade of oil.

Oiliness

It is the degree of ability to form an oily surface on the metal. It is a combined measure of surface-tension and wettability to form a film of lubrication.

Neutralisation number

This number indicates the acidity or alkalinity of lubricants and is mea sured by the amount of potassium hydroxide (KOH) in milligrams required to neutralise one gram of oil.

Saponification number

Its purpose is to identify animal and vegetable oils. It is also tested as the milligrams of 'KOH' required to form soaps, when it is added to l gm of oil. Mineral oil will not help to form soaps and thus it identifies fat or fatty acids.

Carbon residual property

It indicates the amount of carbon residue on evaporating the oil under special conditions. Such tests are generally conducted while using oils for Internal Combustion Engines.

Types of Lubricants

The llubricants may be either solid or liquid or semi-liquid. The solid lubricants are used when oils or greases are unsuitable or not permissible, such as at very low or high temperatures. The solid lubricants may be used in pure or combination or suspended form.

The common solid lubricants are soap, mica, molybdenum-disulphite etc. Graphite is also used in mixing it with oils. The adhesive characteristic of graphite with oil enables it to retain film or oil on the surface. Polytetrafluorethylene (PTFE), being highly resistant to aggressive media, is also popularly used.

The very popular semi-liquid lubricant is grease. It has got a number of grades and depending on the types of lubrication required, the grade is judiciously selected. Grease is basically a mineral or synthetic oil and is thickened with thickening agents. The grease may be of various types, such as, Non-soap greases, Lithium base greases, Calcium base greases, etc.

Additives

To use very common lubricants, different additives are sometimes added with them to use the lubricant purposefully to establish an optimum operating condition. Viscosity-index improvers, corrosion resistants etc. act as additives to allow the lubricants to function at an atmosphere over the temperature range or to retard corrosion and rusting of metals etc.

Different types of additives are discussed below:

Additives may be of three major types. Primarily, it may be 'Antifriction Additives' which are generally fatty acids and natural or synthetic fats. They are used mostly to improve the power transmission capacities as such additives retain on the metal surface . Secondly, it may be' Anti wear Additives'. They are generally derivatives of Phosphorous and Phosphoric acids. They are operable upto a temperature of 200°C. Such additives reduce the wear of rubbing surfaces at moderate temperatures and loads.

Finally, it may be Extreme Pressure Additives. They can withstand high contact temperature and loads. They are generally organic derivatives of sulphur and chlorine. Such additives actually resist the effect of scuffing of the mating surfaces.

Selection of lubricant

The best selection of a lubricant for a particular type will be the simplest and cheapest one, which will of course, meet the requirements.

In general, in any assembly without any feed mechanism, a small quantity of mineral oil is ideal. Again, a small quantity may not be functional when wear debris creates problem or heat generation more. In that case, an oil-feed system may be preferred.

When cooling is not much required but sealing is the criterion, 'grease' is the choice.

But under high temperature range or for special requirements e.g. low flameability or low carbonisation, synthetic oil is preferred.

Generally, oil of lower viscosity is suitable for high-speed mechanisms and for high load and low-speed mechanisms, highly viscous oil or grease is used.

If the temperature becomes too high or too low, no product contamination is acceptable, if extremely long life is required, the best choice may be a solid lubricant.

Modes of lubrication

The modes of lubrications are usually characterised by the friction and wear behaviour of surface layers. It is generally of three types.

(*a*) Boundary lubrication;
(*b*) Fuid film lubrication; and
(*c*) Mixed lubrication.

The above modes can be identified considering the value of 'R', where,

$$R = \frac{\text{Mean fluid film thickness}}{\text{Surface Roughness (cla)}}$$

where, $R \leq 1$, it is boundary lubrication
If $5 \leq R \leq 100$, it is fluid film lubrication.
and when, $1 \leq R \leq 5$, it is mixed lubrication.

(*a*) Boundary lubrication

This type of lubrication mode is generally found in machine tool slide ways and most commonly in door-hinges. The principle is shown in Fig. 9.1. 1 and 2 represent the underlying materials. The area 4 is the boundary lubricant and the area 3 is the protective film which may be present, due to the physical and chemical reactions between the surface layer and the lubricant. It is generally an oxide film.

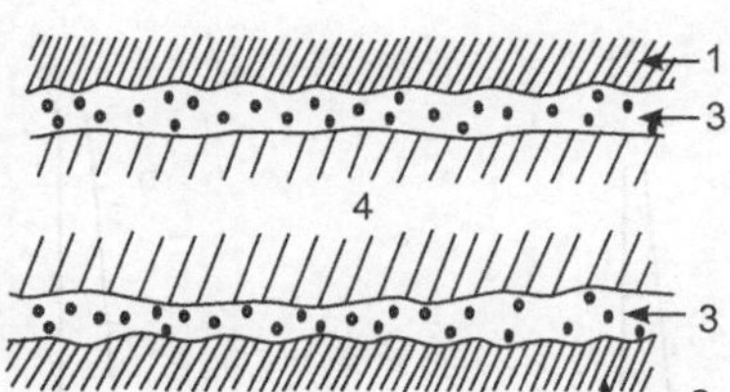

Fig. 9.1. Boundary lubrication.

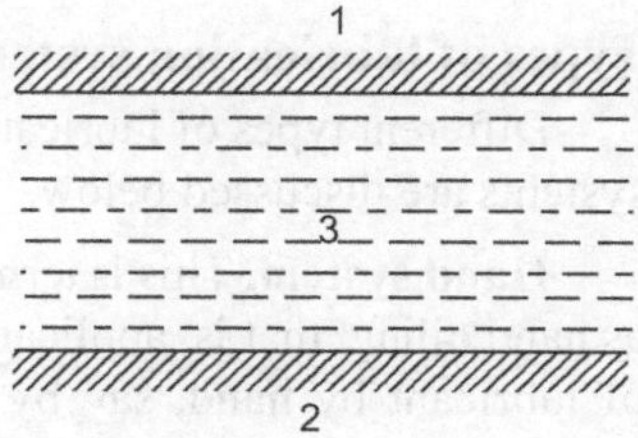

Fig. 9.2. Fluid Film

(*b*) Fluid film lubrication

Surface wear can best be minimised by the formation of a fluid film (3) in between the mating surfaces (1 and 2) as shown in Fig. 9.2. This film can be formed by the motion of solids or by the application of an external pressure. The former one is known as 'self acting' or 'hydrodynamic' type and the later one is 'externally pressurised' or 'hydrostatic type' of lubrication.

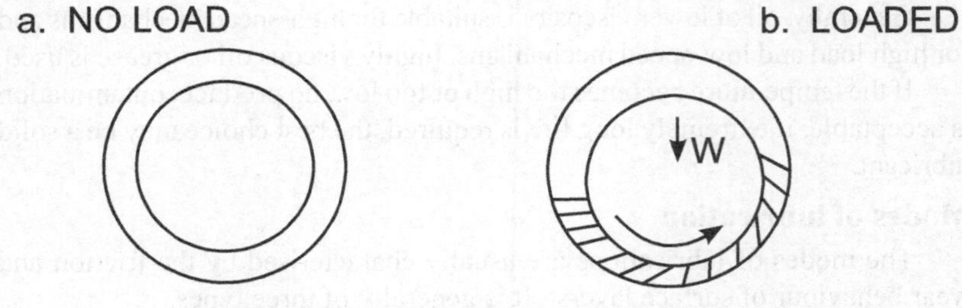

Fig. 9.3. Lubrication on journal bearing.

However, in both cases the main objective is to form a film of lubrication which should be thicker than the dimensions of surface irregularities. The best example of such a lubrication is as in case of a journal bearing (Fig. 9.3). The applied load (W) forces the journal into an eccentric position within the bearing, thus providing a converging film as shown in the diagram (Fig. 9.3*b*). At no load condition the film thickness remain constant (Fig. 9.3*a*).

Mixed lubrication

Although the mode of lubrication is classified as above but practically in most of the machine elements the lubrication mechanism is a mixture of both 'boundary' and 'film lubrication.'

In most cases, the surface interactions and boundary lubrication con tributes to the overall friction and wear characteristics in addition to a significant fluid film lubrication action. Such a phenomenon is termed as 'mixed lubrication', for which both the physical properties and chemical interactions with solids are to be taken into consideration for an efficient lubrication.

Types of lubrication systems

Different types of lubrication systems are discussed below.

Hand system. This is termed as hand oiling, that is, application of lubricant by hand, say by oil can. It is of course very common at a number of circumstances. But it is not at all an efficient system. Too much wastage of lubricant is caused. A chance of applying oil in excessive quantity is always there. Of course, for simple bearings and where an easy access is available,

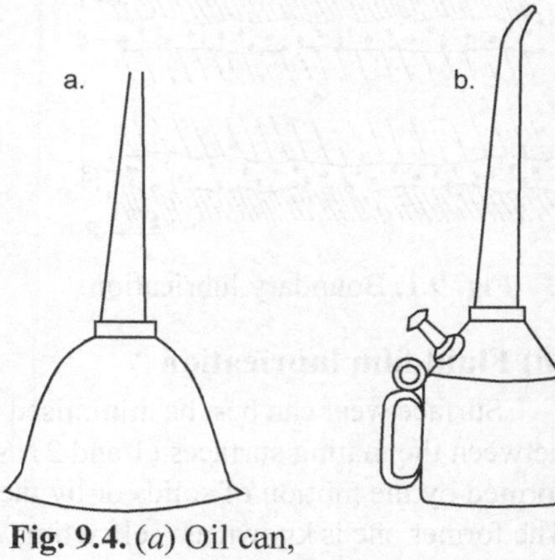

Fig. 9.4. (*a*) Oil can,
(*b*) Force feed lubricator.

this system is followed in good numbers. It is always better to use this system as a temporary measure. The Fig. 9.4 shows two types of oil cans, such as (*a*) a simple bench type and (*b*) a forcefeed type oil can.

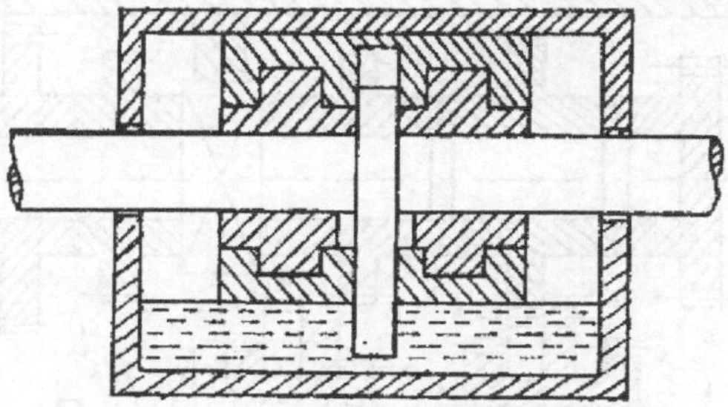

Fig. 9.5. Ring type lubrication.

Dip type. The principle is to keep a ring of chain dipped into the oil and its rotating motion will provide lubrication at the appropriate place (Fig. 9.5). The horizontal bearing can be easily lubricated by this system. The ring may rest on the journal. When the journal rotates, the ring will also rotate at a lower speed and will carry oil from the bath along with it. But, such a system is very much inefficient in high speeds.

Wick type. This system functions with the capillary principles (Fig. 9.6). A capillary tube (2) is inserted in a oil reservoir (3). The top of the capillary tube will certainly remain above the oil level. The wick (1) is inserted through it by insert stick (4) up to the point of lubrication. A continuous flow of oil will take place through the dipped wick. Amount of oil transfer will depend upon the formation of the wick and on the viscosity of the oil. Care should be taken to change the wick when it ceases to function desirably or it becomes contaminated.

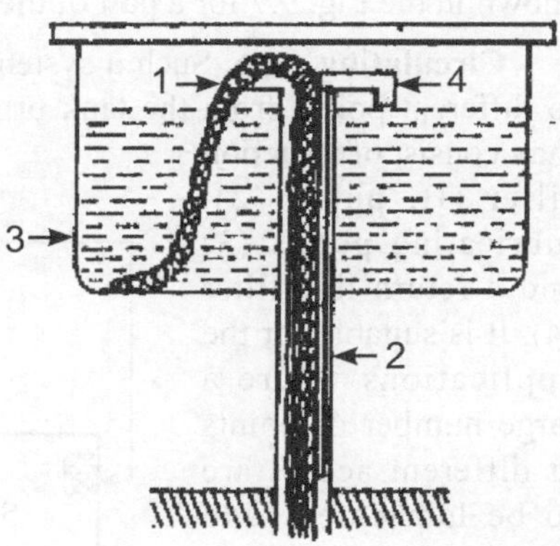

Fig. 9.6. Wick lubrication.

Splash type. Such a system operates when a running component itself is dipped into a oil sump and it splashes oil round the parts requiring it. Splash lubrication is very much common in small power vertical and horizontal high-speed engines with a closed crank case. It is also effectively employed

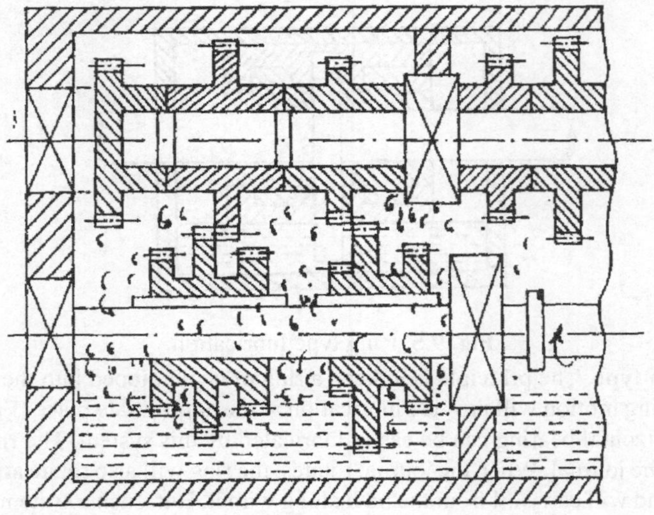

Fig. 9.7. Splash type lubrication.

in the all-gear-drives. The lower part of the gears remain dipping in the oil, as shown in the Fig. 9.7 for a part of the feed box of a lathe.

Circulating type. Such a system provides means to supply a lubricant to different points from the tank or sump by a pressure pump (Fig.9.8). It may consist of a suction filter (1), pump (2), lubricating points (3) and a return-line filter (4). It is suitable for the applications where a large number of points at different access are to be lubricated and the cost is ultimately justified.

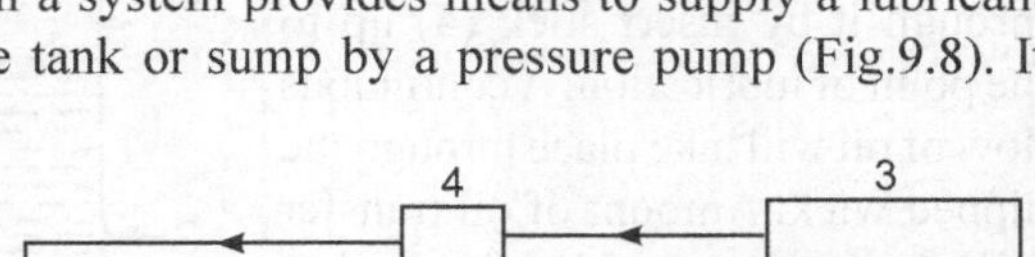

Fig. 9.8. Circulating lubrication.

Mist system. Its principle is to produce an aerosal spray. Lubricating oil is mixed with compressed air·, drop by drop and at the lubricating points the mist is throttled to atmospheric pressure. Thus, the finely divided oil particles make the rubbing surfaces wet and the air is permitted to escape to atmosphere.

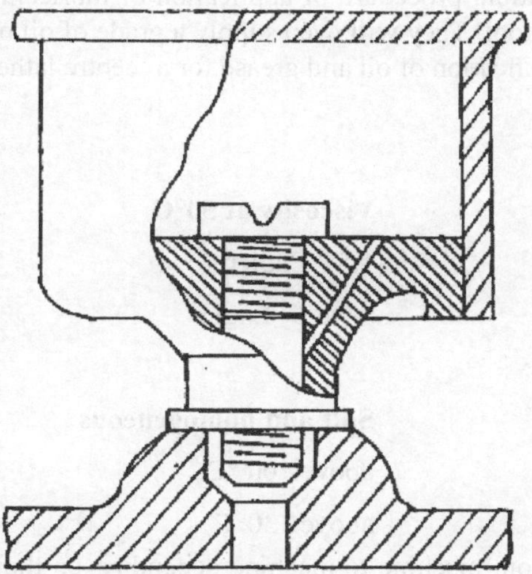

Grease cup. Grease lubrication is a popular type of lubrication and may be applied by the help of a grease cup (Fig. 9.9).

The rotation of the cup presses the grease and forces it to pass through the hole in the base piece.

Fig. 9.9. Grease Cup.

Such an arrangement may be employed only where design has provided the means for lubrication with such procedure.

Lubrication — a regular maintenance

It is needless to emphasise the importance of lubrication any more. It is a regular work being performed by the operators, attendants, during the operation. It will be sufficient to note that each and every system will be supplied by the manufacturer along with their suggestions and instructions for lubrication, regarding the quality of lubricants required, several procedures of lubrication to be followed, key points to be lubricated, intervals of lubrication etc. It is separately discussed along with most of the chapters in this book.

However, a typical example may be mentioned here to discuss the types of instructions, being generally supplied by the manufacturers.

Lubrication of machine tools

A special mention is hereby made on the practice of lubricating schedules of methods for machine tools.

Regular lubrication with properly selected lubricants is no doubt an important factor in the maintenance of high grade machine tools working under continuous load.

The manufacturer of machine tools will specify the type of lubricants to be used, frequency of lubrication, procedure of application of lubricants for a particular type of machine tool. They will even supply a grade of oil or grease to be used. A typical specification of oil and grease for a 'centre lathe' may be as follows —

(*a*) *Specification.of oil* :

Reservoir	**Viscosity at 50°C**
Head stock	1.5 ° Engler
Feed box, Apron	5° to 6° Engler

(*b*) *Specification of grease* :

Consistency at 20°C	**Soft and homogeneous**
Pour point	above 100 °C.
Drop point	above 130°C

This will of course, vary from machine to machine depending on their sizes etc.

Lubricating instructions

Manufacturers will generally prescribe the schedule of lubrication. The instructions, which should be strictly followed as a maintenance measure, will clearly specify the lubricating points, type of lubrication along with the quantity of lubricants to be applied and the frequency.

A typical lubrication instruction sheet for a centre lathe may be as given below –

Sl. No.	*Lubrication points*	*Type of lubrication*	*Frequency*	*Quantity*
1.	Head stock	Oil lubrication	1500 to 2000 running hrs.	18 litres
2.	Feed box	Oil lubrication	1500 to 2000 running hrs.	1.5 litres
3.	Apron	Oil lubrication	1500 to 2000 running hrs.	0.5 litre
4.	Saddle oil cup	Oil lubrication	Daily	Refill full
5.	Oil tray above quadrant	Oil lubrication	Daily	Refill full
6.	Oil nipple in quadrant	Oil lubrication	Daily	4-5 strokes of Oil gun

Sl. No.	*Lubrication points*	*Type of lubrication*	*Frequency*	*Quantity*
7.	Oil nipple in Tailstock	Oil lubrication	Daily	4-5 strokes of Oil gun
8.	Oil nipple in Cross slide	Oil lubrication	Daily	4-5 strokes of Oil gun
9.	Oil nipple in Top slide	Oil lubrication	Daily	4-5 strokes of Oil gun
10.	Oil nipple in End bearing	Oil lubrication	Daily	4-5 strokes of Oil gun
11.	Lead screw and Feed rod	Oil lubrication	Daily	Oil coat with a clean brush
12.	Bed guide ways	Oil lubrication	Daily	Oil coat with a clean brush
13.	Motor bearing	Grease lubrication	4000 running hrs.	Replenish grease

Cutting Fluids : Cutting fluids are frequently referred to as lubricants, applied to the material being cut. It is also known as 'Coolants.' The contact surface between the tool and the work piece or between tool and chips is lubricated in metal cutting operations, by the application of proper cutting fluids or coolants. It reduces the friction and the chances of welding between the contact surfaces.

Besides the above major functions, the cutting fluids also dissipate the heat generated during the cutting process. It also washes away the chips from the cutting zone, reduces the cutting force and improves the surface finish.

Friction is inevitable in sliding contact. But there is a difference between the friction occurs in general engineering practice and in metal cutting. The ratio of apparent and real area contact in case of metal cutting is nearly unity, where as the area of contact in normal practice is too small, may be about one percent of the apparent area. Moreover the contact pressure at the cutting edge is extremely high. So, the mode of lubrication in metal cutting is only boundary lubrication, as the laws of friction do not hold good at the above mentioned conditions. The chemical properties of cutting fluids are, that is why, more important than the physical properties.

Classification of Cutting Fluids :

The cutting fluids may be broadly classified as shown in the following chart.

-----OIL

----------PETROLEUM--------
BASED

------EMULSIFIABLE
OILS

CUTTING FLUIDS-------

-----WATER BASED
FLUIDS

---------NON PETROLEUM-----
BASED

---------SYNTHETIC
AND
CHEMICALS

Of course, it is to be further mentioned that air is also used, such as, air draft is used for grinding or boring operations. The main purpose is to remove the chips from the cutting zone, although some cooling effect is also performed.

Oils mean neat oils. It may be mineral or vegetable oil. Mineral oils are commonly used as additives, when the cutting speed is too high. Oils with extreme pressure additives are used widely. Mineral oils, blended with lard oil, sperm oil etc., commonly known as compounded oil possess high boundary lubricating properties.

Emulsifiable oil is soluble oil emulsified in the ratio of 1 : 20 to 100 parts of water. The proportion of the water is dependent on the type of operation. For example, a ratio of 1:30 is satisfactory for gear hobbing, milling, drilling etc. The concentration of mixture is also dependent on the requirement of level of rust prevention of the metal being operated. This is a low cost cutting fluid and is widely used. For severe cutting operations suitable additives are blended with it.

Synthetics and Chemicals are now a days used as the cutting fluids. A blend of water and different chemical agents, like Phosphates, Nitrites, Nitrates, Chlorine, Sulphur etc. are having industrial uses for specific purposes. Basically such fluids are mostly coolants. The major problem to use such synthetics and chemicals is their compatibility with other lubricants.

Selection of Cutting Fluids :

The selection of cutting fluids depends on various factors, such as, speed of metal cutting, machining process, tool material or work material etc. It may be remembered that the major advantage of a petroleum based cutting fluid is it's low surface tension and that of water based fluid is the high specific heat of water.

For a high speed machining, cooling action becomes more important than the lubrication, where as, for a low speed cutting, the lubricating action is mostly required for the severe cutting pressure, chances of welding of aspirates with the work piece or the tool. So, for a high speed cutting (cutting speed is more than 30 or 35 m/min), like turning, milling, drilling, grinding etc., water based fluids are generally used and for low speed cutting (tapping, broaching etc.), oils are recommended.

The other factors which are also to be considered during the selection may be the compatibility, economic feasibility etc. The chances of health hazards should also be considered before using any cutting fluid.

REVIEW QUESTIONS

1. What are the functions of lubrication ? Discuss the most important properties of lubricants.
2. How may the lubricants be classified ? Mention some of their uses.
3. How may the modes of lubrication be characterised ? Discuss different modes in details.
4. Describe with sketches different types of lubricating systems.
5. What type of lubrication devices you will recommend for the following cases:
    (*a*) For steam turbine - (Splash type/Pressure circulating system).
    (*b*) For Line shaft (Waste pad/Grease cup).
    (*c*) For the bearing of a machine tool (Gravity system/Pressure gun application).
    (*d*) For all gear drives of a centre lathe (Dip type/Splash type).
    (*e*) Multi-cylinder vertical I.C. Engine (Gravity circulation system/ Pressure lubrication system).

10

CORROSION, ITS CONTROL AND CHEMICAL CLEANING

Introduction

The term corrosion indicates electro-chemical reactions on metals with the environment. Metals are basically constituted with some or other chemical compounds and it has an obvious tendency to transform to its original form whenever chances are provided. This particular tendency generates corrosion and converts the metal into metallic oxides, sulphides, hydroxides, carbonates etc. Result is the loss of some amount of material and consequently loss of strength of material.

There may be different types of corrosion. It may be localised or uniform over the entire surface. The corrosion attack may be either direct or in two stages or electro-chemically.

In the direct chemical attack, a uniform reaction takes place throughout the entire surface and the deposits may act sometimes as an insulation for further corrosion.

But in the two-stage corrosion attack, the deposits through primary corrosion act as a second corrosive product.

And in the electro-chemical mechanism, the second stage corrosion products, if deposited at some distance, an electric current flows between the precipitates. The contact between dissimilar metals also result in such corrosion. However, this type of mechanism is due to a transfer of electrons between cathodic and anodic areas.

Types of corrosion

Different common forms of corrosions are discussed as follows:

(*a*) Galvanic Corrosion

Presence of two different materials in contact in a conductive solution produces galvanic corrosion. This corrosion mechanism is similar to the principle in a battery, where a potential difference occurs in between a cathode and an anode metal surfaces. In such a galvanic arrangement the active material deteriorates quickly.

The preventive measure is to keep dissimilar metals well-apart or to use metals which are closer together in the galvanic series.

(*b*) Pitting corrosion

Pitting is caused generally due to a rupture of the protective film on a metallic surface, either by surface irregularities or by local straining or through impingement of foreign materials (Fig. 10.1). Such a corrosion occurs like a pinhole (1) primarily in a localised way but, ultimately it spreads and may form microcracks resulting in failure of material.

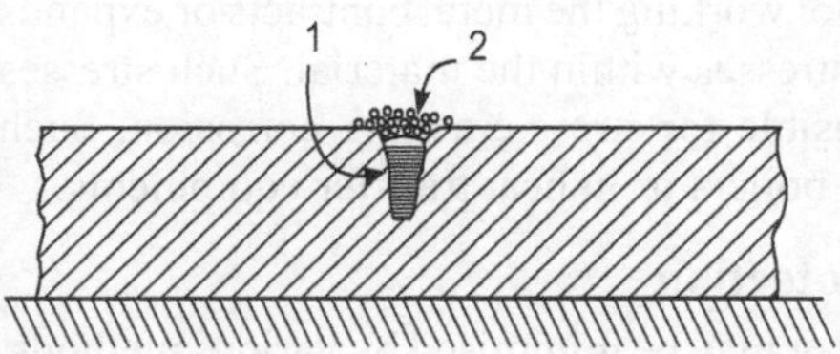

Fig. 10.1. Pitting corrosion.

The metals like brass, copper etc. produce surface oxide readily and the pinholes are generated on the surface. Such corrosions are generally due to change in the acidity of the pit-area. Sometimes the corrosion products (2) act as an insulator to the pit.

Pitting corrosion is very common with stainless steel or aluminium in an aqueous environment containing metallic chloride salts.

(*c*) Intergranular Corrosion

This corrosion occurs mostly on the grain boundaries. Metals are composed of different elemental components. They are supposed to be evenly distributed throughout the surface. But this seldom occurs. After different operations, like welding etc. the components concentrate in different locations forming different pockets and the grain boundaries (1) act as more anodic than of the grain centre (2) in a corroding medium. The result is intergranular corrosion. The mechanism of which is shown in Fig. 10.2.

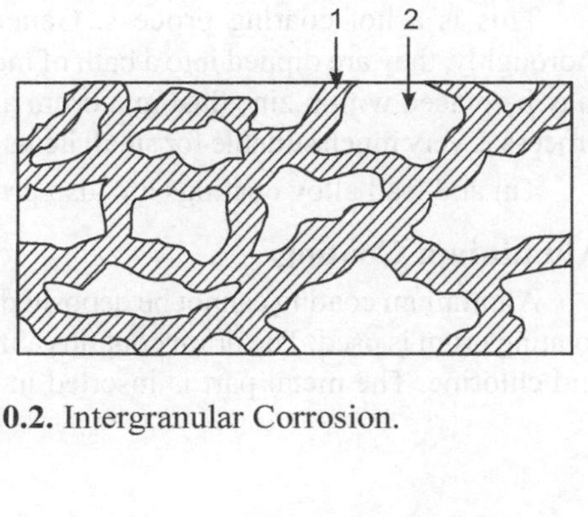

Fig. 10.2. Intergranular Corrosion.

(*d*) Erosion Corrosion

For a worn away metal surface such type of corrosion is very much common. It is not the direct effect of electrochemical reaction but it is a combined result of wear and corrosion.

The protective oxide films are raputured due to wearing of metal and the electrochemical reaction proceeds creating corrosion.

This is very much common in condenser tubes, turbines, pump impellers etc. where some liquid flows with high velocities and strikes the surface.

(*e*) Stress Corrosion

A metal surface may experience repeated stress concentration. When such phenomenon occurs in a corrosive environment, the metal is subjected to stress corrosion. During cold working or hot working the metal contracts or expands creating repeated concentration of stresses within the material. Such stresses create micro-cracks and are responsible for stress cracking corrosion. Such corrosion is very common in steam boilers or in heat transfer equipments.

Corrosion Minimisation or Protection

The corrosion may be protected or may be minimised by various methods, a few of which are discussed below.

Electroplating

Electroplating may be defined as a process in which metals are deposited or plated from solutions containing metallic ions by means of electrolysis. It consists of setting-up the articles (to be coated) as cathodes in an electrolytic bath, the anode being metallic zinc or nickel etc. Generally, zinc, cadmium, tin, nickel, chromium coating can be applied by this method. Zinc coatings are very much common. A light coating of cadmium works similar to a thick coating of zinc. But, to put importance on appearance also, nickel or chromium is used.

Such coatings, if free from pin holes are very much resistant to corrosion.

Galvanising

This is a hot coating process. Generally, after cleaning the articles thoroughly, they are dipped into a bath of molten zinc. Sometimes, the articles may be placed within zinc dust in a drum and is heated up to 300°C. Such a process is very much suitable for small items like screws, bolts, pipe fittings etc.

Tin and lead alloy coatings are also produced by galvanising.

Aluminium Coating

Aluminium coating cannot be deposited by electro deposition. Here also, a rotating drum is used. The drum contains aluminium powder, aluminium oxide and chlorine. The metal part is inserted in the rotating drum. Such coatings are proved to be efficient on duralumin. But on steel or iron the layers of such coatings are to some extent brittle in nature.

Phosphate Coating

A paste like solution is made by mixing H_2PO_4 with iron fillings. The iron and steel articles are to be kept immersed in the boiling solution for about 3 to 4 hours. Application of oxidising agents improves the coating quality.

Of course, the corrosion resistant quality of phosphate coatings is very much limited. They are generally used as antirust coating.

Metalising

Such a process is similar to a metal spraying process. Metal powder is fed on an oxyacetylene flame at a controlled rate, over a surface. The atomized metal will be impinged on the surface forming a thin layer coating. It has been discussed in Chapter 8.

Thermoplastic Coating

Fine plastic powder or a liquid PVC is used to coat the materials in this process. The hot metal piece is to be dipped into the bath to produce the necessary coating. Moreover, tin, nickel, copper or chromium coating on thin sheets produce better results. However, lead-coating on steel is very popular in industrial use.

Anodic Coating

An anodic coating produces a coating of oxides to aluminium, magnesium or to their alloys. The base metal is immersed in an electrolytic solution and is connected as an anode. The film, thus deposited on the base metal enhances the resistance to corrosion. The base metal is to be cleaned in alkali after degreasing and the anodic coating (3 to 10% solution of chromic acid) is applied.

The coating contains aluminium oxides and chrome salts. Sulphuric acid, oxalic acid or boric acid bath process may also be employed to produce anodic coating.

Chemical Cleaning

'Cleaning' indicates the removal of dirt, soil and other impurities from the surfaces of any kind. The cleaning processes should be powerful enough to replace the undesirable films on a surface by acceptable films, that is compatible with the end use of metal cleaning like painting, lacquering, enamelling or finishing with some other operations, if necessary.

The variety of soils encountered in the cleaning of metals is numerous.

They may be generally comprised of debris from previous processing, mineral oils, corrosion inhibitors etc. Cleaning activities do not have any well-defined classification. However, broadly the surface cleaning may be discussed with respect to the type of soil or film to be cleaned. Such as; (*a*) Removal of organic soils and (*b*) removal of inorganic films.

(*a*) Removal of Organic Soils

Organic soils may be classified as mineral oil and animal or vegetable oils. Mineral oils are neither soluble in water nor in acid or alkaline solutions. They are generally cleaned with chlorinated hydrocarbons. So, both the naptha (petroleum solvents) and trichloroethylene may be used for this purpose. They also remove fats, greases besides flushing away the insoluble particles.

Animal oils may be accumulated on the machined surface from lubricating oils. They are also soluble in chlorinated hydrocarbons.

But, thermal decomposition of coolants may create typical problems.

They may be cleaned by paraffin-based or sulphurised mineral oils. Solvent cleaning is also a popular method, which may be discussed in brief in this regard. In most of the cases, emulsifying solvents are used. They contain both penetrating solvent and dissolved emulsifying agents.

The work to be cleaned is dipped into a solution of petroleum based hydrocarbon solution (*e.g.* naptha). The oil on the work gets diluted and the emulsifying agent is spreaded over the surface. Then a water-spray washing is provided to remove the contamination.

Alkaline cleaning is the oldest and popularly used technique to remove organic soil. Caustic soda is a very common ingredient used in such cleaning. Silicates, orthophosphates are also widely used.

(*b*) Removal of inorganic films

Inorganic films may be scale or smut, rust and tarnish, dirt or cleaning residues. The removal of scales and smuts are very much difficult. Mechanical means are sometimes employed to remove them. Rust and tarnish are formed in the corrosive environment. These films react readily with acid and cyanide. Inhibited muriatic acid or sulphuric acids may be used for dissolving iron oxides and calcium containing scales. Nitric acids are also used for stainless steel or for brightening aluminium surfaces.

The films like cleaning residues may also be removed in acid dipping operation. Specially phosphoric acid-dips may serve the purpose of removal of both organic and inorganic cells.

The miscellaneous title for all other inorganic residues is referred to 'dirt' which may be cleaned by wiping by hand or machine and brushing or by some other mechanical procedures.

Removal of films by Mechanical means

The chemical cleaning organisations also employ mechanical procedures like hydraulic spraying, steam gun spraying, polishing and other abrasion methods to remove a contaminated surface from metals. Polishing wheels commonly use aluminium oxide, silicon carbide, emery and corundum as abrasive heads. Buffing is another mechanical technique to provide lustrous or brushed finishes on metals. Such mechanical means clean the surface, removing the undesirable films which may have been formed due to the metallurgy of the work-material.

REVIEW QUESTIONS

1. What is meant by corrosion? Explain various corrosion control procedures.

2. Describe in brief the different methods of improving corrosion resistance in metals.

3. What are different types of corrosion? What are their preventive measures?

4. What is cleaning? In what cases the chemical cleanings are preferred? Describe different types of chemical cleaning.

5. Identify the correct statements and rewrite the false statements correctly from the following:

(*a*) Pitting corrosion is very common on metals like brass, copper etc.

(*b*) Intergrannular corrosion is seen to be localised on a surface.

(*c*) Aluminium coating is very efficient on iron.

(*d*) Electroplating is very much suitable for small items like screws, bolts etc.

(*e*) Galvanising is a hot coating process.

11

PUMPS AND AIR COMPRESSORS

PUMPS

Introduction

The rank of pump may be considered next to electric motors among the equipments being used in large quantities in industries.

Maintenance of different types of pumps is a common feature in industrial sectors. However, here only the centrifugal and reciprocating pumps have been discussed with respect to their installation or maintenance for a trouble-free operation.

Foundation (The procedural steps are given below)

A rigid foundation into the ground is normally required for installing a pumpset, unless it is to be mounted on a rigid structural platform. It should be suitable to provide a permanent rigid support for the baseplate for taking the weight of the set and should have enough mass to absorb any vibrations. The surface area of the foundation should be of sufficient dimensions so as to leave adequate space of sufficient width, around the pump-base when installed on it, to allow for grouting the foundation bolts, etc. The depth of the foundation depends on the nature of the soil in which it is to be laid.

Continuous concreting should be done as far as possible for laying the entire block of foundation, leaving provisions for grouting and pockets for locating foundation bolts in the proper position.

The size and number of foundation bolts for each pumpset depend on its kW rating and type of prime mover used.

Before placing the pumpset with its monoframe base (in case of monosets) or base plate (in case of direct coupled sets), on the concrete foundation, it is recommended to chip off its upper surface for ensuring perfect adhesion of the grout. Moreover, water should be sprayed on the foundation before pouring the grout. The pumpset should be placed on the foundation with its foundation bolts in position, making it rest on shims and/or wedges placed at frequent intervals (close to foundation bolts position) between the baseplate and concrete block, to provide sufficient space for grouting.

When the pumpset is installed on a platform, constructed inside an open dug well for reducing pump suction lift or for some other reasons, it should be ensured that main supporting wooden beams or steel structures are of suitable size for the required span between their supports and rigid and strong enough to take the weight of the pumpset and two/three men, as also for absorbing vibrations, if any, likely to be caused while running the pumpset.

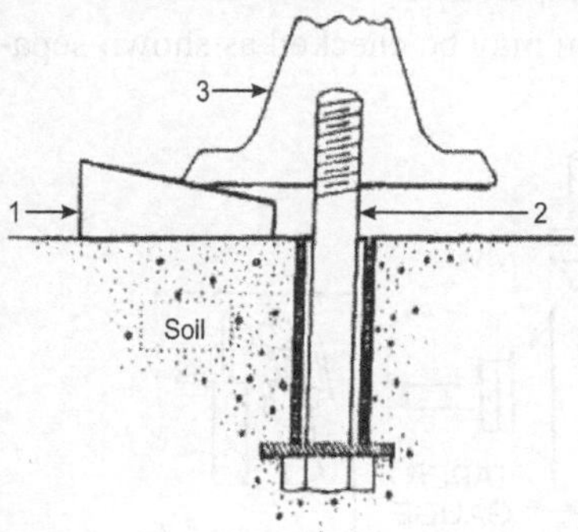

Fig. 11.1. Base plate on foundation.

The Fig. 11.1 shows the special care to be taken to place the base-plate. The base plate (3) is to be placed above the floor leaving a space (about 2-3 cm) to accommodate wedges (1) for levelling. The size of foundation bolts (2) may be selected depending on the rating (kW) of the pump as told earlier.

Grouting

The importance of grouting has been discussed a number of times previously. In this case also, the grouting will arrest the possibility of any lateral movement of the pumpset providing a rigid connection with the foundation block. The levelling of the pump-set is an essential task to be performed efficiently before the grouting, adjusting the blocks or wedges etc. on which the pumpset has been made to rest. When the levelling is over, the foundation bolts are placed in position and the nuts are tightened primarily by hand.

The mixture for grouting is generally recommended to prepare with 1 : 2 ratio of cement and sand. Sufficient water should be added to facilitate the grout to flow evenly beneath the base.

A wooden wall of proper height is made all-around the pump to avoid leakages of grout and the mixture is poured continuously. Care is to be taken to avoid formation of any air pocket within the grout. A sufficient time is then to be allowed for hardening and setting. Of course, a rapid drying may generate cracks and that is why a common practice is to sprinkle water on it, or to keep it covered with wet burlap until it is sufficiently set.

The wooden walls may then be removed and the foundation alongwith the grout may be finished. The foundation bolts are tightened properly after sufficient time.

Alignment

Alignment of the shafts of the pump and its prime mover is the vital job during erection. A flexible coupling is generally used to take care of the end movements of the shafts which may occur during the transmission of power or due to the change of temperature.

However, it is essential to keep and maintain the surfaces of the couplings perfectly parallel. Their axes should remain parallel.

Both the angularity and the parallelism may be checked as shown separately in the Fig. 11.2 *a* and *b*.

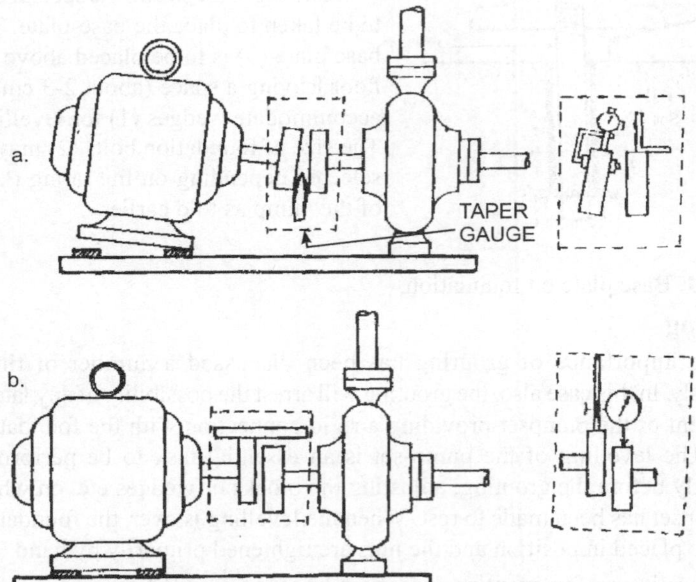

Fig. 11.2. (*a*) Alignment testing of pump and driver shaft (Angular)
(*b*) Alignment testing of pump and driver shaft (Parallel).

To measure the clearances between the surfaces, feeler gauges are used. Measurement is taken on all four sides, say at an interval of 90°.

The parallelism is inspected usually by straight edges. The diameter of coupling faces may be different but the coupling face should remain aligned with the main axis.

Both the tests may accurately be done with the help of a dial gauge as shown separately for both the cases.

Angular and parallel misalignments which are checked by inserting taper gauge or placing straight edge as shown in Fig. 11.2*a* and 11.2*b* respectively, are corrected by means of shims under the motor mounting feet. After each change, it is necessary to recheck the alignment of the coupling halves. Adjustment in one direction may disturb adjustments already made in another direction. It should not be necessary to adjust the shims under the pump.

Fittings etc

The installation work may be completed keeping the following points in mind:

(*i*) A suction pipe should always preferably be one or two sizes larger and never smaller than the pump nozzle.

Eccentric reducer (1) from a larger suction pipe is usually fitted to the pump to prevent the formation of air pockets, as shown in Fig. 11.3*a*. Similar increasers are fitted on the discharge line.

On the contrary, the tapered pipe (2) shown in Fig. 11.3*b*, should always be avoided as it forms air pockets (3).

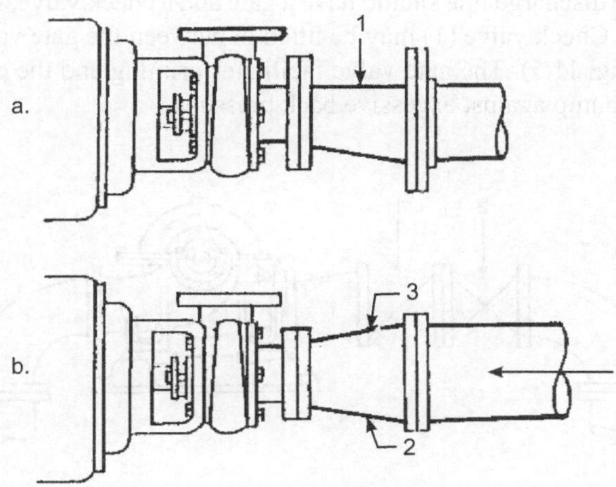

Fig. 11.3. (*a*) Eccentric reducer in suction line.
(*b*) Tapered reducer in suction line.

(*ii*) A foot valve is better to use to the suction line to keep the section line full of liquid when the pump is idle.

(*iii*) The suction line should be as short and direct as possible (Fig. 11.4*a*). It should never be allowed to run with several bends and ups and downs with several joints. Such installation develops air pockets (1) and hamper proper functioning of pumps (Fig. 11.4*b*).

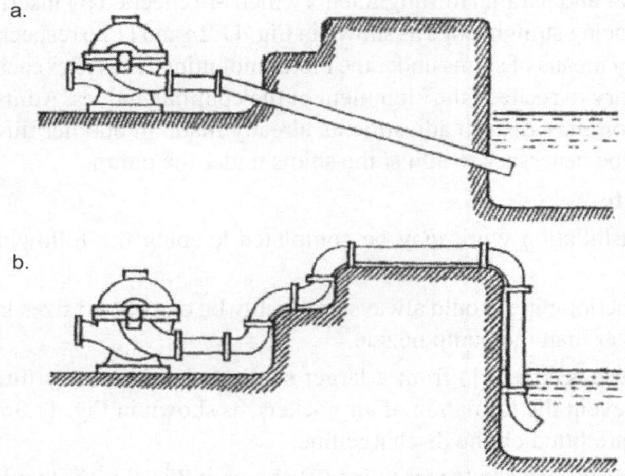

Fig. 11.4. (*a*) Direct suction line (*b*) Suction line with joints and bends.

(*iv*) The discharge line should have a gate and a check valve, fitted closer to the pump. Check valve (1) may be fitted in between the gate valve (2) and the pump (Fig. 11.5). The gate valve facilitates priming and the check valve protects the pump against excessive back pressure.

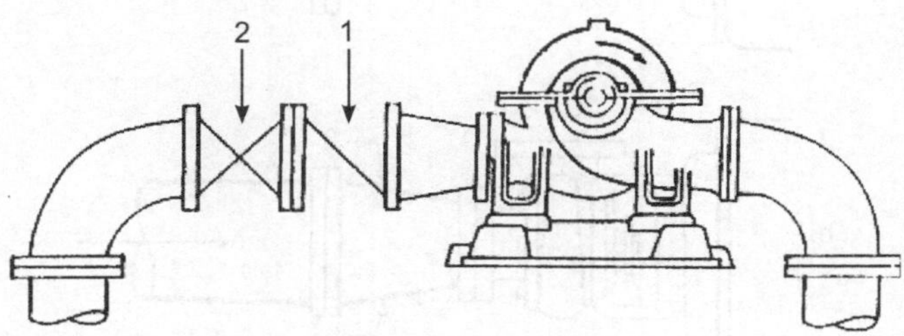

Fig. 11.5. Discharge line.

(*v*) A suitable strainer, much larger in diameter (may be four times) than the suction pipe should be connected at the end of the suction line to avoid foreign matters to enter into the pump. Otherwise, the impeller will be clogged' and will create trouble.

(*vi*) (*a*) While connecting the pipe lines, care should be taken to make it true. The suction pipe should never run above the axis of the pump (2). It should always be fitted either along to be axis (l) or below it (Fig. 11.6).

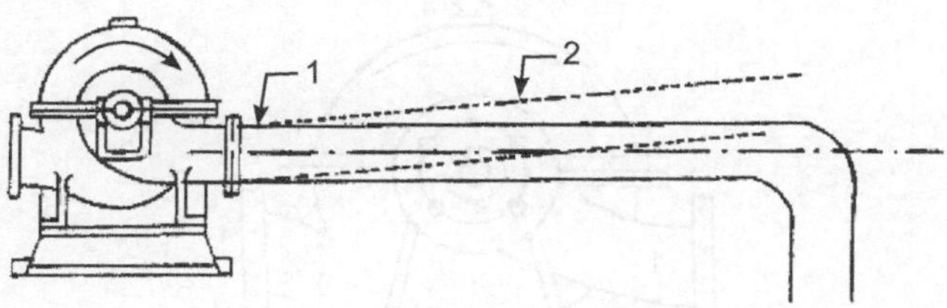

Fig. 11.6. Axis of suction pipe.

(*b*) Pipes should never be forced or pulled to connect suction or discharge flanges. This will cause strain on the joints and will create noise.

(*c*) When the length of the pipe line is too long or when it operates with hot liquid, provision for expansion and contraction is to be properly made in the pipe lines.

(*d*) All joints in the pipe lines should be thoroughly checked to avoid air leakage.

Direction of rotation of pump

It is of primary importance, while installation of a pump, to note that it should never run in the reverse direction. Centrifugal pumps are commonly designated either as right hand or as a left hand, commercially.

An easy procedure may be followed to identify them. One should stand on the driving end and will look over the motor to the pump. If the rotation is seen as clockwise, it is a right hand pump or otherwise (if the rotation is counterclockwise) it is a left hand pump.

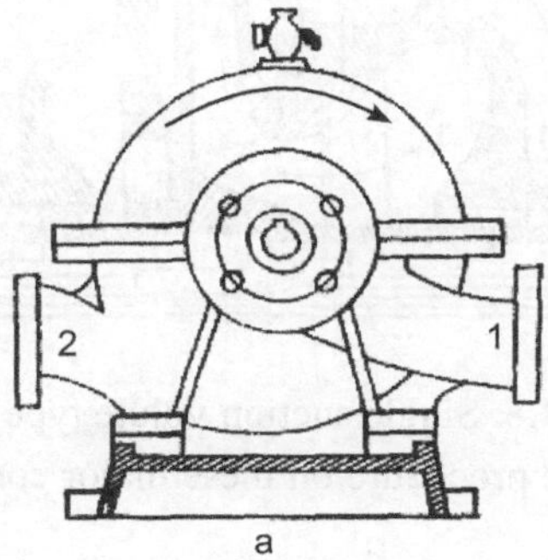

Fig. 11.7. (*a*) Right hand pump rotation.

For a right hand pump (Fig. 11.7*a*), the suction (l) is at the right and the delivery end (2) is at the left and for a left hand pump (Fig. 11.7*b*) the suction (1) is at the left and the discharge (2) is at the right.

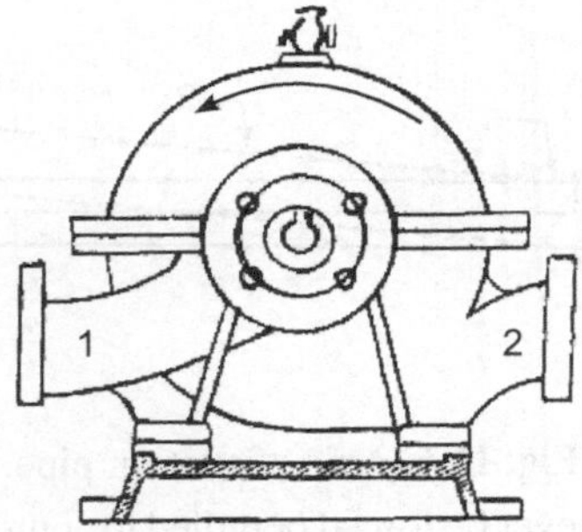

Fig. 11.7. (*b*) Left hand pump rotation

Maintenance

The most important items, which need major maintenance work are shown in the part figure (Fig. 11.8) of single suction volute type pump. The impeller (4) is generally made of bronze and it is keyed (7) to the shaft. On both sides of the impeller the shaft possesses sleeve (5), which is held by nut. To prevent leakage of water, wearing rings (3) are provided and the casing is mounted on the shaft over a stuffing box (8) or brass packing.

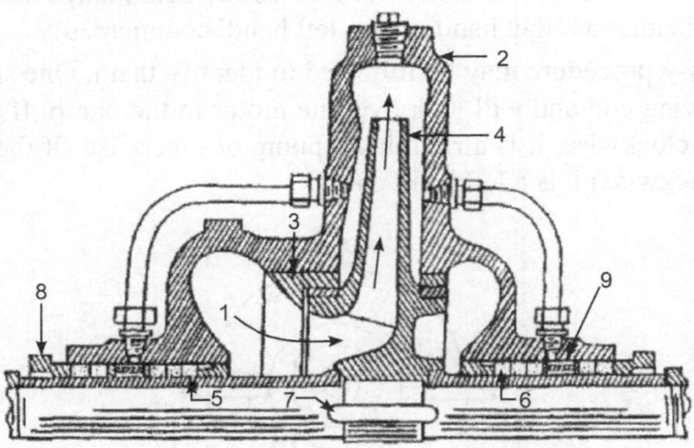

Fig. 11.8. Single suction volute type pump.

Close maintenance procedure on these major components are discussed ahead.

Wearing Ring

The purpose of a wearing ring is to prevent leakage from the exist side to the suction side. The percentage of water that will be leaked, will cause to reduce the efficiency of the pump in similar proportion alongwith enhancing the requirement of power input.

So, wearing rings should be checked time to time. The inspection should be frequent, which may be even monthly when the pump handles gritty or corrosive materials.

If the wearing ring clearance increases alarmingly (say double the original clearance), it should immediately be replaced to restore the capacity of the pump.

Shaft Sleeve

It is generally made of bronze, it may be of alloy steels also. The function is to protect the shaft from the liquid, the pump handles. The wear by the packings of the stuffing box is also prevented by the sleeve.

If it is seen that an excessive tightening of gland is being required to control leakages, it indicates that the shaft sleeve should be replaced immediately.

Packing

The stuffing box prevents leakage of liquid and it should also prevent air leakage in the suction side. To install the stuffing box, the gland is to be taken out and it should be positioned correctly. It generally consists of three square rings and rows of compressed packing. Once it is positioned by hand the gland is to be tightened. It may be remembered that the efficient fitting of packing is a job of experience. Leakage of a few drops of water per minute is permitted and it helps in lubricating the joints of shaft sleeve and packing.

The packing is to be changed when score marks appear on the shaft or the packing materials become hard.

Bearing

Axial hydraulic pressure on the impeller shaft is very much obvious in centrifugal pumps. It is therefore, a must, to provide thrust bearings on all pumps to control the thrust along the shaft. Generally, deep grooved radial ball bearings are mounted in housings and it is to be fitted in the bracket which is an integral casting with the lower half of the casing. Roller bearings are also extensively used today. However, it is to be noted that antifriction bearings commonly receive injury due to over lubrication. Application of too much lubricant is to be avoided, as the main function in lubrication of such a highly polished surface is to form a coating of lubricants to protect it from corrosion. Presence of too much lubricants may harmfully produce heating zones.

It is sufficient to apply lubricants two or three times a year in such bearings. It is to be overhauled once a year by flashing the bearing surface thoroughly with gasoline or kerosene and applying fresh lubricants.

Mineral oils may be used in case of oil lubricated ball bearings. For grease lubricated bearings, a clean and neutral, good quality general grease may be used.

Cleanliness of the bearing housing is of great importance for the long life span of the bearings.

The inspection of bearings for wear should be conducted at frequent intervals and dirty oil or grease should always be changed.

Troubles with centrifugal pump and their causes

1. Pump is not getting started

(*a*) Pump may not be primed. It is to be primed opening all the vents to secure a steady stream of water.

(*b*) Total static head may be too high. Operating conditions are to be checked; both suction and discharge heads should not exceed the specified limit.

(*c*) Impeller may be clogged or otherwise the strainer of suction line may also be clogged. Proper inspection and cleaning is required.

(*d*) Impeller may rotate in a wrong direction. It should be corrected. Generally, an arrow on the pump case indicates the direction of rotation.

2. Pump is getting started but stops after a while

(*a*) Improper priming.

(*b*) Air leakage in exit line.

(*c*) Water seal line may be clogged.

(*d*) Suction head may be too high.

Inspection may be done and corrective measures may be taken accordingly.

3. Pump capacity is below the specification

(*a*) Air leakage may be in the suction line.

(*b*) Air leakage may occur through stuffing box.

(*c*) Wearing rings may be worn. The clearance may be checked and the ring may be replaced if necessary.

(*d*) Suction or discharge head may be too high.

(*e*) Impeller may be partially clogged.

(*f*) Speed may be low.

4. Pump consumes too much power

(*a*) Misalignment may be caused.

(*b*) Speed may be beyond rating. ·

(*c*) Mechanical defects may occur, such as, casing may have been distorted, shafts may be bent, rotating elements may be binding, stuffing box may be too tight etc. Casing may be repaired, run out of shaft may be checked, wearing ring, stuffing box, bearings may be inspected.

5. Pump is getting overheated

(*a*) Bearings may be worn.

(*b*) Rotor may have gone unbalanced.

(*c*) Rotating parts rubbing somewhere.

(*d*) Either low voltage or too high speed may make the motor hot.

6. Pump produces too much noise

(*a*) Water supply may be insufficient.

(*b*) Air leakage may be there in the suction line.

(*c*) Faulty foundation may be a source of vibration and noise.

7. Miscellaneous

(*a*) The wearing of bearing is frequent. This may be due to a bent shaft, misalignment or lubrication failure.

(*b*) An undue vibration of the pump. It may be due to a foundation defect or again it may be due to partial clogging of impeller.

(*c*) Coupling bushes are getting worn rapidly. It is generally due to misalignment.

Reciprocating pumps

Reciprocating pumps are mostly driven by steam wherein available and is known as direct acting pump. It may also be a motor driven power pump. However, reciprocating pumps are positive displacement pumps.

Installation

It may primarily be noted that the reciprocating parts have got inertia forces and that is why the foundation should be heavy and rigid following the instructions of the manufacturer.

Before installation the pump is to be thoroughly checked removing all slush and lacquer from the pump components by suitable solvents. The cylinders, valve passages, should possess no dirt of foreign matters. Then the pump may be placed on the foundation and the levelling is obviously an essential job. The foundation should be at least 250 mm above the ground.

Generally the reciprocating pumps are delivered from the companies without fixing the packings at a proper place. So, the rings are to be inserted separately with a moderate force. After fixing them firmly the gland may be mounted and tightening of nuts should be done carefully. Leakage of a few drops of liquid at all times is preferred for cooling and lubricating the packing.

The following points may be taken into consideration while installing a reciprocating pump.

(*i*) Suction pipe should always be larger than the pump suction connection. It should be as short and direct as possible and should be free from any leakage.

(*ii*) The suction lift in general, should not be more than 6 or 7 metres.

(*iii*) When a suction lift is in existence a foot valve in the suction end is to be provided to keep the pump primed.

(*iv*) A relief valve should be installed at the discharge end as it enables to provide safety in case of any obstruction.

(*v*) A strainer should be used at the end of the suction line to prevent the entry of any foreign substance.

(*vi*) Whenever feasible, regular 90 degree fittings may be avoided during bending. Generally, 45 degree or long sweep 90 degree fittings are used in the pipe lines.

Caution on Operating the Pump

A few essential points may be noted for using a reciprocating pump.

(*i*) If not extremely necessary, the pump should never run at the maximum speed.

(*ii*) Specially to pump heavy liquids or to operate against high pressure, a slow piston speed is desirable.

(*iii*) The slide valves are to be occasionally checked and efficiently adjusted.

(*iv*) To stop the pump, follow the sequence of closing the 'steam valve', 'exhaust valve', 'suction', and 'discharge valve' one after another. Finally the drain valves should be kept opened.

AIR COMPRESSOR

Introduction

Air compressors are being used as a medium of power to operate mechanical equipments for a long time. It is now a common practice in the industries to run the automatic machines, material handling equipments, air tools etc. with the stored energy of the compressed air. Remarkable developments have also taken place in this regard in recent days. To-day's sophisticated compressors are precision built machines and they deserve similar care and precision for their installation and maintenance.

The proper selection of a compressor for a definite purpose is very much important for the users. The decision making on the selection of a site to install the air compressor is also important besides maintenance work.

Types of Compressors

Before discussing the selection, let us be conversant with the types of the compressors. It may be classified in various ways. Primarily, it may be classified as a reciprocating type or a rotary type. The reciprocating type of compressor as commonly used, may again be classified taking into account the number of cylinders along with other considerations as illustrated and as shown in Fig. 11.9 as per Indian standard (IS : 6206 - 1985).

The general consideration of classifying reciprocating compressors is dependent on the position of the axes of the cylinders. They may be horizontal, vertical, inclined etc.

However, the chart is given on page 168 to identify different types of reciprocating compressors.

Chart Identifying different types of reciprocating compressors

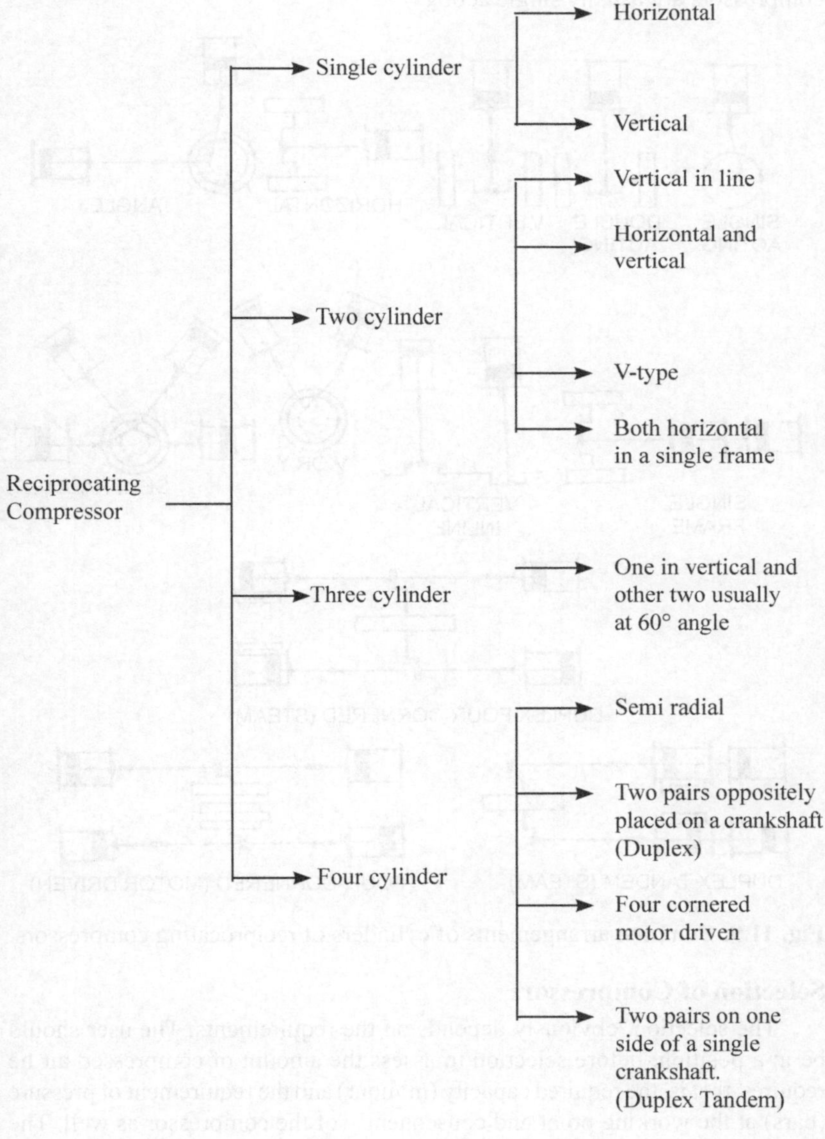

Mostly, each type may be either single or double acting. Of course, horizontal and vertical two cylinder compressors, or horizontal single cylinder compressors are usually single acting.

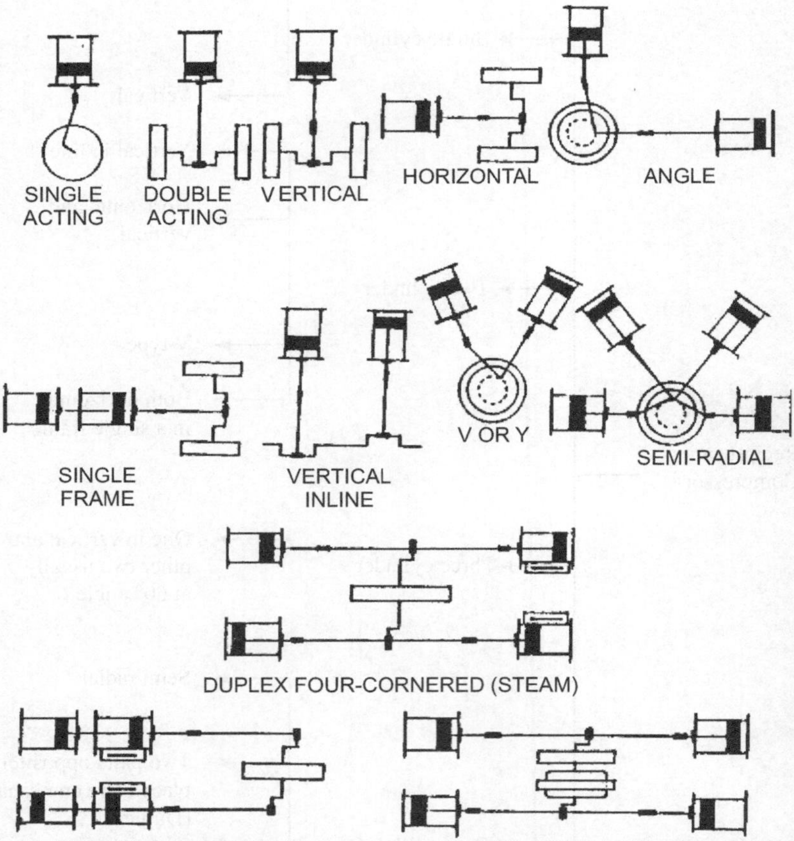

Fig. 11.9. Common arrangements of cylinders of reciprocating compressors.

Selection of Compressors

The selection, obviously depends on the requirements. The user should be in a position, before selection to assess the amount of compressed air he requires, that is, the required capacity (m^3/hour) and the requirement of pressure (bars) at the working point and consequently of the compressor as well. The efficiency and the power consumption may also be compared among the alternatives for final selection.

However, as per the Indian standards, a selection chart, as shown in Fig. 11.10, may also be consulted to select an air compressor.

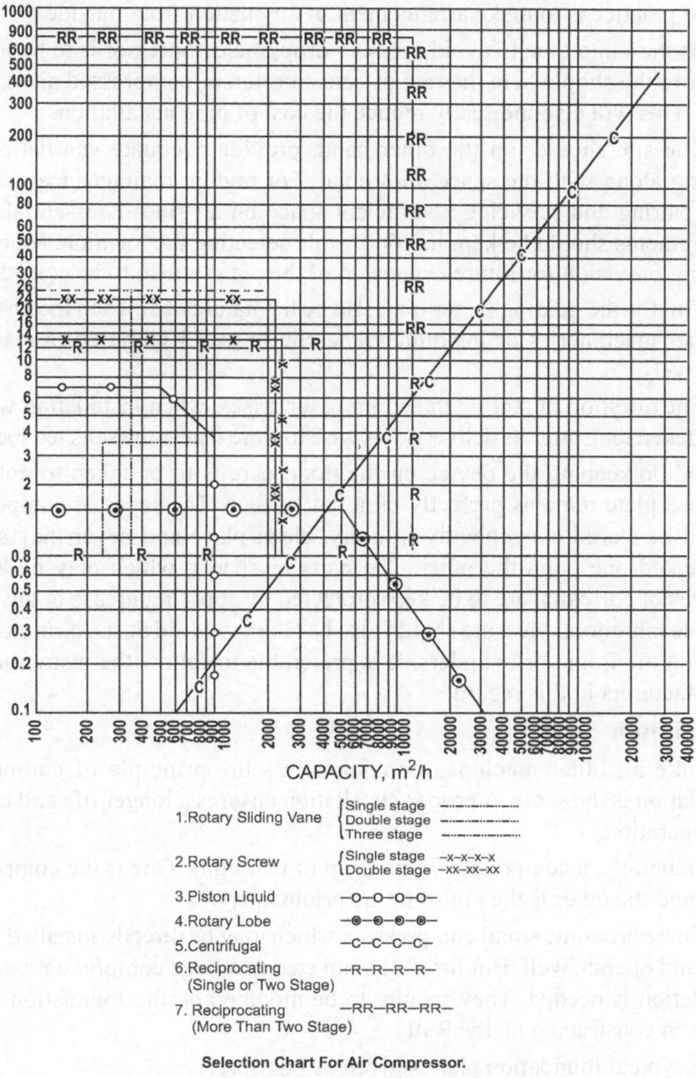

Selection Chart For Air Compressor.

Fig. 11.10. Selection chart for Air Compressor.

Site selection and storing of compressor

The major factors, which act as the guidelines to finalise a site for the installation of a compressor are generally, availability of clean air, nature of end uses, necessary space for fittings, maintenance, dismantling etc.

The most vital factor is to have suction air clean. The dirt or foreign matters of the surroundings may be avoided using suitable filters. Of course, a good practice is to make arrangement of air suction from outside the room.

At the same time to avoid pressure drop, the compressor is to be located nearer to the shop where the rate of consumption of compressed air is maximum. This will also indirectly reduce the cost of pipe installation.

The site should, on the other hand, provide adequate ventilation and lighting along with the space all-around. For regular maintenance work, or for repairing and servicing a sufficient space on all sides is essential. Such requirements should be kept in mind while selecting the location. Moreover, keeping provision for future expansion of the plant is also to be considered.

Finally, the guiding factor may also be the nature of soil and the positions of other machineries or instruments nearby to avoid vibration propagation effectively.

The question of storing the compressor arises, when installation work is undertaken long after its delivery. It is wise to store it near the selected location.

While keeping the device on the floor, care is to be taken to note that the base plate remains perfectly on a flat surface. The separate components should be stored conveniently in a dry, clean place and never they should be stacked one upon the other. The parts or spares, which may be having chances of corrosion are to be kept protected by applying grease or any other suitable inhibitors. Measure should also be taken to avoid dusts, dirts etc. from entering the frame or cylinders. It is preferable to follow the instructions of manufacturers in this regard.

Installation

Like all other machines or equipments the principle of compressor installation is the same. A proper installation ensures a longer life and trouble free operation.

Generally, a compressor is made up of two units. One is the compressor itself and the other is the motor or the prime mover.

There are some small compressors which may be directly installed on the floor and operate well. But for a medium size or a large compressor a definite foundation is needed. They require to be mounted on the foundation which has been constructed on the Raft.

A typical foundation plan is given in Fig. 11.11.

Lowering down the unit, fixation of foundation bolts with the plate, and grouting etc., which are invariably to be done carefully, are not discussed in detail. However, it is mention worthy, that before grouting, levelling of the base plate of both the compressor and the motor is to be checked efficiently in both the directions, longitudinally and crosswise.

Moreover, the total foundation, levelling and grouting work are to be done continuously at a time for both the compressor and the prime mover. Before erection, the top of the foundation may be made roughened and it should be made properly cleaned.

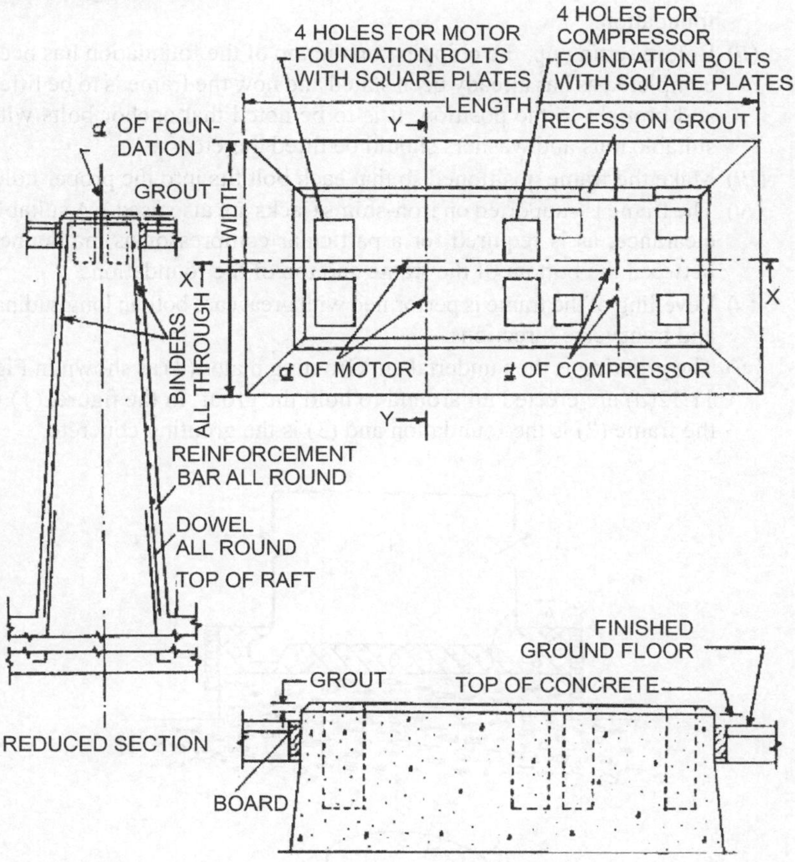

Fig. 11.11. Foundation plan of an Air Compressor.

To design the foundation plan, special care is needed to consider the weight of the compressor along with all other accessories and the unbalanced force. Such data are always being supplied by the manufacturer. It may be noted that a complete balancing of a reciprocating compressor is practically very difficult, but the line of unbalanced forces are to be calculated and it should at least be absorbed by the foundation.

The procedural steps are described below in brief.

(*i*) The foundation block, as described above, should be made ready in advance with regard to the date of setting up of the compressor. Proper time (say 5 to 7 days)should be allowed for setting up of the foundation.

(*ii*) Before setting up, the chipping on the top of the foundation has been completed as has already been stated and now the frame is to be lifted and brought to the position. It is to be noted that anchor bolts with suitable nuts and washers should be fitted thereto.

(*iii*) Make the frame positioned so that each bolt fits into the proper hole.

(*iv*) The frame is supported on iron-shims (jacks are also used). A suitable clearance, as is required for a particular compressor, is maintained between the bottom of the frame and top of the foundations.

(*v*) Levelling of the frame is performed with great care both in longitudinal and transverse directions.

(*vi*) The grouting is then undertaken. Grouting plates (4) as shown in Fig. 11.12(*a*) are erected all around to hold the grout. In the figure, (1) is the frame (2) is the foundation and (3) is the grouting concrete.

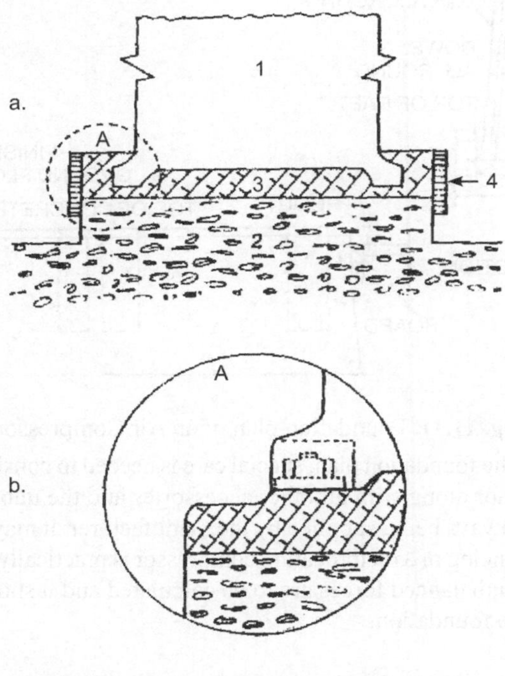

Fig. 11.12. Grouting of frame on foundation.

(*vii*) When grouting is set after a few days, the anchor bolts are tightened. Make sure that the supporting shims or the jacks have been removed just after the cement has been cast.

(*viii*) The levelling of the frame is again checked.

(*ix*) Assembly of compressor cylinders to the frame is then taken up. Generally the cylinders are supplied fully assembled with valves, pistons, packing glands etc.

(*x*) So, the cylinders are now to be connected to the cross heads.

(*xi*) The piston rod is fitted to the crosshead while the later is positioned to its dead centre. This part of installation is very much important, as, a poor execution of position-connection will lead to seizing up of the cross head after few minutes operation.

(*xii*) To check the assembly, the shaft may be rotated three or four turns and its free turning should be ascertained.

(*xiii*) After completion of installation of cylinder, its level is checked and the fitting of auxiliary equipments are started.

(*xiv*) While relevant accessories like coolers, condensate separators are installed, attention should be paid, so that pipings and accessory units are duly supported to prevent loadings and expansions from bringing about unforeseen stresses.

(*xv*) The pipings and each of the units, before being fitted should be properly cleaned. The machine can suffer serious damages if the cleaning of all the components, being fitted, are not properly done.

In general, it may be mentioned that the installation of a large size compressor is always made isolated from rest of the building and a large size compressor is better to install on reinforced bars as shown in the foundation plan in Fig. 11.11.

Finally, it may mentioned that the base of the foundation block is to be finished with the layer of washable oil resistant paint so as to prevent possible dripping from being absorbed by concrete, damaging the same.

Important accessories and components

A few important accessories etc. are discussed below with respect to their fittings and positioning etc.

Air Receiver

Air receivers should always be properly sized and placed at a proper location.

A horizontal receiver should be mounted levelled as for as practicable. The drain cocks are placed generally at the bottom of the receiver.

For a vertical receiver, the air pipe should be connected at the midway up the side and should penetrate to some extent. The end of the line should be made directed downwards to divert any moisture towards the wall of the receiver.

It is preferable to have the discharge line of minimum length so that no pressure drop occurs in between the compressor and the receiver. So the air receivers are recommended to locate as nearer to the compressor as possible.

The outlet (take-off) pipes are connected from the top of the receivers.

The safety valves may be placed below the receiver with opening down. It will also serve the purpose of water leakage. Of course, the water drain cock should separately be connected and should be opened frequently during working hours.

Air Cleaners

Each and every compressor should be fitted with an air cleaner or intake filter. The air entering the compressor should be cool, clean, acid free for the durability and the higher efficiency of the system.

An air cleaner is better to place as close as possible to the compressor. The requirements of an air cleaner may be summarised as follows :

(*i*) There should be no pressure drop in it.

(*ii*) It should possess high separating capacity.

(*iii*) It should be robust in design as it is to be subjected to continuous air pulsation in the air suction line.

The air cleaners should be cleaned from dirt etc. at a regular interval.

After coolers and inter coolers

These are very common parts of a compressor. They are simply heat exchangers. With high water vapour and temperature, the efficiency of a compressor come down and so an intercooler serves the purpose of condensing the water vapour and cooling the air between the stages.

The purpose of after coolers are also same and they are fitted in between the compressor and the air receiver along with a separator to collect the condensed moisture. It will cool and dry the air enhancing efficiency and providing safety.

Coolers should be drained off regularly.

Compressor Valves

A compressor valve may be of disk type or plate type. The lift of the valve play a major role on the efficiency of the compressor. So a careful maintenance of such valves are very essential. Low air delivery, rise of temperature on the valve compartments, blow back noise etc. are the common indications to identify the faults in the valve operation. The life of a valve depends on the efficiency of air filter and the quality of lubricating oil besides overall condition of the compressor.

Checking the valve at a definite interval, is an essential job.

Piston Rings

Wear of piston rings may also be responsible for the loss of efficiency of a compressor. Piston rings, having good lubricating facilities may serve effectively in a clean cylinder for about 20 years. However, a small amount of unattended wear may gradually increase and some of the rings may fail to offer the valuing action.

To identify the wear of the piston ring, air pressure may be applied on the top of the piston. A blow-by sound through the piston will indicate the wearing feature of the piston rings.

If such blow-by noise occurs, both the cylinder wear and piston ring-wear should be checked. When cylinder gets scoured or worn, it may be re-bored and finished, and new pistons and piston rings may be used.

Maintenance of Compressor

The maintenance of compressors are generally performed during periodical overhauling or servicing. All repair operations are performed with the machine stopped and the system unloaded. The routine overhauling consists of the following major checking.

Crank Shaft

Whenever the shaft is removed the following activities are done.

(*a*) Thorough cleaning of oil passages.

(*b*) Crank pins and journals are checked. The surface of all pins should remain polished and free from seizing traces.

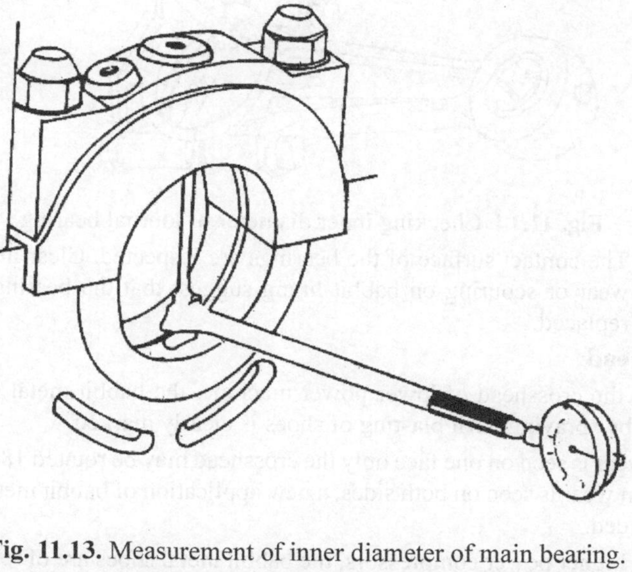

Fig. 11.13. Measurement of inner diameter of main bearing.

(*c*) Surface scouring may be eliminated with fine grain emery cloth or by grinding, if necessary.

(*d*) The clearance between the crank pin and main bearing is compared with prescribed data.

(*e*) The out of roundness of pins are checked.

(*f*) The inner diameter of the main bearing is measured with bearing fitted on the frame. Checking is to be made at different locations to ensure the roundness (Fig. 11.13).

Main and connecting rod bearings

Both the main and journal bearings are to be checked essentially.

(*i*) The smoothness of the lining surface is thoroughly inspected.

(*ii*) If any spots are found all over the lining, it is to be replaced.

Connecting Road

After cleaning the connecting rod properly, the small end bearing and the journal bearing are checked.

(*i*) The inner diameters of both the journal bearing and small end bearing are checked.

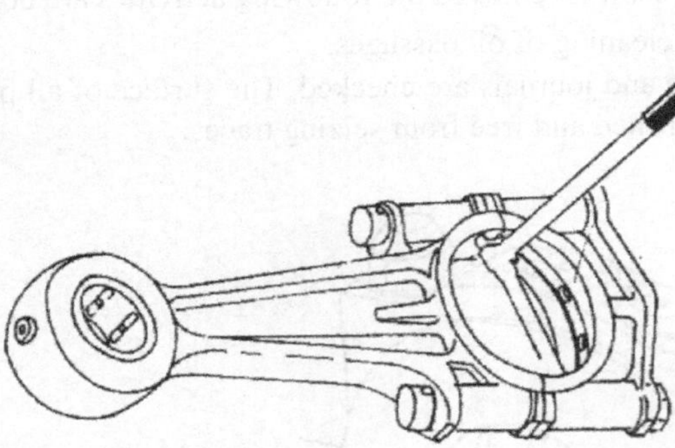

Fig. 11.14. Checking inner diameter of journal bearing.

(*ii*) The contact surface of the bearings are inspected. Clear marking of wear or scouring on babbit lining suggest that the bearing is to be replaced.

Crosshead

For the crosshead of lower power machine, the babbit metal shoes are applied by spraying. Grit blasting of shoes is clearly marked.

If wear is seen on one face only the crosshead may be rotated 180 degrees but when wear is seen on both sides, a new application of babbit metal should be provided.

For higher power compressors, the babbit metal shoes are of replaceable types. A micrometric gauge may be used to inspect the dimensions and they may be replaced if necessary.

Piston

The piston along with the piston rod is to be removed for checking.

(*i*) The traces of seizing, if any is to be checked. Normally the wear seen in the lower bearing surface.

(*ii*) The play between the piston and cylinder liner is checked by feeler gauges.

(*iii*) The gap of the piston ring and the side play of the same is checked.

(*iv*) The piston rings are removed and the relevant grooves are checked for proper condition.

(*v*) When the piston or the cylinder liner is changed, it is advisable to change the piston rings as well.

Cylinder

(*i*) The surface condition of the cylinder liner or the cylinders are checked.

(*ii*) If deep scouring is found, out of roundness of the cylinder is located, the liner is to be replaced.

(*iii*) For integral cylinders, without liner, reaming is done and the piston along with the rings are replaced suitably.

(*iv*) Very fine grain emery cloth, abrasive cloths soaked with kerosene are sometimes used to reface the cylinder.

(*v*) Ultimately the cylinder and the inner parts are properly cleaned and are dried. Greasing of cylinders with a light film of lubricating oil should be performed after operation.

Packing gland

The components of the packing gland require periodical cleaning. It is generally done, when leakages are seen. For a severe leakage, it is to be disassembled and the rings are replaced.

When it will be disassembled, the gaps of the rings are checked. If the end surfaces of the gap are in contact, the ring is to be replaced.

If a packing gland is found to be replaced, the diameter and the out of roundness of the respective surface of the piston rods are checked. Necessary steps are to be taken to get it repaired, if necessary.

Rotary Compressor

In the rotary compressors, there are valves in the housing in lieu of ports and there are rotors. Such compressors are having less number of wearing parts. The vibration is also less.

They may be of different types such as (*a*) Dry screw compressors, (*b*) Oil injected rotary screw compressors, (*c*) Rotary vane compressors etc.

These compressors are reliable and they are subjected to low maintenance costs. They may be of single stage or of two stages. For higher pressures up to 21 bars, they may be in three stages. However, multistages are also available for higher pressure requirement.

Major attention as maintenance

An air compressor, while in smooth running condition, is supposed to supply compressed air at rated pressure and capacity and the power consumption should also be reasonable. For a smooth running, the following major points may be attended regularly.

(*i*) The machine is to be kept always clean.
(*ii*) Oil level in the sump and its temperature is to be checked.
(*iii*) Cooling water temperature is to be constantly monitored.
(*iv*) The valve assemblies are to be inspected and the temperature of the bearings and the valves are to be properly checked.
(*v*) Flow of oil, oil pressure and grease points are to be inspected.
(*vi*) Air leakage from any valve or pipelines are to be checked.
(*vii*) Any unusal vibration or sound are to be immediately attended.
(*viii*) The oil filter, the intake air filter are to be kept clean.
(*ix*) The nuts on the cylinder flanges, as well as other bolts and nuts are to be checked and tightened whenever necessary.

However, a preventive maintenance schedule is invariably required for a compressor plant and should be maintained as per the manufacturers, suggestions.

Maintenance Schedule

For the guidance of the user, a schedule for the maintenance of compressor installation is given below. However, it is not intended to constrain any of the recommendations given by the manufacturer in the maintenance manual.

Reciprocating compressor

(*a*) Daily:

(*i*) Check oil level in crankcase;
(*ii*) Drain receivers, aftercoolers and intercoolers and any other cooling system;
(*iii*) Safety valve to be checked by manual operation;
(*iv*) Check the lubricating oil pressure;
(*v*) Drain water from lubricating oil sump and top up with lubrication oil;
(*vi*) Turn machine by hand with inter stage drain open. At the same time give the cylinder oil pump a few strokes by hand, and
(*vii*) Grease barrel of water pump.

(*b*) After 50 h of running:

(*i*) Bolts to be tightened and adjusted;

(*ii*) Check suction filters;

(*iii*) Check cooling water pump gland;

(*iv*) Check stage pressure and temperature;

(*v*) Check lubricating oil filter; and

(*vi*) Change lubricating oil required for 50 h of running.

(*c*) After 150 h of running :

(*i*) Clean air inlet filters;

(*ii*) Remove and clean lubricating oil strainer in sump;

(*iii*) Check engine lubricating oil system for leaks;

(*iv*) Open engine crankcase cover and examine title screwing arrangement and slackness of big end, cross-head and main bearings, inspect screwing arrangement of all bearings, lubricating pipes and connections. Examine lubrication oil pumps;

(*v*) Examine samples of lurbicating oil from sump for metal particles and emulsification (if the examination indicates severe discoloration or emulsification it should be changed);

(*vi*) Clean water pump suction weed trap; and

(*vii*) Clean circulating water suction strainer (if fitted).

(*d*) After 300 h of running :

(*i*) Clean valve assembly;

(*ii*) Check air governor for its functioning;

(*iii*) Check safety valve setting for its correctness of pressure at which it should be blown;

(*iv*) Clean lubricating oil sump and renew oil;

(*v*) Examine engine big end, cross-head and main bearings and adjust oil clearances; and

(*vi*) Check that the freedom of movement of a compressor, if mounted on shock and vibration mountings, is not restricted.

(*e*) After 1000 h of running:

(*i*) Check crankshaft alignment;

(*ii*) Check crankshaft on moving parts, connecting rod bolts, nuts, split pins, etc;

(*iii*) Examine for wear and adjust compressor big-end bearing, gudgeon pin and bush;

(*iv*) Check aftercooler and intercooler for scaling if any;

(*v*) On shock/vibration mountings, inspect condition of necol lacquer. Clean and reseat (if it is to be used);

(*vi*) Remove and clean auto-clean strainer cartridges;

(*vii*) Clean out and inspect intercoolers and water spaces on cylinder, cylinder head, etc.;

(*viii*) Hydraulic test of all cylinders, intercoolers, oil coolers and water spaces;

(*ix*) Examine samples of lubricating oil from sump for metal particles and emulsification (if found necessary, lubricating oil should be changed);

(*x*) Open crankcase cover and examine the screwing arrangement and slackness of big end, cross-head and main bearings, connecting rod bolts, balance weight bolts. Examine lubricating oil pump and lubricating pipe connections;

(*xi*) Examine for all stages, suction, delivery valves for wear and damage. It is important that valves and seatings should have clean and unpitted surfaces. Valve springs to be inspected for damage and tension and replaced, if necessary;

(*xii*) Examine big end bearing bolt for elongation; and

(*xiii*) Examine and gauge for wear compressor plunger and cylinder.

(*f*) After 2000 h of running :

(*i*) Check all moving parts for wear;

(*ii*) Examine cylinder lubricating oil pump for wear, bell valves to be examined for pitting or corrosion and renewed. Check all valve springs for condition and tension;

(*iii*) Open the main big end and cross-head bearings and examine, adjust bearings oil clearances;

(*iv*) Examine connecting rod and crank for straightness;

(*v*) Pressure gauge for air and oil to be recalibrated; and

(*vi*) Clean the sump and renew oil.

(*g*) After 5000 h or running :

(*i*) Dismantle and clean all parts and replace worn out parts as per recommendations from manufacturers. An approved cleaning chemical is to be used for the parts and care is also to be taken to dry completely the parts before assembling. If compressor shall be kept idle for longer periods, an anti-rust oil shall be used along with lubricating oil and compressor is rotated 5 to 6 times after ever 2 to 3 days to lubricate all sliding or moving parts; and

(*ii*) Replace connecting rod bolts and nuts.

Rotary Compressors

(*a*) Daily:

(*i*) Check levels of fuel, oil in sumps and filters;

(*ii*) Drain the moisture separator;

(*iii*) Drain any other cooling system provided;

(*iv*) Drain water from lubricating oil sump, check level of oil in sump;

(*v*) Give one turn to grease cups fitted to water pump and drive; and

(*vi*) Tum compressor by hand or motor.

(*b*) After 50 h of running:

(*i*) Run compressor upto working pressure. Check performances;

(*ii*) Clean and oil all joints;

(*iii*) Drain and clean compressor sump. Clean lubricating oil strainer. Refill sump; and

(*iv*) Check for oil leakage and the security of all keeps and fastenings.

(*c*) After every 200 h of running :

(*i*) Check suction filter elements and clean;

(*ii*) Drain the receiver for any water collected at the bottom;

(*iii*) Examine suction and delivery valves for wear or damage and refit as necessary;

(*iv*) Check lifting of air reservoir relief valve at designed pressure ;

(*v*) Clean intercooler fins;

(*vi*) Inspect condition of necol lacquar on shock/vibration mountings, clean and recoat if necessary (if shock/vibration mountings are not fitted, check fastenings and holding-down bolts for tightness);

(*vii*) Examine and refit the air compressor;

(*viii*) Strip, examine the bearings, adjust the clearances of gears, repack glands of circulating water pump; and

(*ix*) Open and refit the circulating water pump discharge valve.

(*d*) After every 500 h of running :

(*i*) Drain compressor oil and refill with fresh oil recommended by the manufacturer;

(*ii*) Change filter elements in oil filter; and

(*iii*) Drain and clean compressor sump, clean lubricating oil strainer and refill sump.

(*e*) After every 1000 h of running:

(*i*) Clean and check all hoses provided on compressor;

(*ii*) Attend to moisture separator;

(*iii*) Check separator elements and replace, if necessary;

(*iv*) Water discharge to cooler, check control valve; and

(*v*) Examine water side and clean tubes of water cooler.

(*f*) After every 5000 h of running :

(*i*) Dismantle and clean the compressor;

(*ii*) Non-metallic blades be replated and metallic plates be replaced/ dressed, if worn-out; and

(*iii*) Replace the damaged parts and also the parts recommended by manufacturers.

REVIEW QUESTIONS

1. State the important aspects of installing a Reciprocating pump. What precautions arc to be taken to prevent leakage at pipe joints and air pocket formation in suction pipes.

2. State the sequence of operations to be followed for installing a medium capacity centrifugal pump for domestic purpose in a multistory building. Name different types of pipe fittings and their respective purposes.

3. State the troubles that may generally be encountered in the operation of centrifugal pump and mention causes and remedies of those troubles.

4. State the important aspects of installing of reciprocating compressor.

5. Write explanatory notes on maintenance of

(*i*) Air cleaner,

(*ii*) Compressor valves,

(*iii*) Piston rings and

(*iv*) Packing gland of a compressor.

6. Point out the work for which major attention is needed for a compressor.

7. Identify the probable causes and remedies for the following in case of a centrifugal pump :

(*a*) Pump consumes too much power.

(*b*) Pump is getting overheated.

(*c*) Too much noise is produced.

(*d*) Bearings are getting worn too quickly.

8. Identify the causes and remedies for the following troubles of a compressor :

(*a*) Efficiency is getting reduced.

(*b*) Pressure drop occurs in between the compressor and the receiver.

(*c*) Vibration is being transmitted to adjacent device.

(*d*) A blow-by sound occurs in the cylinder.

12

BOILER MAINTENANCE

Introduction

The main function of a boiler is to transmit heat to the water by burning the fuels (coal, oil, gas) to produce steam which acts as the energy to spin the turbine generators for electricity generation or to use desired steam pressures or quality of steam for various industrial purposes.

The following basic features generally differentiate the boilers in several classifications:

(*a*) *Position of furnace*: Externally fired or internally fired boilers *e.g.* 'Babcock and Wilcox' and 'Lancashire boilers.'

(*b*) *According to use:* Stationary, portable, locomotive or marine.

(*c*) *Relative position of hot gases:* Water-tube and fire-tube boilers *e.g.* Sterling and 'Cornish' or 'Lancashire' boiler.

(*d*) *Direction of axis of the boiler:* Horizontal, inclined or vertical boilers.

(*e*) *Circulation of Water:* Most boilers are operated on natural circulation, *e.g.* Lancashire, Babcock Wilcox boilers etc. The other type of circulation is forced circulation *e.g.* Velox boiler, Lamont boiler etc.

Installation of Boiler

The installation work usually involves laying out, foundation, erection and fittings of mountings and accessories.

We will have a discussion mainly on the foundation work. The boilers, naturally require a foundation upon which it will rest or within which a steel frame work is to be constructed to support the boiler independently.

For a hard soil, digging of a trench may serve the purpose for a small boiler. But, to have rigid support, trenches are to be made widen and at the bottom a concrete raft reinforced with steel or iron bars is to be constructed. Pile driven rafts may also be necessary considering the condition of the soil.

It may be mentioned that boiler foundations are not subjected to excessive vibration. It only supports heavy loads. So a medium quality mixture of concrete may be used to construct foundation blocks.

Once the foundation has been set in, the ashpit may be laid alongwith the combustion chamber and should be allowed to be hardened.

Actually the procedural steps of erection depend on the types of boiler. The boilers may be supported by brackets, the brackets may be supported on side walls or against boiler-supporting columns, and crossbar erection may have to be done according to the requirement. As an example, the horizontal return-tubular boilers are erected with steel in-cased settings.

However, brick work may be constructed as high as possible.

A typical brickwork construction for a Lancashire boiler is shown in the Fig. 12.1.

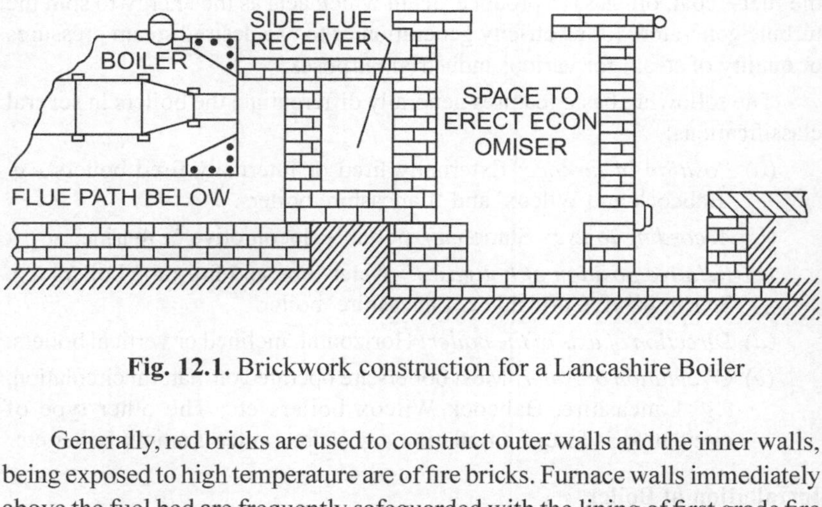

Fig. 12.1. Brickwork construction for a Lancashire Boiler

Generally, red bricks are used to construct outer walls and the inner walls, being exposed to high temperature are of fire bricks. Furnace walls immediately above the fuel bed are frequently safeguarded with the lining of first grade fire brick or a plastic refractory material, designed for this purpose.

The provision of ventilation is also there to preserve the lining. The Fig. 12.2 shows a few examples of modern boiler front wall (Fig. 12.2*a*) or side wall (Fig. 12.2*b*)construction, whereas, an efficient inner fire brick wall separated from outer steel casing is shown in (Fig. 12.2*c*). The air space acts as an insulation and air flows constantly through this air space.

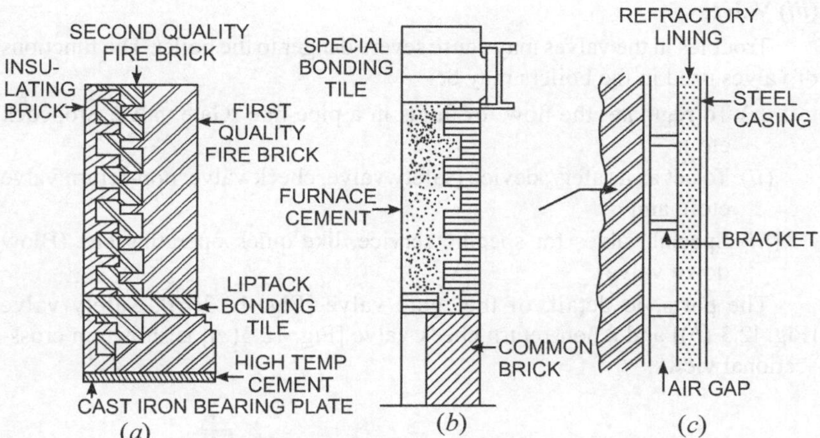

Fig. 12.2 (*a*) Front wall, (*b*) Side wall and (*c*) Fire brick wall construction

Maintenance of Boiler

Boiler maintenance is a massive job and a schedule should be prepared for the maintenance of the entire plant. However, the maintenance procedure of important mountings, accessories etc. are discussed below.

Mountings

(*i*) Water Level Indicator

The indicator should be placed in such a position that it remains always well-lighted and the level is easily seen at any times. The operator should keep himself assured that both the water and steam passages are remaining clear at all times. As a test, the water tap may be shut to check whether steam is blowing through the drain and again shutting the steam trap if the water tap is opened, water should blow through. Finally, the drain should be closed and steam trap is to be opened. The water should rapidly rise up to working level in the gauge glass.

(*ii*) Steam Pressure Gauge

The pressure gauge should be graduated to at least $1\frac{1}{2}$ times the set pressure. This may be periodically tested by comparing the calibration and range with the inspector's test gauge or through other testers *e.g.* screwed plunger pump or dead weight plunger pump.

(*iii*) Valves

Troubles in the valves may cause severe danger to the boiler. The functions of valves used in the boilers may be:

(*i*) To regulate the flow of fluids in a pipe line (Gate valve, Stopcock etc.);

(*ii*) To act as a safety device (safety valve, check valve, non-return valve etc.); and

(*iii*) Special valves for specific service, like quick operating etc. (Blow down valve).

The parts, in detail, of the gate valve [Fig. 12.3(*a*)]. Safety valve [Fig. l2.3 (*b*)] and a non-return check valve [Fig. 12.3(*c*)] is shown in cross-sectional views.

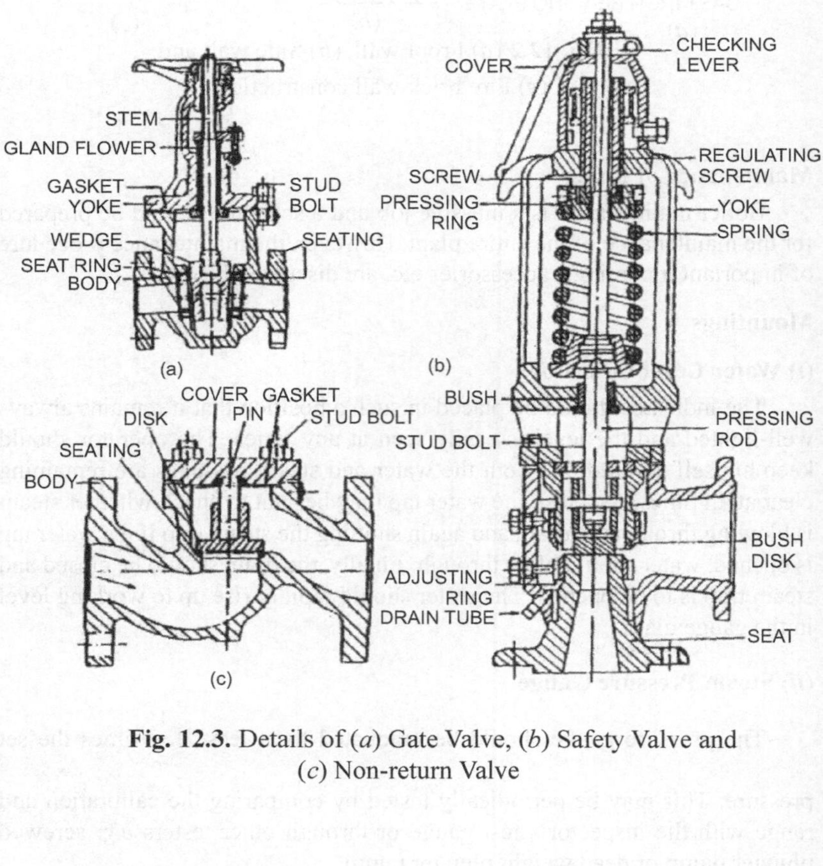

Fig. 12.3. Details of (*a*) Gate Valve, (*b*) Safety Valve and (*c*) Non-return Valve

The construction of the components will convey an idea about the major points of maintenance.

As an example, the maintenance of a gate valve is much difficult as the two tapered seats in the wedge are to be matched perfectly with the seat of the body [Fig. 12.2(*a*)]. The gland packing is also shown in the figure. While tightening the gland covers, it should be noted that they are able to apply sufficient pressure on the stem to prevent leakage. The packing materials should withstand the operating temperature.

In case of a safety valve [Fig. 12.3(*b*)], it is evident that corrosion may occur in the area of a valve seat if it is not operated for a long span of time. So, to maintain such a valve a regular blow off at certain intervals is an essential task. Its capacity should be tested before commissioning to avoid any accident.

A non-return valve [Fig. 12.3(*c*)] allows steam to flow from a boiler to the using end by raising the disk with the steam pressure.

But it prevents the flow of steam in the reverse direction if at any time the boiler pressure falls below the system pressure due to any reason. So, it can cut-off the boiler from the system.

The perfectness of the cushioning effect of the disk and the pin assembly should be checked obviously during screwing period.

A number of types of other valves are also used in the boiler house for air, steam and water.

The major factors to be considered for connecting the valves with the pipelines in general, may be discussed in brief.

The flanged joints are to some extent disadvantageous, as they are tightened with gaskets and the failure of the gasket, while the system is in operation, may create accidents. Screwed joints are popular but they are generally adopted for low pressure lines and smaller sizes. For a medium and high pressure line, socket welded joints are used. But, welded joints are also widely used.

However, the general difficulties encountered during the operation of various valves are given below. The type of problems, the probable causes and the remedies in general are mentioned in tabular form.

PROBLEMS WITH VALVES

Problem	*Probable causes*	*Remedies*
Seat passing.	(*a*) Gland bolts are over tightened.	Gland nuts may be loosened slightly.
	(*b*) Lubrication may be insufficient.	The yoke, stem etc. are lubricated as per specification.
	(*c*) Damage may be in the disc, or wedge seat	Check and rectify the cause by machining, grinding, etc.

Problem	*Probable causes*	*Remedies*
Leakage through cover joint.	(*a*) Improper bolt tightening. (*b*) Gasket may be damaged. (*c*) Seating surface is defective.	Make it proper. Check and replace if necessary. Right type of gasket is to be used. Check and rectify by machining.
Leakage through gland packing	(*a*) Gland pressure is not sufficient.	The gland bolts may be tightened accordingly.
	(*b*) Packing rings are not in sufficient number. They may even be defective.	Required number of packing rings are to be inserted. If the rings are defective a few numbers of rings or the total set may be replaced.
	(*c*) Stem has gone worn.	The pitting, scouring or wearing of stem is checked and rectified. The stem may be replaced if necessary.
Operational Problem	(*a*) Too much tightening of gland packing.	Tightening should be up to an extent only to arrest leakage.
	(*b*) Lubrication is not proper.	Stem threads, bearings are lubricated with proper grade and types of lubricants as has been instructed by the supplier.
	(*c*) Yoke bush or trapezoidal threads have been damaged.	To be verified carefully. Bush may be replaced. Component may also be replaced, if necessary.

Accessories

The maintenance of a few major accessories are discussed here. The major accessories like Economiser, Air Pre-heater, Feed pump and Injector have been discussed in detail. Maintenance of super heaters are mostly similar to that of an economiser.

Economiser

Economiser uses a large portion of the waste heat carried away with the flue gases to the chimney to heat the feed water being passed through a set of pipes. The principle of economiser is shown in the schematic diagram (Fig. 12.4).

In modern boilers, higher feed water temperature is available due to the pulverised fuel firing and so the presence of bonded deposits and deterioration of the metal internally are seldom encountered. However, external corrosion owing to the condensation of moisture, the presence of damp soot or chemical action of waste gases are found on economiser tubes.

As 'on-load cleaning', shot cleaners at flue gas inlet to economisers may be used. Water washing can be done to remove bonded deposits if any. This is a 'off-load' cleaning procedure. In severe cases, soda-ash solution may also be employed.

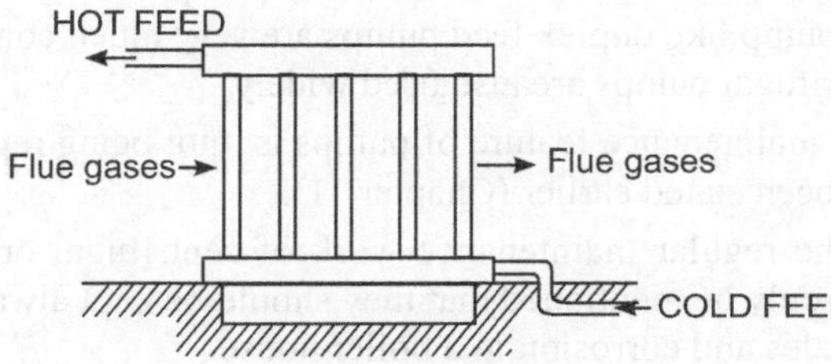

Fig. 12.4. Principle of Economiser

Generally, an welded type economiser requires little maintenance apart from routine cleaning. However, all tubes can be examined internally and externally, once a year, as, leakage from one tube may cause damage to other tubes, adjacent to it, by steam impingement and erosion in the case of an economiser, superheater or air pre-heater.

Air Pre-heater

Air pre-heater is generally arranged at the back of the boiler along with the economiser. It is a normal accessory to a large size water tube boiler. Air is pre-heated before it enters the boiler furnace, passing the air through the air heater tubes, which are exposed to the hot flue gas passing to the chimney. It enables to have a higher furnace temperature, and to increase the percentage of 'CO_2' in the flue gas resulting in a larger amount of steam.

Fouling is a common problem in air heaters. The degree of fouling during operation can be identified on the pressure drop across the air heater. 'On-load-cleaning', such as, air bypass, hot air re-circulation etc. can be occasionally performed if the provision is there, or otherwise 'off-load cleaning' is to be done at a planned shut down period. The fouling occurs generally due to plugging corrosion, or erosion. If necessary the tubes may be replaced. During annual servicing, the air heaters should be thoroughly cleaned and washed with water.

Leakage of the air heaters can be checked analysing the drop of percentage of CO_2 of the flue gas across the air heater. The leakage may be through tube joints, corrosion or erosion holes.

A regenerative air heater is of rotary type. A preventive maintenance is necessary to check the drives, bearings, oil circulation system etc.

On installing air pre-heater the draught intensity will be severely impaired due to the loss of temperature of the flue gas. To overcome this difficulty an induced draught fan is always necessary and a force draught fan is also to be installed to force the air through the air heater and ducts into the furnace chamber.

The proper maintenance of these fans are also an integral part of the maintenance of the air heater and economiser for proper functioning.

Feed Pump and Injector

To deliver feed water into the boiler, feed pumps are commonly used. A reciprocating pump like duplex feed pumps are very much common for such purpose. Centrifugal pumps are also used widely.

However, maintenance feature of pumps is not being repeated here, as it has already been stated earlier (Chapter 11).

Besides the regular maintenance work of centrifugal or reciprocating pumps, it may only be mentioned that they should be kept always clean from dusts, dirts, scales and corrosion in a boiler house.

The solvent action of the liquid passing through the pumps may corrode the pumps. Moreover, any leakage of air in the pump or the air which may exist within the liquid being pumped, accelerate corrosion. So the pumps should be kept vented and should be kept fully filled with the liquid.

The use of injectors are mostly limited to the smaller plants. It has got practically no moving parts and consequently requires less maintenance cost. A set of nozzles are arranged in it to expand the steam. The steam pressure drops but the velocity is highly increased. A vacuum is created when the steam passes across the space between the steam nozzle and suction nozzle. As a result water comes in into the suction chamber and the water is forced against the boiler pressure by the 'pressure energy' which is a result of the conversion of 'kinetic energy' of the steam jet.

However, the problems in the injectors may be (*i*) plugging of tube or nozzle, due to dirt or scale (*ii*) leakage in the suction pipe (*iii*) fouling in the suction or the discharge line, etc.

The nozzles are to be made cleaned after dismantling it. If scale is found in the tubes, they may be cleaned by dilute muriatic acid. Suction and delivery lines are to be kept clean for efficient operation of the injectors.

It may be remembered that injectors cannot handle too hot water. So if the temperature is too high, cold water may be used as feed water.

Major Maintenance work and troubles

Major maintenance work in the boiler require the service of a well experienced group of people. The service of maintenance personnel may be necessary due to various reasons, a few of which are discussed below.

(*a*) Cleaning and Scaling

Generally, in a boiler plant each and every opportunity to clean the boiler from accumulation of grease, scale, sludge and soot is readily utilised.

The tubular boilers require more frequent cleaning than other types of boilers.

However, each and every boiler components, accessories should be kept cleaned from dirt, coak or coal dust. The salts of calcium and magnesium in the feed water are mostly responsible to form scales. Formation of scales which occurs mostly at the hottest part of the tubes, obstruct the transmission of heat, lowering down the capacity of the boiler and ultimately causes a failure of pressure parts. The rate and amount of scale formation depends on the quality of feed water, type of boiler, pressure etc.

Scales are removed by wire-brushes, hand or mechanical scaling tools. Power driven cleaners, soot blowers etc. are used now-a-days to remove scale from the tubes.

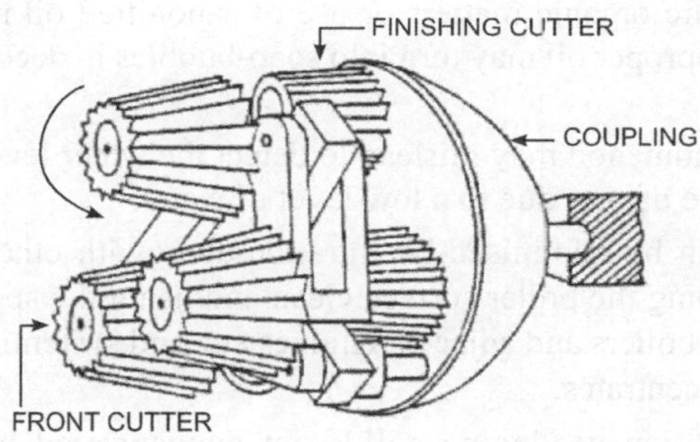

Fig. 12.5. Cutter type tube cleaner

The boiler should have a required number of hand holes or washout plugs for inspection and as well as for inserting hose pipes and cleaning tools.

A cutter type mechanical cleaner is shown in the Fig. 12.5. The top end is only shown in the figure. This is a hydraulically driven device. The cutter head is rotated at high speeds. The cutters remove the scale into small chips. Arrangement of cooling the cutter is done by a stream of air which simultaneously washes away the removed scales from the tube.

Corrosion

The corrosion of internal parts of the boiler is mostly due to the non-treated quality of feed water. Corrosion generally occurs as cavities and pits are formed in localised areas and ultimately penetrate into the metal. The dissolved gases in the 'make up water' or in the condensate are also equally responsible for severe corrosion. So, it may be summarised that presence of air and carbon dioxide or the vegetable and mineral acids in the feed water causes internal corrosion. Proper feed water treatment, chemical coatings, installation of a deaerator may minimise internal corrosion. Moreover, oils or fatty substances, being carried from the adjacent machineries may also cause corrosion. Attention should be specially given to all seating and flue covers and all parts in contact with brick work.

The external corrosion is mostly due to carbon dioxide, which in combination with moisture form carbonic acid. It may also be caused by leakage from seams, cocks and valves.

It may be mentioned that superheaters' corrosion is generally due to the contact of metal and steam at high temperature. The economiser suffers due to the fact that the oxygen dissolved in the feed water comes first in contact with the high temperature surface of an economiser.

Foaming and Priming

Formation of undesirable steam bubbles in the boiler is termed as foaming.

Some specific organic matters or use of saponified oil may cause such problems. An improper oil may turn into soap-bubbles in decomposition with alkali.

Such a phenomenon may mislead to detect the water level perfectly and finally may cause danger due to a low level of water.

Foaming can be minimised by filtration alongwith other treatments of feed water, keeping the boiler surface clean and using the specified grade of lubricants in the boilers and adjacent engines etc. and systemic blow down of boiler water concentrates.

When the foam producing solids get concentrated in the boiler, it accelerates carrying over impurities and water slugs alongwith the steam. This harmful procedure is called priming.

Priming mostly occurs when the rate of evaporation is forced beyond the capacity of the boiler. Such carrying over will cause deposit formation in superheaters and may damage steam turbines also.

The remedy of priming is to use well-treated feed water, clean grease and other suitable lubricating oils in the system.

Water Hammer

When the steam is admitted to a cold pipe, containing water or condensed steam, the steam comes into contact· with the water at low temperature. Naturally, a large amount of steam condensation occurs creating a vacuum there. In this situation, the water may run through the pipe at high velocity causing the effect of water hammering.

Water hammering may be eliminated if the concentrate removal rate and the steam entry rate in a pipe is kept balanced.

Moreover, the design of steam pipe fittings should not allow the accumulation of water at any part to avoid water hammering.

Feed Water Characteristics

An ample supply of feed water is needed to run the boiler house. But pure water is seldom found in nature. All natural water contains more or less foreign matters. They may remain in suspension or as mixture or being soluble. Moreover, the elements like air, oil, grease particles etc. may enter from other equipments of the plant.

The impurities, whether they are soluble or insoluble or belong in suspension, create hazards in the boiler plant.

The unsuitable or impure feed water causes the following problems:

(*i*) Tube failure.

(*ii*) Embrittlement and corrosion of boiler steel.

(*iii*) Loss of heat due to the deposit of scale, dirt and oil on heating surfaces.

(*iv*) Enhance span of servicing period for inspection, cleaning and repair.

(*v*) Enhance maintenance cost due to the investment in spare components.

(*vi*) Frequent blow down of boilers, heaters etc. causing loss of heat.

(*vii*) Increase the steam consumption in the prime mover. This may be due to accumulation of dirt, scale on valves, nozzles etc.

(*viii*) Priming and foaming.

So, it is a common practice to employ definite treatment of feed water to eliminate or at least minimise the above difficulties.

Feed Water Impurities

Organic material in the feed water is responsible for foaming. The inorganic compounds may be there either in suspension or in soluble form. The silica, iron, alumina etc. may be present in suspension whereas calcium, magnesium, potassium or sodium in the form of carbonates, sulphates, nitrates and chlorides may be present in the water being soluble. They cause corrosion etc.

Rain water also contains gases likeCO_2 and very often hydrogen sulphide.

The solid impurities whether in suspension or soluble ultimately remain in the boiler as evaporation is continued. Of course, a small portion may be carried over to the connected machine along with the steam. But, most of the impurities accumulate on the heating surface. The dissolved gases, being free in course of evaporation are mostly discharged with steam. The remaining portion generates different combinations in contact with iron and causes scale, erosion and pitting corrosion on the boiler wall.

The 'hardness' is again a major factor. It may be a temporary hardness due to presence of limes, magnesia etc. or it may be a permanent hardness if the water contains sulphates, chlorides of lime, iron and magnesia.

Thus, a treatment on boiler feed water is an essential job in the steam power plant.

Feed Water Treatment

The feed water treatment of a boiler is essential to remove soluble and insoluble impurities and the dissolved gases to soften the water. A proper chemical analysis of water will regulate the selection of process to be adopted.

The following techniques are generally used:

(*a*) Filtration

The suspended solids are removed through filtration. Both raw and treated water may be passed through a filtration plant which removes a large part of the impurities.

The supplied water may be passed through the bed of silica, gravels etc. The mud, sand and other small particles may even settle down automatically.

However, the filtration bed should have a large area and it should be washed frequently to avoid choking and formation of leakages. The precipitated impurities are discharged at a regular interval.

(*b*) Lime soda procedure

It is primarily a precipitation type of purification. Of course, it involves chemical treatment also to remove hardness. The re-agents used are hydrated lime (calcium hydrate). Soda ash (sodium carbonate) and caustic soda (sodium hydroxide).

The chemical compounds should be prepared in definite proportion. Automatic mixing gear is generally employed in the industries.

The supplied water is taken to the water receiver and is passed to the tank below (Fig. 12.6), along with a definite amount of chemical compound. The tank is compartmentalised and after repeated precipitation in different portions, the water comes up through a bed of filter and is stored in the end tank as purified water. It may be pumped as feed water to the boiler.

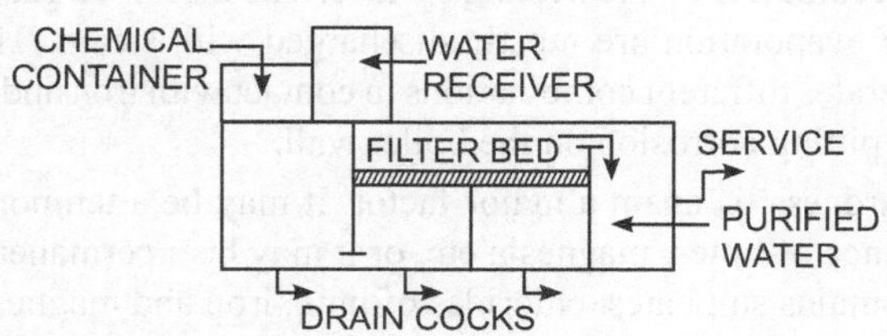

Fig. 12.6. Feed water treatment by Lime-Soda-procedure

In modern plants, two separate reaction "tanks are used. The water is treated with lime in the first tank and then with soda in the next tank.

But, in both the processes some hardness (being un-reacted) will remain in the water. If the chemical reactions are carried out at just near to the boiling point of water, a better result is obtained.

In this process the lime reduces the hardness due to carbonate.

$$MgSO_4 + Ca(OH)_2 + Na_2CO_3 = Mg(OH)_2 + CaCO_3 + Na_2SO_4$$

The magnesium hydroxide $Mg(OH)_2$ and calcium carbonate $CaCO_3$ are not soluble in water. They will precipitate as a sludge.

The permanent hardness, that may be caused due to presence of sulphates of lime (as Na_2SO_4 in the above equation) etc. is reduced by the application of soda ash. This will decompose soluble sulphates into solid carbonates and will precipitate, purifying the water.

(*c*) Base Exchange Process

Some minerals like Sodium, Aluminium silicate or some commercially produced synthetic resins like phenolic type materials have got the ability to change sodium ions for calcium and magnesium ions, when they are present in the water.

Zeolite method of softening feed water functions with the above principle.

The softening tanks are having beds of crushed marble, zeolite and gravel within it (Fig. 12.7).

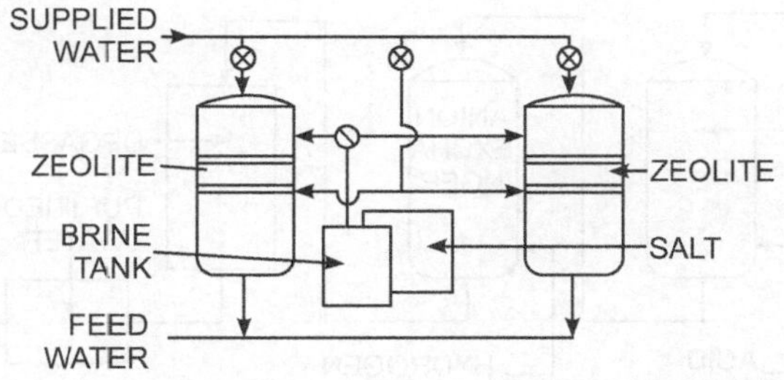

Fig. 12.7. Feed water treatment by base exchange process

Now-a-days artificial porous zeolites are manufactured for this purpose. The water is delivered at the top of the bed and is allowed to percolate through the layers of bed. The calcium and the magnesium ions will be retained by the zeolite portion removing the hardness. Of course, the water will be rich in sodium salts, which are non-scale forming in nature.

When the sodium of the zeolite will be about to be exhausted a backwash process is to be employed to restore the softening capacity of the zeolite.

After isolating the bed from the regular cycle line, water is to be forced to flow upwards firstly for cleaning of deposits from the bed and then brine water (water mixed with common salt in definite ratio) is to be allowed to flow downwards.

Modern plants employ water meter, time switch, float switch etc. to backwash the plant after softening a definite amount of water automatically.

The amount of common salt brine, required to flow through the bed automatically accumulates in the brine tank from the salt-tank.

Thus, the zeolite bed may be restored fully for further cycles of softening feed water.

The main disadvantage is that the method cannot be applied for the water which contain acids unless the acid is previously neutralised.

(*d*) Demineralisation Method

In this process also different synthetic organic resins are used. Hydrogen or Hydroxyl ions of these resins are replaced with calcium, sodium or other metallic cations and the hydroxyl ions can exchange chloride, sulphate or other anions from the water.

With this principle, separate tanks of cation and anion exchanger may be arranged serially (Fig. 12.8) first to remove calcium, sodium etc. and then to release chlorides, sulphates etc. as described above. The carbonic acid in the pure water may then be eliminated by setting a de-gasser in the line to obtain demineralised water to supply as feed water to the boiler.

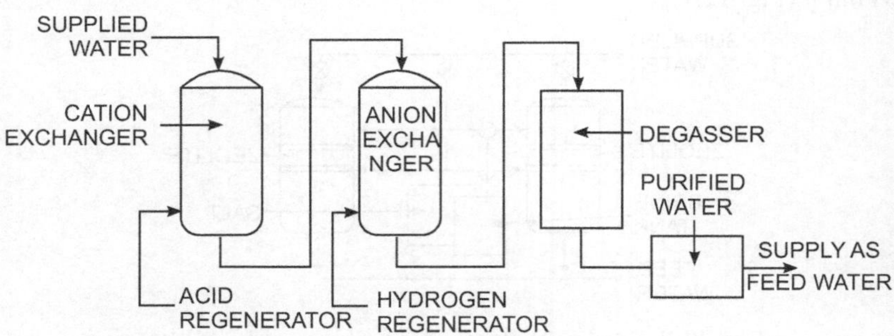

Fig. 12.8. Feed water treatment by demineralisation method

Both the exchanger will loss their capacity of exchanging after handling a definite amount of water. The capacity may be restored regenerating the cation exchanger by acid and anion exchanger by caustic soda to regain hydrogen and hydroxil ions respectively.

(*e*) Methods of applying Boiler Compounds

The treatment by a boiler compound involves introduction of reagents directly into the boiler and the reaction is allowed to take place inside the boiler.

Generally, soda ash, sodium silicate, sodium aluminate, sodium phosphates etc. are used in definite proportion. Different boiler compounds are commercially available. They may be of various nature, such as, liquid, powdered, or solid in form. The solution may be introduced through the injector or it may be fed to the suction side of the boiler pump by means of a side feed lubricator. The ingredients of the compound should be selected after the feed water analysis.

Such a process is generally applicable for a medium sized plant and where the feed water contains minimum scale forming elements.

The most disadvantage of such a low cost, simple method is that, the scale forming substances accumulate in the boiler itself. A frequent blow off and efficient supervision is required for such a system.

A few Repairing Work Stay bolt

Stay bolts are generally used to support flat surfaces at a short distance. The stay bolts (Fig. 12.9) are threaded in the surfaces. If they are to be replaced, the old one may be removed by cutting it off. Then a suitable size of drilling may be done, so that a thin layer of metal exists above the threads of the flat surfaces. A round nose chisel may be used to remove the thin metal and restoring the threads. If the threads are in good condition, a similar size of stay bolt may be fitted.

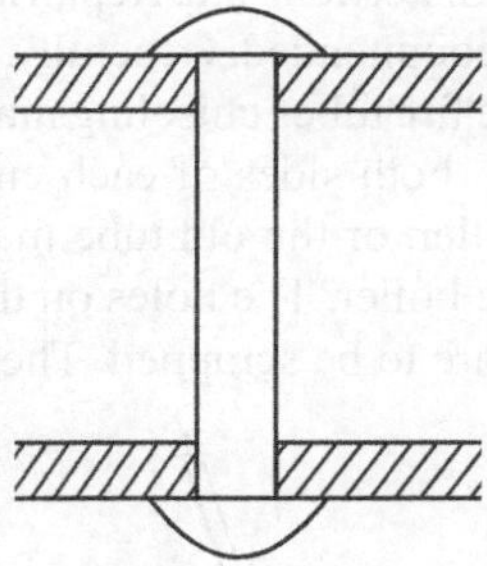

Fig. 12.9. Stay bolt

If the threads become damaged, a larger size tapping will enable to fit a suitable stay bolt there. The rivetting of ends are to be done carefully.

Manholes and Hand holes

Such openings are oval in shape. They are cut in the shell plate. The doors are fitted from the inside of the manhole. The leakages through such doors should be immediately attended.

The major cause of leakage is the softening of asbestos gasket due to a rise in temperature. Usual practice is to tighten the respective nuts as the boiler pressure goes up. If the leakage continues even after this, the gasket is to be replaced.

Valves

A good number of valves for various purposes are installed as the boiler mountings. The general defects may be due to 'Spoilage of packings or gaskets', the valve seat may create problem due to the corrosion of the valve seating surface. Moreover, cracks, low scratching resistance as well as corrosion and erosion of metals are the main causes of failure of valves.

Hard facing, if required, is to be done depositing high alloy metal (chromium-nickel stainless steel or alloys having chromium about 17% and carbon upto 2%) upto the proper thickness. After the hard facing it may be machined suitably. It may be noted that satellite hard facing requires pre-heating and post heating to avoid cracks in the base metal. Finally, lapping is to be performed for a required surface finish. The accuracy depends on the selection of a lapping compound, skill of the operator, the quality of surface before lapping etc. It is not true that the longer the time taken for lapping, better will be the finish, choice and selection or proper lapping compound may minimise the machining time.

Tube Repairs

Boiler tubes should be thoroughly inspected to identify leakages, bulging or distortions etc. Repairing by welding or some other means are generally not recommended. Generally, boiler tubes are replaced by a new one. For removing the fire tubes chiseling may be done [Fig. 12.10 (*a*)) to open up the old beading on both sides of each end. The tube may then be pulled out by a suitable puller, or the old tube may be cut on both sides [Fig. 12.10 (*b*)] from inside the boiler. The holes on the tube sheet is to be cleaned and the projections on it are to be scrapped. The edges are to be made round off as well.

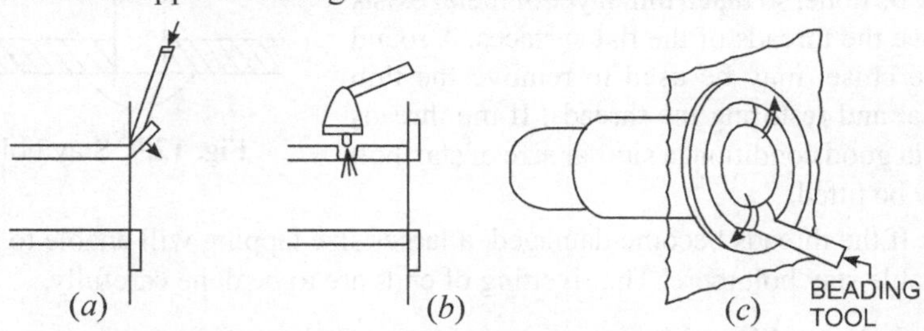

Fig. 12.10. (*a*) Removing old beading by chisel (*b*) Cutting of tube, (*c*) Tube end beading with beading tool

The ends of the new tube are to be simultaneously cleaned, and the tube is placed through the tube sheet in such a way that, equal projection of tube ends are available on both the sides. The ends of the tube are expanded by a suitable expander and the tube ends are beaded by the help of a hammer and beading tool. The ends of the tubes are usually annealed before installation to avoid cracks etc. during hammering and beading. The new tube [Fig. 12.10 (*c*)) joint may finally be checked and put into service.

Cracks

Cracks on boiler surfaces are created mostly due to stresses, caused by a repeated difference of temperature. Cracks may be available in the unstayed or in the stayed areas.

Welding is not always permissible to repair cracks. Generally, cracks of smaller length may be welded, if they have not been spreaded on various sides. The welds should always be relieved of stresses, caused due to high temperature of welding. The crack on a lapped area should not be welded. Such cracks generally progress upto the failure. It is always better to take suggestions from boiler inspection authorities to repair cracks. Cracks in some areas like the bottom of the shell, water legs etc. may be repaired using patches. For the bottom of the shell, the affected portion may be cut-off and a plate of equal thickness, shape and material is to be riveted taking care to obtain previous strength as far as practicable. Welding is not recommended in this case, as the shell bottom is to withstand the furnace heat. Welding may by done for the water leg portion. The patches are welded from the inner side.

After any welding work it is better to test through X-ray or other testing equipments.

On the other hand, for any riveted work, the edges of the plates are to be caulked. Caulking presses the entire edge downwards tightly on the plate at the bottom to achieve a water or steam tight joint.

Mechanical Stokers

Consistent plentiful availability of good quality of coal is simply impossible to-day. Now-a-days the common hand fired boilers are working below their economic capacity and moreover, the efficiency of the boiler rapidly goes down when cheap and low grade fuels are used.

Besides the elimination of hard labour of hand firing, the mechanical stoker has got the advantage of using cheaper grades of fuel with greater regularity and efficiency.

So, the essential requirements for efficient and smokeless combustion, such as continuous feeding of the fuel and uniform distillation of the volatile matter in the presence of oxygen, are available with the mechanical firing.

As per the working principles of the mechanical stokers they may be classified into four categories:

(*i*) Sprinkling stoker;
(*ii*) Chain grate stokers or Travelling grate stokers;
(*iii*) Overfeed stokers;
(*iv*) Underfeed stokers.

(*i*) Sprinkling stoker

Sprinkling stokers throw variable quantities of fuel over the fire bed covering the stationary and moving fire grate at regular intervals as shown in Fig. 12.11. The spreading of fuel can be done evenly using a shaking grate. Most of the sprinkling stokers work with sliding and shaking self-cleaning firegrates. The ash pit is arranged at the bottom. Both natural or mechanical drought can be used with this type of stokers.

***(ii)* Chain grate stokers**

The travelling type chain grate stokers are highly successful in burning all bituminous coals without imparting any agitation and are very popularly used in industries.

An endless chain of grate bars is mounted on sprockets at both the front and rear end and is supported by guide rails or slides. It may be connected to a motor and should have a variable speed transmission device.

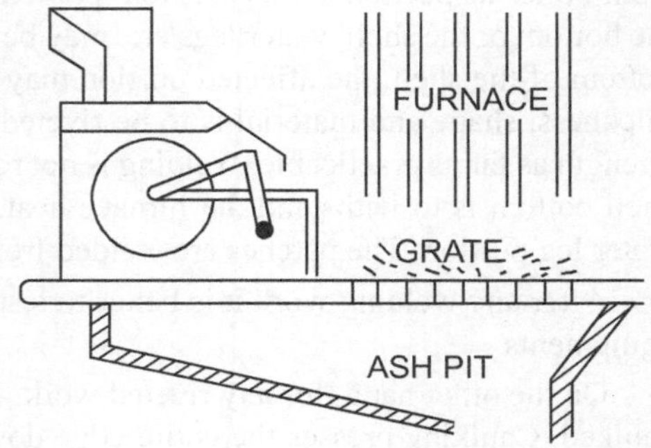

Fig. 12.11. Outline of a Sprinkling stoker

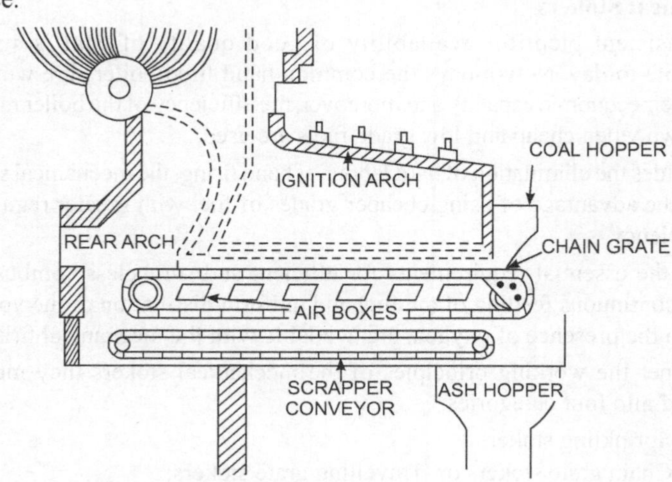

Fig. 12.12. Chain grate stoker

Coal is fed from the hopper on the front part of the grate through an adjustable door to regulate the thickness of the fuel bed (Fig. 12.12). Such stoker should have an ignition arch for igniting the incoming fuel. Both natural and forced draught systems are available with such stokers. The speed of the chain grate travel is to be actuated according to the requirement of time for complete combustion of the fuel on the bed.

(*iii*) Overfeed stokers

It consists of hopper, distribution chambers just like sprinkling stoker, but here fuel comes on a dead plate from the hopper and then is pushed in variable quantities with the help of a reciprocating arm on the grate. The grate is inclined downwards and travels slowly (Fig. 12.13). Ultimately the burnt materials are accumulated on the dumping grate and finally discharged to the ash pit.

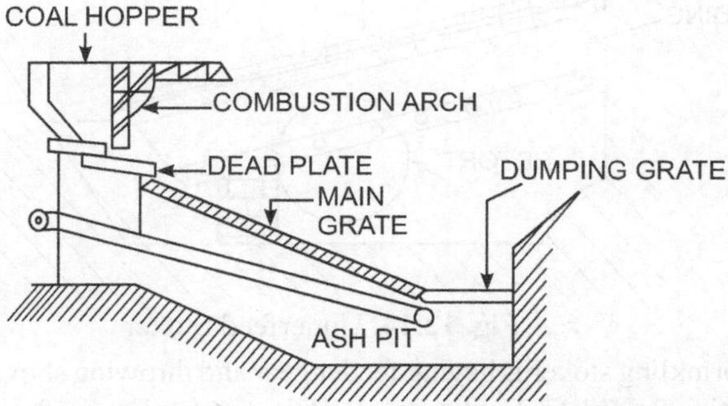

Fig. 12.13. Line drawing of Overfeed Stoker

There are two basic types of overfeed stokers. They may be front feed where the inclination of a grate is from front to rear and side feed, where the inclination of a grate is from side to the centre.

The overfeed stokers are generally of natural draught type and are used for small boiler plants.

Underfeed Stokers

The only difference of an underfeed stoker with the overfeed stoker is that in this type of stoker, the feed is from the underneath instead of feeding fuel on the surface of the grate. Underfeed stoker uses the gas producer principle.

The reciprocating ram (not shown in the figure) supplies fuel to the retort where it is subjected to coaking action. The coal spills over the inclined lower side of the grate. After liberation of volatile gases the coal is pushed towards the top of the fire. Such a system facilitates complete combustion as actual firing occurs from the coal which is partly coaked. Air is supplied for combustion horizontally into the-bed at the point of distillation of the gases, that is, where the fresh fuels are fed as shown in Fig. 12.14.

Maintenance of mechanical stokers

As the name implies the stokers are consisted of a large number of mechanical components and mechanisms. They are to withstand high temperature, undergo variable speeds and carry heavy loads. As a result, the wear and tear and possibility of breakage is more in mechanical stokers. So the stokers need to have a regular maintenance schedule.

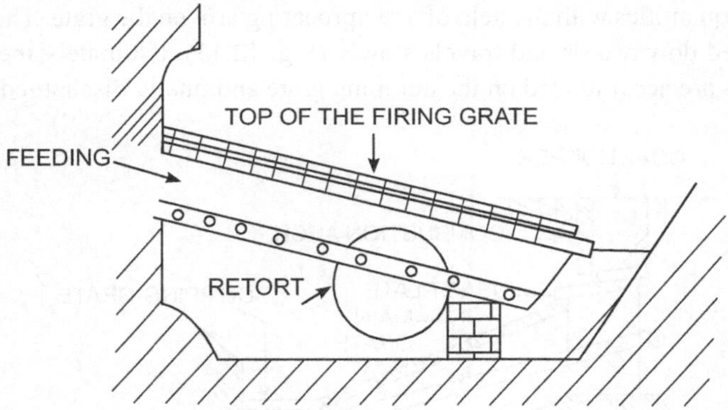

Fig. 12.14. Underfeed Stoker

The sprinkling stoker is having feeder gear and throwing shovel along with a reciprocating feed plate. The time scale of the mechanism is often regulated by renewable camfaces. These faces wear too much and spare faces should be kept ready. Similarly, chain links for a chain grate stoker is an essential inventory.

For a chain grate stoker, the important feature is to regulate speed. Accuracy of variable speed transmission devices should be checked at regular Intermittent checking of supplied size of coal, required bed thickness and combustion efficiency are the important tasks of the operator.

Intermittent checking of supplied size of coal, required bed thickness and combustion efficiency is an important task of the operator.

Draught control is a vital criterion for proper functioning. Insufficiency of air will cause smoke and supply of air in excess will enhance fuel consumption unnecessarily. At the same time overloading of fuel should be avoided. Frequent checking of temperature of vital parts of stoker is a part of effective maintenance. Overall cleaning and lubricating should be done at a predetermined interval.

Oil Fuel Burners

Liquid fuels possesses a superior heating value. The main principle is to atomise the oil through some means and spraying the atomised liquid into the furnace. The major requirements for this are to have oil fuel installation, an arrangement for lightening it up, to preheat the oil if necessary, both suction and discharge filters and air distributers.

The burners generally used may be of three types:

Pressure type, Steam atomising type and Air pressure type.

Pressure type

The oil is drawn from the oil tank and is supplied through a horizontal or vertical fuel pump. The oil may be pre-heated to rise its temperature upto the flash point and the burner receives it from the discharge of a oil filter. The oil is atomised by the tangential slots of burner and is thrown on the furnace.

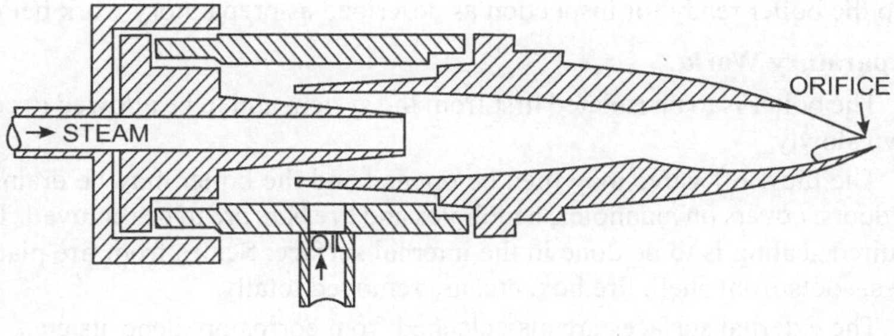

Fig. 12.15. Oil fuel burner

Steam type

In a atomised steam type fuel burner, the steam and the oil enter the burner through different passages as shown in Fig. 12.15.

The high velocity steam imparts a whirling motion to the atomised fuel and the mixture is carried by the air draught.

Air Pressure type

The principle is the same as that of a steam type system, the only difference is that the oil mixes with the air, supplied to the burners by blower or air compressor.

Such type of burners are mostly used in industries. It is to be noted that oil should be of low viscosity when it comes to the burner.

Maintenance of Burners

The major maintenance work with the oil-burners is to keep them clean. The tip is to be checked and cleaned from deposit at regular intervals. The size of the tip will regulate the proportion of mixture of fuel and air or steam. Much care is to be taken to clean the tips. Breakage of tip may occur during cleaning.

While in operation, attention should be given on rate of admission and exhaust of fuel and those of steam or air.

It may be mentioned that the fire, if occurs accidentally, will not be extinguished by water. So the chemical extinguishers, sands etc. should be kept stored as precautionary measure.

Boiler Inspection

It is a routine job as per the boiler act. Generally, competent authority will visit the plant annually and will inspect the boiler thoroughly and will prepare a report on the overall condition of the boiler.

Such inspection work should be assisted by the users. It is the duty of the owner of the boiler to provide all required helps to the boiler inspectors and to keep the boiler ready for inspection as described as preparatory work below.

Preparatory Work

The boiler is to be isolated first from the system and to be allowed to cool down slowly.

The blow off valve may then be opened and the boiler may be drained. All doors, covers on manholes, handholes and fire bars are to be removed. The required scaling is to be done in the internal surface. Scales from fire plates, tubes, soots from shell, fire box, etc. are removed totally.

The external surfaces are also cleaned from corrosion, deposits etc.

In brief, the washing and cleaning should be such that the inspector can inspect the surface metals.

The accessories like superheaters, economisers etc. will also be inspected simultaneously. They are also to be kept clean.

It is also the duty of the owner to point out any known defect during inspection.

Inspection Job

The boiler inspector will check all the mountings, accessories as well as internal and external surfaces of the boiler.

He has got the right to suspend the operation of a boiler until any pre scribed repair, if any defect is found by him.

The inspection can be termed as satisfactory when the inspection is carried out under sufficient lighting and all essential parts are found to be in satisfactory condition.

Hydraulic Test

The objective of the hydraulic test is to check the pressure withstanding ability of the boiler. Water is forced to the boiler by the feed pump or otherwise by a hand pump until the water comes out of the air vent.

All the valves are kept closed. Safety valve is made gagged to prevent any release.

The hydrostatic pressure is raised upto at least one and half times of the working pressure. The leakages through valve settings, hole covers etc. are checked thoroughly holding the raised pressure for a sufficient time.

The weak points, if any, may thus be detected. Light hammer blows are also applied during hydraulic testing of a boiler.

Boiler Starting up

The following are the important points to be taken care of during starting of a boiler.

(*a*) Purge the boiler with adequate air for a few minutes.

(*b*) Starting up a cold boiler may cause stresses in the boiler pressure parts, particularly in the boiler drum (as the metal thickness is more). The specified rate of heating should be strictly maintained.

(*c*) The feed pump, the valves in the feeding line, blow off valve, safety valve and water level indicators are thoroughly checked.

(*d*) Be assured that a draught is available to remove the products of combustion, specially for liquid and gaseous fuel boilers.

(*e*) Fire may be lighted with the help of oil soaked waste jutes, a few amount of coals or kindling wood.

(*f*) The fire doors are closed and dampers are adjusted to provide a small air flow.

(*g*) The stoker may be allowed to function at its lowest speed.

(*h*) When the warming up will take place, the water in the boiler at atmospheric pressure, will attain its boiling point. The passing out of steam or air through the air cock at the top of the boiler will clearly indicate this.

(*i*) When all the airs will be supposed to have gone out, the air cock is closed.

(*j*) A flow of steam may be permitted to the superheater to raise the temperature of its tubes, after opening a drain valve.

(*k*) Pressure may then be allowed to rise maintaining the specified rate of rise of temperature per minute.

(*l*) During the warming period, all condensate should be removed from the lower parts of the pipes.

(*m*) Firing rate should be in full control when the boiler will approach to its operating pressure. A steam flow may also be established.

(*n*) Once the steady steam flow is established, the outlet drain of the superheater is shutdown and steam is allowed to flow from the boiler as per the preplanned requirement.

REVIEW QUESTIONS

1. Distinguish between 'Inspection' and 'Test' of a boiler. Explain clearly how a boiler is to be prepared for Hydraulic test and mention the purpose of this test.
2. What are the technical duties of Boiler Owners prior to Annual Inspection? State the salient features of Inspection of an installed boiler.

3. Describe in brief the different methods of feed water treatment in a steam power plant.
4. Why is feed water treatment for a boiler necessary? State the maintenance procedure required for a feed water treatment plant.
5. What are mechanical stokers? Classify them. Write down the maintenance features of Mechanical stokers.
6. Write short notes on:
   (*a*) Steam test of a boiler
   (*b*) Boiler starting procedures
   (*c*) Oil fuel burner
   (*d*) Replacing of a boiler tube
7. What corrective action is to be taken if,
   (*a*) There is leakage through gland packing of a valve.
   (*b*) Fouling is caused in the air heater.
   (*c*) Steam bubbles are formed in the boiler.
   (*d*) A crack is identified at the bottom of the boiler shell.
   (*e*) Feed water pump is getting corroded.
8. Fill up the gaps with appropriate word:
   (*a*) Lime soda procedure of feed water treatment is a _________ (chemical/precipitation) type of purification.
   (*b*) _________ (Sprinkling/Chain grate) stokers are very popularly used in industries.
   (*c*) The objective of the hydraulic test is to check the _________ (steam raising/pressure withstanding) ability of a boiler.
   (*d*) The permanent hardness of feed water may be reduced by _________ (lime/soda ash).
   (*e*) The purpose of a safety valve is to drain out _________ (steam/water) from the boiler.

13

HEAT ENGINE

Introduction

Heat engines are the devices which utilise heat to transform it to useful work. Heat is generated by the combustion of the fuel, which may be solid, liquid or gaseous.

The heat engines are mostly utilised as the prime movers. We will discuss the installation procedure, maintenance of major components etc. in this chapter with regard to the prime movers, like steam engines, turbines and I.C. Engines.

STEAM ENGINE

So far as the mechanical design is concerned, steam engine is comparable to I.C. Engine. It is also having Cylinders, Piston, Crankshaft etc. in addition to the controlling valves of steam entrance and exit. But it is mention worthy that the installation work of a steam engine is much complicated than that of an I.C. Engine, as the foundation of a steam engine should be basically robust. The foundation design of a steam engine is better to be performed by the contractors who are having much exposure in this area.

Foundation

A typical foundation plan for a steam engine foundation is shown in Fig. 13.1.

It may be noted that the foundation block of the engine (2) will have provision of foundation bolts (l) as per the requirements of the supplied engine. The cylinder foundation blocks (3 and 4) should also be constructed simultaneously to have the alignment. The provision for the foundation of the alternator (7) should be made along the crankshaft along with the construction of fly wheel pit (5) and the pedestal blocks (6) to erect the bearing.

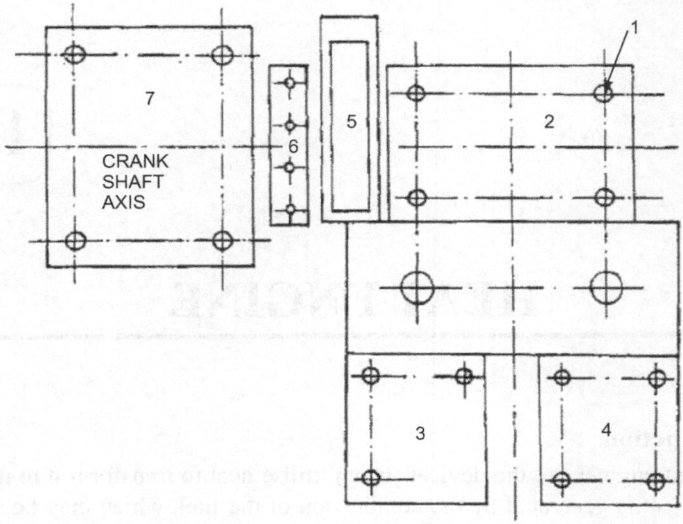

Fig. 13.1. Foundation plan.

The steam engine foundation is always constructed over a concrete raft of required thickness. For a vertical engine the depth of the foundation is made sufficiently larger to withstand static and dynamic loads.

Of course, the size of the foundation block depends on the quality of the subsoil.

An efficient levelling and grouting is naturally the essential features of the installation.

Maintenance of Steam Engine

The maintenance of an engine may be illustrated taking the important individual parts separately. The Fig. 13.2 shows a vertical non-condensing steam engine.

Steam comes to the chamber within which the slide valve (9) is functioning. It-allows the steam to enter or exit through the steam passages. The piston (2) moves up and down within the cylinder. The movement of piston takes place against the load communicated through the piston rod (3). The other end of the piston rod is connected to the cross-head with which one end of the connecting rod (5) is attached by a pin as shown in the figure. The flat base (4) of the crosshead acts as guide to move the piston rod in the axis of the cylinder. Another end of the connecting rod is connected to the crank pin (6) which moves in a circular motion. The crankshaft (11) also gets a rotary motion along with the eccentric (7), and the flywheel (12). The eccentric rod (8) simultaneously actuates the slide valve through the valve spindle guide (l0).

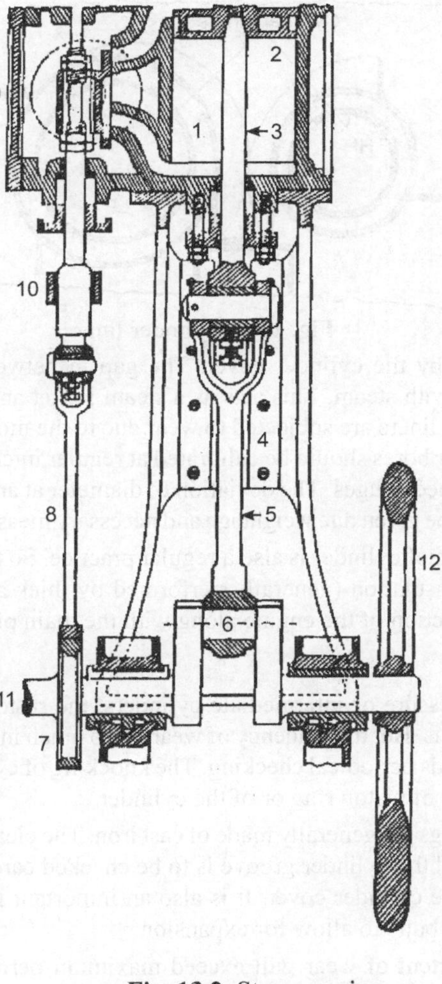

Fig. 13.2. Steam engine

The brief working procedure of a steam engine establishes the need of accuracy of the functioning of different components and obviously indicates the key points to be considered for major maintenance work, a few of which are described ahead.

Cylinder

Most of the steam engine cylinders possess liners with it. It may be in the form of a barrel as shown in Fig. 13.3 screwed at the bottom and the other

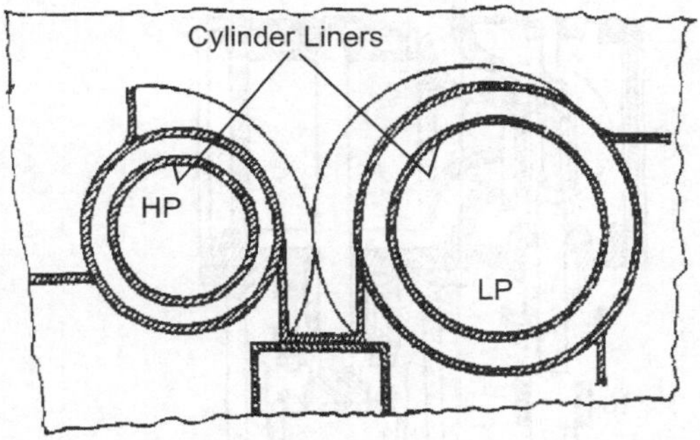

Fig. 13.3. Cylinder liner.

end is supported by the cylinder cover. The gap in between the liner and cylinder is filled with steam. This acts as a steam jacket and reduces loss of heat. The cylinder liners are subjected to wear due to the movement of piston rings. The cylinder bores should be calibrated at regular interval with the help of specially designed gauges. The deviation of diameter at any point along the cylinder axis is to be given due weightage and necessary measure is to be taken.

The insulation of cylinder is also a regular practice. So the lagging of the cylinders or the insulation (generally performed by thick asbestos) is to be made after the erection of the engine along with the main pipe lines.

Piston Rings

For a low pressure or intermediate cylinders, the piston rings last long without adjustments. But the tendency of wear is too much in the high pressure cylinder and it needs periodical checking. The knocking of cylinder is an usual indication of wear of piston ring or of the cylinder.

The piston rings are generally made of cast iron. The clearance in between the piston ring and the cylinder groove is to be checked carefully with feeler gauge, opening the cylinder cover. It is also an important feature to inspect the opening at the butts to allow for expansion.

Whenever, extent of wear will exceed maximum permissible limit the piston rings are to be replaced. It may also be necessary to rebore the cylinder groove after a long years of service.

Main Bearing

Main bearing is mounted on the frame and positioned at opposite to the cylinder. The crankshaft journal is supported on the bearing.

While installing, the exact bedding of the journal on the bearing should be assured by smearing technique.

And the wearing, if any, takes place while in operation, should be attended readily. Generally thin plates of steel or brass, usually called shims, are placed in between the two halves of the bearings. Addition or alternation of number of shims sometimes provide easier means for adjustments.

A regular attention should be paid to check if the bearing has got unusually heated. If the adjustments are all perfect, such phenomenon may occur due to failure of lubrication.

If such heating of bearing occurs, temporary measure may be taken using castor oil as a lubricant to cool down the bearing. But it is safer to stop the engine and to remove the causes of such undesirable heating.

Glands

Different types of glands are used in a steam engine assembly. As an example, the piston rod passes through the cover of the cylinder. It should be made steam tight and a stuffing box is usually used (Fig. 13.4). It is so made that a small space around the rod is filled with some form of flexible material to achieve steam joint round the piston rod and a gland is used as a cover pressed down on the rod, with two screwed bolts. Stuffing box and gland is mounted to the slide valve rod, pump rods, steam valves and for other similar purposes.

Glands deserve periodical attention. Say for pump glands a greasy packing is used. The packing should be properly bedded and whenever gun metal is there (*e.g.* pump plunger) the packing is kept slightly slackened to avoid corrosion.

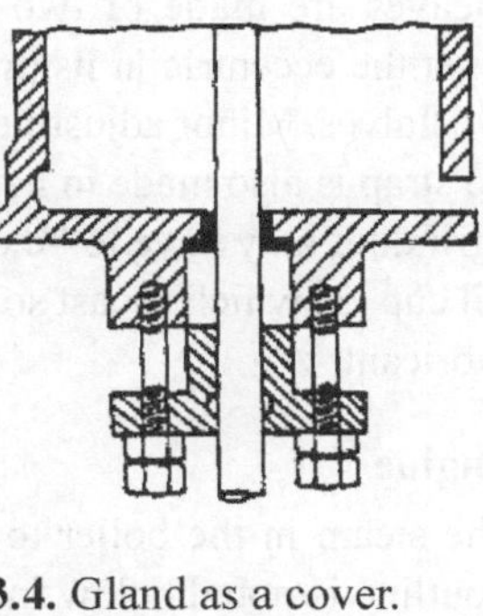

Fig. 13.4. Gland as a cover.

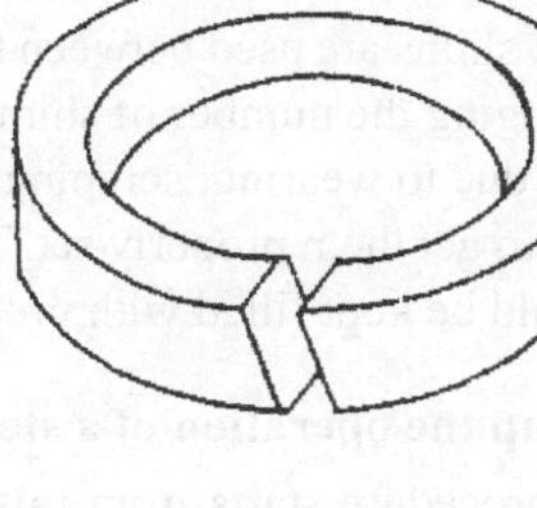

Fig. 13.5. Gland ring.

It is a bad practice to allow the packing rings to be completely unusable, without attending it at proper time. The gland rings as shown in Fig. 13.5 are having a beveled cut on it. Such rings facilitate easy withdrawal and fitting. During periodical inspection, one or two rings may be replaced if necessary. If the situation deserves, complete set may have to be changed.

While placing ring inside, it should be placed smoothly and it should not nip between the side of the bore and the gland. Such nipping will cause undue wear and breakage.

Necessary care should also be taken in this regard to check the trueness of the rod for proper functioning of stuffing box. If the rod is found to be worn out of shape, proper grinding should be performed and new segments may then be added.

Eccentric Strap

Eccentrics are used to connect rotary motion into a reciprocating motion such as in the steam engine slide walves or in the pumps. The eccentric strap is subjected to repeated application of loads. Wear in this juncture is very much obvious to occur. But care should be taken to avoid it to have a trouble free operation of the system. The Fig. 13.6 shows a typical eccentric strap. The sheave (1), a circular solid disc is surrounded by a metal band, called strap (2).

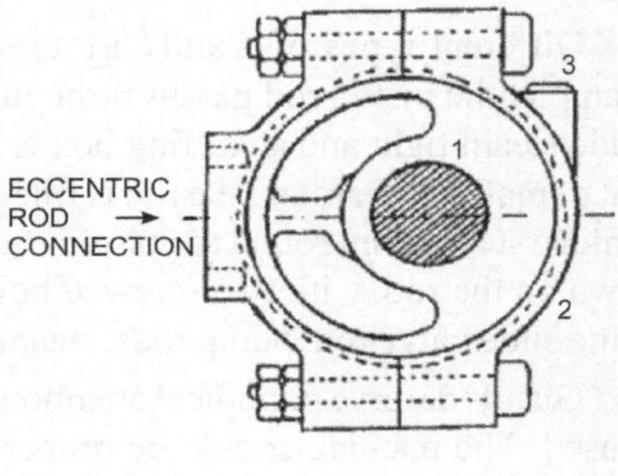

Fig. 13.6. Eccentric Strap.

The sheave rotates about the centre of the shaft and a to and fro motion is resulted after transmitting this revolving motion through the strap and the eccentric rod. The sheaves are made of two halves. So during erection, care should be taken to get the eccentric in its proper place. Moreover, shims are used between the two halves. Minor adjustments may be made changing the number of shims. The strap is also made in halves. If it is necessary due to wearing, scraping and grinding may have to be done on the two faces to get them properly set. The oil cup (3) which is cast solid with the strap should be kept filled with proper lubricant.

Starting up the operation of a steam engine

The procedure starts from raising the steam in the boiler to allow it to enter the valve chest and so on. A brief outline is noted below in this regard stepwise.

(*a*) The manhole, mudhole, gauges of the boiler is checked.

(*b*) Water is allowed to fill the boiler.

(*c*) The boiler is fired, being sure that all valves are turned off.

(*d*) The water level is checked frequently.

(*e*) Being assumed that the feed water supply is satisfactory, the pressure is allowed to rise.

(*f*) The engine cylinders should be primarily heated to avoid higher condensation rate inside the cold engine.

(*g*) The drain taps are kept opened.

(*h*) The stop valve is opened slowly primarily to allow steam through the drain tap to one end of the cylinder then it is to be stopped. Turning the engine the stop valve should again be opened to allow steam to blow to the other end.

(*i*) For a single cylinder engine, the crank is placed over the dead centre and in case of a compound engine the sampling valve which allows high pressure steam to enter the low pressure cylinder may be opened slowly and should immediately be closed as soon as the engine starts running.

(*j*) When the engine comes to the working speed and the governor takes the control of engine, the load may be gradually applied. But all the drain taps are to be kept closed and the stop valve is to be opened fully.

Lubrication

A special mention is required for the lubrication of the steam engine. This is an essential but complicated job. The problem in the steam engine assembly exists in manifold so far as the lubrication procedure is concerned.

The engine cylinder and the piston, the valve and the valve rod, the piston rod etc are subjected to high temperature. So a special oil, termed as cylinder oil may be used for such parts.

The parts like crosshead bearing, crankshaft main bearing or the crank pin are not directly in contact of stream. So mineral oil may be selected for the lubrication of these items.

Besides the high temperature of super heated steam, the moisture content in the saturated steam also creates problem. Lubricants, if applied in excess quantity, may be carried by the steam to the condenser and then run to the boiler along with the feed water. If the lubricants get precipitated in the boiler tube there is a chance of rupture. To avoid such entrance of lubricating oil to the boiler, an oil separator is installed in the return line.

Generally, hydrostatic or mechanically actuated lubricating systems are used for proper lubrication. Wick lubricator or drop feed oilers are also used for slow-speed engines. For an enclosed engine splash lubrication is mostly used.

The procedure of a typical central system of lubrication for a large station is shown in Fig. 13.7.

Finally, it should be mentioned that sometimes 'atomiser' is used to introduce lubricants in the steam line. The lubricant gets broken into mist and reaches all over the internal surface of the cylinder. The most efficient location of the atomiser is in between the throttle valve and the valve chest to introduce the oil mist in the steam line.

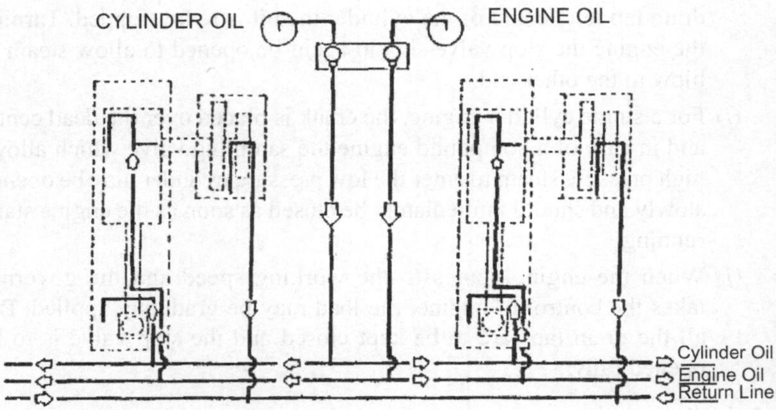

Fig. 13.7. Lubrication line of a Steam engine.

STEAM TURBINE

Steam Turbine has occupied a key position as a prime mover, specially in a power generating station.

It is a continuous process to take all possible measure to increase the efficiency of a turbine and this objective may be given due weightage from the very foundation work.

Foundation

The installation of turbine is a job of care and experience along with skill. In the very beginning it may be noted that in case of a turbine, the inertia forces are very much insignificant and that is why turbine foundation does not call for a massive foundation as essential for a steam engine. But of course turbine foundation must be rigid and capable of providing firm support and also should be heavy to take care of unbalanced forces while the turbine is in operation. Mostly in a steam power plant, the condenser is placed just below the turbine.

Generally the main structure for foundation (1) is constructed on the raft as shown in a typical turbine foundation in Fig. 13.8.

The concrete raft is first constructed on the floor. The depth of its construction depends on the nature of the subsoil.

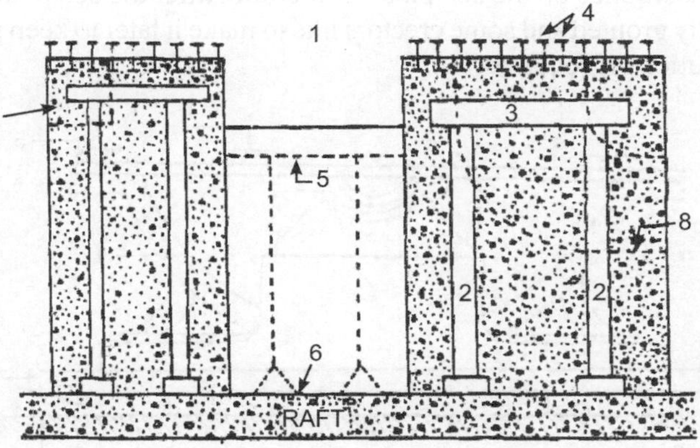

Fig. 13.8. Foundation of turbine.

If the subsoil is considered to be soft or marshy, raft may be constructed after driving piles in the site. Such piles, driven in the ground will restrict any sinking possibility. Main structure is generally concrete blocks (1) with an internal steel work (2) to support the main joints. The surface of the foundation should be made rough for grouting. It may be noted that the sole plates (4) are shown above the surface which enables to have bondage between the main foundation block and the grouting. Such sole plates are connected on the main joints (3).

The girders may also be built up in the concrete foundation. The upper girder (S) is to support the turbine and the lower girders (6) the condenser.

The provision for the recess for steam pipes (7), airduct (8) under the alternator are also to be arranged during foundation work.

The total job is to be performed as per the drawings supplied and suggestions made by the manufacturers of turbine.

Erection of Turbine

Erection work of turbine should start from the finalisation of datum points to facilitate the assembly of different components at perfect positions. The finished floor level, the walls of the building and the centre line of any adjacent sets may be taken as datum points to take horizontal or vertical dimensions.

The condenser may either be erected or may be placed at proper place along with other heavy parts underneath the turbine.

The erection work for bed plate (1) of the turbine may be taken up now. The bed plate is placed at proper position as shown in Fig. 13.9 and is levelled with packing pieces (2), wedges (3) and then lower half of the cylinder (4) may be positioned on the bed plate. Sometimes after the bed plate erection, it is finally grouted and some erectors like to make it later to keep provisions for alterations if required.

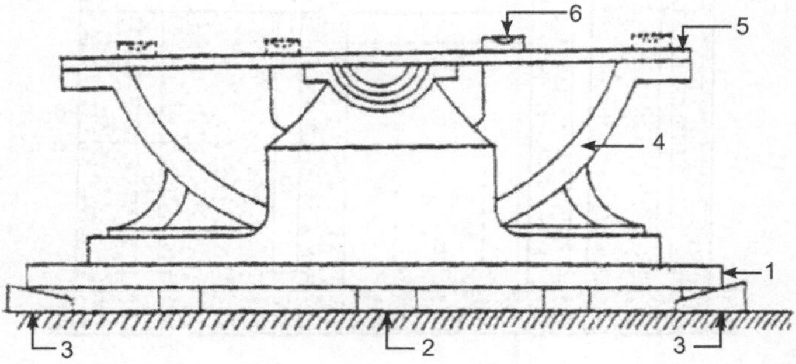

Fig. 13.9. Erection of cylinder half.

However, the lower part of the turbine is to be levelled longitudinally or transversely with the help of straight edges and spirit level.

All the oilways, ports, steam inlet branches are carefully checked and cleaned and the next step is to position turbine rotor.

The bearings used are generally made of two shells. The lower shell (Fig.13.10) is provided with three spherically seated pads (1). A fourth pad is attached to the upper half. Liners (2) of varying thickness are placed between the pads and the bearing shell (3). The position of the 'centre of bearing' setting for its alignment may be adjusted upto a small limit by adjusting the position of the liners.

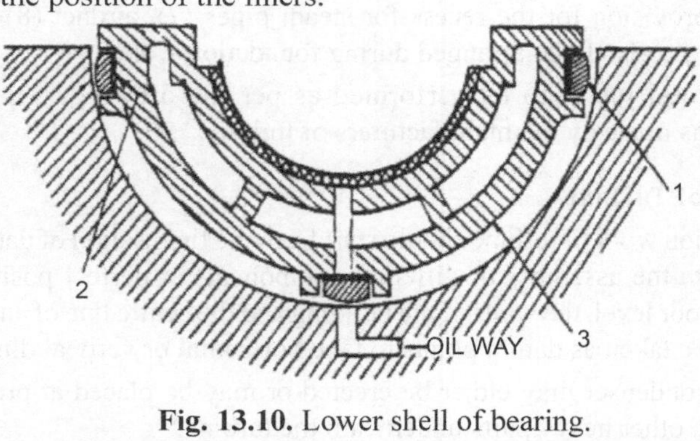

Fig. 13.10. Lower shell of bearing.

Once it is placed with the properly positioned centre line, the rotor is to be placed in position. To lift the turbine spindle the manufacturers generally supply guiding bars (l). The length of the slings (2) are so adjusted that the rotor (3) comes down perfectly horizontal as shown in Fig. 13.11. The rotor is placed and 'aligned and a final checking is made whether the rotor is fouling the diaphragms or any other part of the cylinder. Once the alignment is perfectly alright, the alternator stator and as well as the rotor may be erected for efficient coupling.

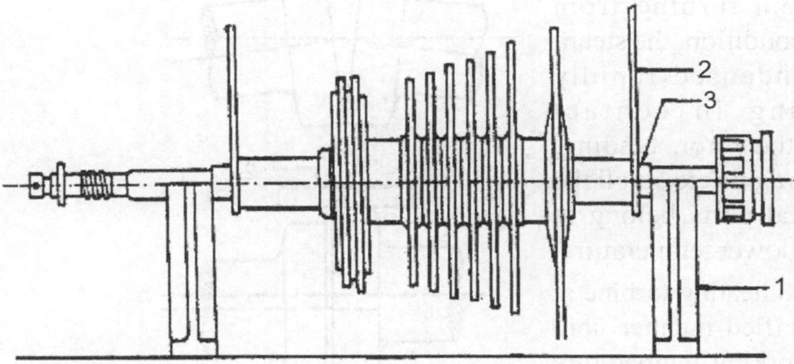

Fig. 13.11. Positioning of rotor.

The sole plates may then be grouted, if it has not been done previously and then the top half of the cylinder and the alternator may be erected and the alignment of bearings are finally checked.

Strictly speaking, the main aim of the erection work is to make the shafts perfectly aligned and levelled.

Shaft alignment checking

Misalignment of shaft, if any, will ultimately spoil all the attempts to achieve an optimum result. Different types of lack of alignment may take place. A few procedures to check the alignment are given here. If the couplings are touching each other at the bottom (Fig. 13.12*a*) the feeler gauge will show the clearance (1) and rotating the assembly, it is to be seen whether the clearance is permissible and whether it remains same throughout.

The alignment of two axes may also be checked by straight edge (2) as shown in Fig. 13.12 (*b*). Such tests classify whether the shafts are in the same horizontal and vertical plane or not.

Consider the Fig. 13.12(*c*). Here the Fig. 13.12(*a*) has actually been repeated to identify the types of adjustments to be done to align the shafts.

While rotating the shaft if it is seen that the clearance at position P_1 and P_3 is different (Fig. 13.12*d*) but same at P_2 and P_4, it indicates that the level of the bearing is to be adjusted. But in a reverse case, when the clearances are same at the points P_1 and P_3 but different at P_2 and P_4, it reflects that a sideway adjustment is required for the bearing. Adjusting the liners beneath the bearing pads, such misalignments may be rectified.

Heating of Turbines

While starting up a steam turbine from cold condition, the steam is condensed rapidly coming in contact with the rotor, blading, diaphragms etc. as all the internal parts belong at much lower temperature.

So heating turbine at a specified manner upto the working temperature is very much essential. Moreover, the heating should be performed symetrically as far as it is practicable. It is evident that while the turbine will be started, the various parts of it, being heated, will expand. Distortion is bound to occur, if they are not being heated symmetrically. In most of the cases, while starting, the steam is allowed to admit all on a sudden through the stop valve upto certain extent, so to put the machine in rotation. The valve is then partially closed until the rotor continues its rotation but at a very low speed.

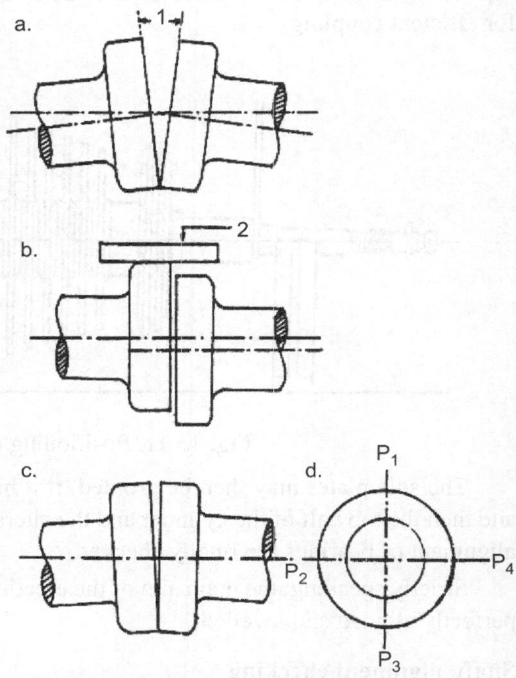

Fig. 13.12. Shaft alignment.

Critical speeds are to be carefully avoided. Otherwise a dangerous vibration will take place. Normally, the rotor is allowed to rotate with 10% of its working speeds for adequate time and then the load may be applied gradually. Practically, a calculated vacuum is maintained in the condenser to admit the steam to keep the machine running in the primary stage.

It is mention worthy that care should invariably be taken for the effective drainage of the steam turbine. Specially, the rapid condensation will produce enough water which should be drained off in such a way so that no part of the turbine is immersed in water.

Starting a Steam Turbine

A systematic general approach is given as follows for the said purpose.

(*i*) Start oil pump and check the pressure of lubrication at bearing.

(*ii*) Start circulating pump to circulate water through the condenser.

(*iii*) Start air pump at low speed.

(*iv*) Open all the drain valves of cylinder for warming up.

(*v*) Admit steam quickly to start the rotor moving and adjust the air pump simultaneously to maintain a suitable vacuum for the purpose of warming up.

(*vi*) Turn on the water supply to the oil cooler.

(*vii*) Try to listen for any internal noises, such as would be caused by rubbing etc.

(*viii*) If abnormal vibration occurs and persists the machine is advised to shut down.

(*ix*) If everything goes normal and the turbine is properly heated, rate of admission of steam may be increased gradually.

(*x*) When the turbine runs at working speed, the governor should be allowed to take the control.

(*xi*) Check all the seals etc.

(*xii*) Close the drains and increase the vacuum to its maximum value.

If upto this stage, the turbine works satisfactorily, the turbine is ready to take the load. The load should be applied and increased gradually.

Common troubles in turbines and remedies

A few common troubles to which a steam turbine is often to be subjected, are discussed in brief. The steps to be followed to avoid such troubles are also discussed. The troubles encountered during operation are also discussed later.

In the very beginning, the turbine failure may be broadly classified into two groups. One is due to faulty design or lack of quality materials used during construction etc., another may be related to the operating procedures and conditions.

(*a*) Shaft Vibration

Every elastic body such as the rotor of the turbine is having a natural frequency. When a forced vibration is synchronised with it, the resonance develops. Such phenomenon of resonance will take place when the rotational speed of the shaft coincides with in natural frequency. The result is extreme vibration creating noise.

Such speeds are called critical or whirling speeds. The range of critical speeds must be passed through too quickly in bringing the turbine upto its running speed to avoid serious vibration.

Shaft vibration may also be caused due to an unbalanced rotor or by badly aligned shaft.

The alignment may be checked and the bearing may be inspected and proper steps may be taken to avoid shaft vibrations.

(*b*) Disc Vibration

The disc of Impulse turbines are also elastic body and moreover, there are some obvious periodic forces, being experienced by discs. So possibility of vibration of disc is very natural. The critical speeds should be avoided in this case also.

When the disc rotates, a recurring impulse is applied to each part of the disc periphery as it passes in front of the nozzles. The impulse may also be due to asymmetry of the blades. However, such impulse forces are naturally responsible for disc vibrations.

However, the changes of load on a turbine bring about changes in the steam temperature. The temperature distribution in the disc will also rapidly alter due to sudden change of loads as the relatively light rim will be either colder or hotter than the heavier hub and the centre portion. This change of temperature may also be responsible for disc vibration.

Slightest vibration of disc should be identified and taken into account carefully or otherwise it will develop fatigue in metals reducing its life span.

(*c*) Diaphragm failure

Diaphragm may fail due to excessive deflection. The major cause of such deflection is the difference of steam pressure on the two sides. But as the diaphragms are of complex structure, it is practically impossible to measure the deflection.

Care is to be taken during construction and designing to take measures to avoid too much deflection.

Besides deflection, distortion may also occur on diaphragms. It is generally a temperature effect. Rapid change of temperature due to sudden change of load is also a cause of diaphragm distortion as it is to be responded to the changed operating condition at equally rapid rate. This may even cause buckling of guide vanes if enough allowance is not provided for expansion.

(*d*) Erosion of blading

Erosion is a mechanical phenomenon and it is usually produced by the striking feature of small water particle suspended in the steam. Any solid matter (say lime) if is being carried over from the boiler, will also cause erosion.

Moderate blade speed, relatively low steam speed may avoid the erosion of blading.

(*e*) Deposits on turbine blades

The deposits on the turbine blades may be due to two major reasons :

(*i*) If exhaust steam from steam engines are admitted to the turbine (in case of low pressure turbine), oil and foreign matters being carried with exhaust steam may produce a sticky greasy layer on the blades.

(*ii*) For high pressure boilers, turbines are liable to become choked up, due to carry over of various substances (soluble or insoluble in water) from the boilers.

Blade fouling, reduction of load carrying capacity of the turbine may be the indications of blade deposits. Necessary measures are to be taken at suitable time to clean and overhaul the turbine.

It has been seen that mostly, the deposits contain silica and iron oxides. The washing procedure is to remove the rotor and to place it outside in a specially designed rotor stand. It may be cleaned by the forced application of a mixture of water and some abrasive materials like fused alumina, bauxites etc. at high pressure.

Corrosion of blades

The corrosion of blades, as well as that of diaphragm may be initiated due to the impure feed water of the boiler. It may also be caused in consequence of the dissolved gases like oxygen or carbon dioxide in the feed water of the boiler.

The impurities are carried to the turbine by priming. Moreover, for an open feed system, the water, being in contact of air dissolves oxygen and ultimately creates problems.

Corrosion, due to such causes may be minimised by employing a closed feed system and using de-aerators for the make-up feed water.

Servicing of Turbines

The overhauling of turbine is a scheduled work. The interval between the two overhauls depend on the type of turbines and instruction of manufacturers. It involves complete dismantling, inspecting, repairing, replacing of parts, if required.

The important points to be checked along with other inspections are briefly outlined as follows :

(*i*) Diaphragms are examined thoroughly. Any indication of corrosion or erosion in the guide vanes to be checked. If the blades are seen to be ragged at the outside edge–it should be fixed properly.

(*ii*) The bladings are checked. The axial clearance between the blades are measured and compared with standards. Tests should be conducted to identify the signs of crack, if any. Blade deposits should be removed by abrasive cleaning or any other befitting procedure. They may be replaced by new blades if required.

(*iii*) The alignment of the rotor is to be assured. Proper action is to be taken if any sign of rubbing is detected.

(*iv*) The bearings are to be inspected whether there is any mark of wear on the bearings. The 'fit' and 'bedding' is carefully checked. The clearance is measured by the feeler gauge and is compared with the requirement, that is, at which it was set.

(*v*) Gland packings are to be cleaned and examined. The conditions of the rings and the shaft sleeve are inspected.

(*vi*) The lubrication system is thoroughly checked. Grease ways are cleaned and examined.

The overhauling work involves a massive task and for any complicated findings, expert opinion should be taken. After the verifications and inspections, necessary measures are adopted to rectify the faults and then the set may be returned for operating.

After overhauling, when again it will be operated, the performance of the turbine should be carefully observed.

Operational troubles and remedies

Major troubles or difficulties, generally being experienced during the operation of turbine are tabulated below, along with the probable causes and remedies with regard to different components.

Component	*Difficulties*	*Probable causes*	*Remedies*
1	2	3	4
(*i*) Steam supply valve.	(*a*) The rod is getting hardened or sticky.	Foregin materials being carried by the steam.	The hub and the rod is cleaned and the steam seal is checked. To identify oil leakages,oil pressure may also be checked.
	(*b*) Sealing is not perfect.	The seat may have got worn or the stationary seat may also get may also get loosened.	Wearing may be rectified by grinding, the seat is also to be verified and to be reset properly.
(*ii*) Admissior valve	(*a*) May be jammed or hardened.	Substances may be carried with steam.	The rod and hub is to be cleaned.
	Sealing is seen deficient	The wearing of seat may have taken place or the seat may have been loosened.	Grinding may be done. The seat is to be checked and reset.
	(*b*) Vibration on the rod.	The dynamic force of the steam may be responsible.	The rod may have to be changed.
(*iii*) Shaft	(*a*) Condinuous vibration	Misalignment of rotor.	To be checked and aligned.
		Improper clearance of the bearings.	Clearance is checked.
		Wearing out of the coupling between the turbine and the alternator.	Coupling is checked and rapaired.

1	2	3	4
		The moment of inertia may be unsymmetrical or introduction of torsional vibration.	Vibration tests are conducted. Manufacturer may be contacted.
		Thermal or dynamical deformations of rotor.	Rotor may be checked. Balancing may be verified and remade. Better to contact manufacturer.
	(*b*) Momentary vibration	Foreign materials may be present on the blade. The steam sealing small rings may have jammed and foreign substances may be present there.	The turbine should be stopped. Exact cause is to be identified.
		Water may be present due to faulty drainage. Pipe drainage may also be faulty.	Drain the machine or piping as is identified.
(*iv*) Blades	(*a*) Seating of blades or bruises on blades.	Foreign particles dragged by the steam.	To get it confirmed steam filter may be checked. The blades are to be made cleaned properly after rectifying the cause.
	(*b*) Corrosion	Drainage of turbine is imperfect and protection from humidity is not properly arranged.	Clean drainages. Adopt proper protection procedures.
	(*c*) Breakage	Due to friction.	The manufacturer may be consulted.
		Severe corrosion by saline deposits.	Continous drainage of boiler is assured and the blades are periodically cleaned. The damaged blades may be replaced.
		Severe vibration and resonance.	Better to contact manufacturer.
(*v*) Gears	(*a*) Noise	Wearing out of teeth or improper lubrication.	Teeth wearing is checked and lubrication is made perfect.
	(*b*) Teeth breakage	Experiencing sudden overloads.	May be during first periods of operation. The surfaces to be cleaned.

1	2	3	4
(*vi*) Coupling	Vibration is noted	Misalignment.	Alignment is checked. If the misalignment is due to wearing out of the bearing, necessary checking and replacement, if necessary, is to be done.
(*vii*) Safety Regulator	Operation is faulty	May be due to friction.	Better to contact manufacturer.
		Displacement of sliding planes.	Contact the manufacturer.
(*viii*) Bearings	Rapid wearing of lining.	Oil supply may not be sufficient.	The quality of the oil is assured and the obstruction if any in the oil supply may be removed.
		Excessive thrust due to clogging of blades (in case of thrust bearing).	The blading is cleaned carefully
(*ix*) Lubrication system	(*a*) Oil pump is producing noise.	May be due to wearing out.	For the cause of wearing out it is better to contact manufacturer.
		Oil thrusts.	Oil ducts re checked to identify obstructions, if any.
		Lubrication is not sufficient.	Leakage if any, may be identified.
	(*b*) Capacity of the oil pump is not sufficient.	Air inlet may be there.	Air is to be eliminated.
		The pump and the bearings may have enough clearance between them.	It is checked and readjusted properly.
	(*c*) Oil supply pressure is not sufficient.	Losses in the piston.	Leakages through flanges etc. are checked.
	(*d*) Excessive oil supply pressure.	Obstruction of tubes.	To be verified and action to be taken.
	(*e*) Excessive loss of oil.	Losses from the flanges of sealing planes.	Check properly. The clearances are verified.
	(*f*) Temperature increase of a bearing.	Quality of oil is not proper.	Check the quality and proper lubricants are used.

INTERNAL COMBUSTION ENGINE

Internal combustion engine is a heat engine. The combustion produces chemical energy and this is released inside the engine cylinder.

There are no standard methods of classification. However, it may be done in many ways. Let us at least mention that according to the method of ignition, it may be of two types:

(*i*) Spark ignition engine (S.I. Engine), and

(*ii*) Combustion ignition engine (C.I. engine)

In S.I. Engine a mixture of air and petrol vapour is drawn into the cylinder in the suction stroke itself. A carburettor performs the function of atomising or vapourising the fuel to mix it with air. At the end of the compression stroke a spark plug ignites the compressed charge to have external work.

But in the C.I. engine no spark plug is used. The ignition is due to the heat of compression. Moreover, the vaporization of fuel does not take place in C.I. engine. The fuel is injected through an atomiser or nozzle.

Diesel Engine

A brief outline of the installation procedure of a stationary, low or medium speed diesel engine is given here.

Foundation

The foundation details naturally depend on the quality of the sub-soil. However a schematic foundation plan (Fig.13.13*a*) (with only line diagram), showing foundation edge (l) and a typical foundation for the engine (2), is given here. The fly wheel race (3) and the recess for the alternator (4) are also shown alongwith the 'Exhaust water pipe' (6) and 'cable' (S) trenches. It is customary to keep provisions and to make plans for auxiliary structures for fuel tanks, water tanks etc. at the same time.

To take up the work, it is always advisable to follow the manufacturers instructions in this regard.

The elevation of a part of the typical wall construction is also shown in the Fig.13.13 (*b*), taking section at A-A.

It is also a general practice to prevent the transmission of foundation vibration and noise. It is done providing vibration isolating materials (2) in between the foundation and the concrete base or raft (1). A small air gap (3) say about 70-80 mm may be kept all around the foundation. These gaps should always be kept clean. The gap may be covered (4) by some boards etc. at the ground level.. [Fig. 13.14].

Use of wooden boxes

It may be mentioned that before taking up concreting, a template may be framed with wooden planks and boxes, representing floor level, the engine, the alternator etc. They may be made and positioned to check the centre line and other details of the bearing pedestals also.

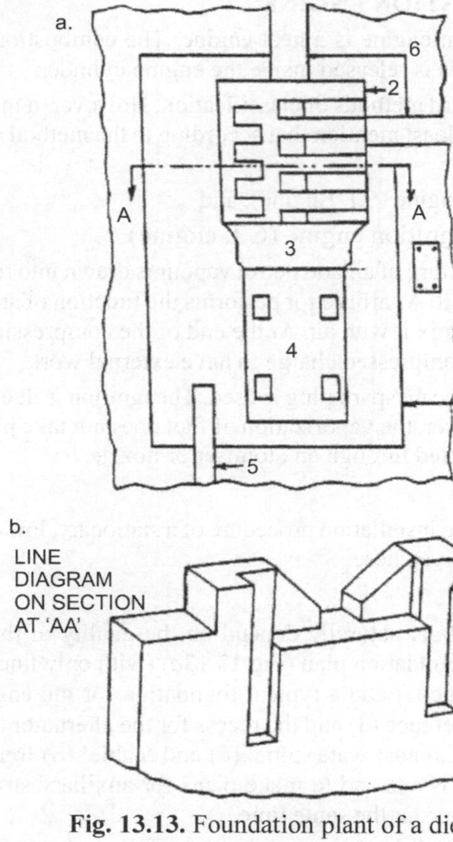

Fig. 13.13. Foundation plant of a diesel engine.

Special care

Special care should be taken to maintain gaps between the 'engine' and the auxiliary items from either the side walls or from other machineries.

Similar care is needed for maintaining recesses under the engine or for fly wheel races.

Special care is needed to mix the concrete at proper proportion and with specified materials.

Fig. 13.14. Vibration isolating foundation.

Lastly, it should be remembered that the total work should be completed.

in continuous operation. So sufficiency of the materials should be verified before starting the work. The position of the wooden boxes should not be disturbed at all during the foundation work.

Erection

The erection procedure is discussed step by step:

(*i*) The engine should be brought to the site smoothly and the engine is to be made cleaned thoroughly. Entry of any foreign particle to any mating parts will cut down the life span of the engine too rapidly.

(*ii*) Before lowering the engine, the foundation bolt is pulled upwards to position the bed plate. After positioning the bed plate it is levelled by steel plates, wedges and nuts are tightened.

(*iii*) After each step, the alignment and positioning of engine is to be checked. At this stage, specially the height of the centre line of the crankshaft from the finished floor level is checked and adjusted with necessary wedges etc.

(*iv*) If the provision for belt or chain drive is there, their alignment should be inspected at this stage, so that the belt or the chain runs truly.

(*v*) The coupling alignment in between the engine and say, the alternator is assured. For this, the coupling halves should be mounted and taken in position. The weight-factor of couplings, flywheel etc. on the crankshaft is also taken into consideration while checking the alignment of coupling. The alignment must be ascertained even when the flexible coupling is used.

(*vi*) The foundation bolts are then to be grouted into the foundation block removing the wedges etc.

(*vii*) The other works like erection of tanks, piping works etc. may be undertaken at this juncture. Of course, final setting of tank should be performed after finishing the pipe lines erection.

(*viii*) The bed plate is finally levelled with the help of straight edge and spirit level. The lining up of the engine is also undertaken. It is better to position the crank first and lining up from it. For effective levelling of the bed also, it is advisable to take the crankshaft in the bearing. The perfect bedding of crankshaft on all the bearings will indicate even levelling of the bed.

(*ix*) The alignment of crankshaft bearings may be checked practically by some easy methods. Taking the crankshaft at the top position as shown in the line diagram [Fig. 13.15 (*a*)] the distance between the crank webs (T) is measured. Again after turning, it is measured when the crankshaft is at the bottom-most position (B), [Fig.13.15 (*b*)]. If B is greater than T, the outer bearing is supposed to be high. But if B is less than T, it indicates that the outer bearing is low.

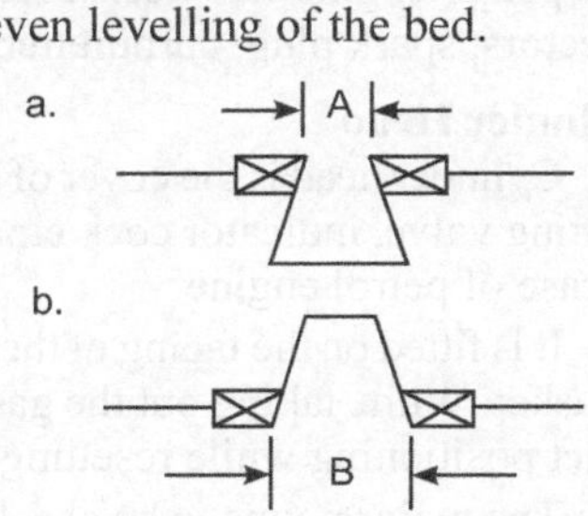

Fig. 13.15. Testing of alignment of crankshaft.

The test may be conducted in a similar way taking the crank in the front and at the back position also.

When it will be confirmed that the bearings are in perfect alignment, the caps can then be put on being assured that the journal is having the' bearing throughout its whole length. The nuts of the bed plate may then be tightened carefully to make it ready for grouting.

(*x*) Mounting of flywheel is then undertaken. The straining of the crankshaft which may be caused due to heavy weight of the flywheel is to be avoided. The support of the extended shaft by the outer bearing should be provided first and the packing support of the flywheel may then be removed to release the full weight on the crankshaft.

(*xi*) After checking the alignment of the crankshaft once more, the grouting work under the bed may be started. A wooden or clay wall may be made all around the bed to hold the grout. The outer bearing is also grouted in a similar way.

(*xii*) When the grout has been satisfactorily hardened, the remaining works like fitting of large end bearing to crankshaft, fitting connecting rod, fitting of piston rings and lowering down pistons in respective cylinders may be completed gradually.

Maintenance features

Maintenance, repairing and over haulting of an I.C. Engine is a vast job and they involve a lot of techniques.

It is simply impossible to discuss them within the scope of this chapter. We will concentrate on a few major points only.

Here is a generalised discussion regarding cylinder heads, cylinder blocks, pistons etc. has only been done.

The fuel pump, fuel filter, cooling systems, governors, water circulating systems etc. are not discussed. Although the maintenance and overhauling of these accessories are equally important.

Of course, at the end of the discussion the combustion technique of diesel and petrol engine has been taken into account and maintenance feature of injectors, spark plug, curburetter have been discussed in brief.

Cylinder Head

Cylinder head is the cover of the cylinder block and it is fitted with injector, starting valve, indicator cock etc. in case of diesel engine and spark plugs etc. in case of petrol engine.

It is fitted on the facing of the cylinder block with the nut and studs through a gasket. While taking out the gasket, it should be marked to facilitate to have exact positioning while resetting.

The major points to be checked, if the cylinder heads are opened, are the gasket, all the studs and nuts, the facing of both the cylinders and the head itself. The deposition of carbon on the head or on the piston head and liner are also to be checked and cleaned.

In case of a large engine each cylinder may have separate cylinder head while for a smaller engine one or more cylinder heads may take care of all the cylinders.

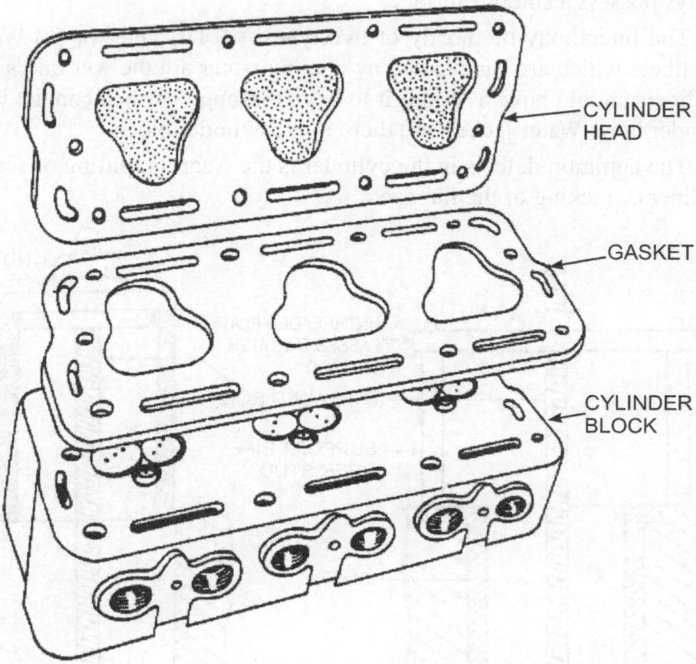

Fig. 13.16. Cylinder head.

The major defect which may occur on cylinder heads is crack generation. Over heating, 'coming in contact with cold water' when it is hot or misalignment of facing of the cylinder block and the head may be the probable causes to generate crack. Formation of scales in the water passages may also generate cracks. However, the common practice to identify the crack or cracks is to hold the piston at top-dead centre and to run a stream of compressed air through the indicator cock. A bubbling sound may indicate that the air is forcing out through a crack in the cylinder head.

If such cracks are identified the cylinder head is to be replaced by a new one.

To open a cylinder head, the water is to drain out first. After removing the cylinder head cover, the injector may be removed. All inlet, exhaust, water pipes are to be detached and then cylinder head nuts may be unscrewed, removing the rocker assembly etc.

Cylinder block and cylinder liner

Most of the cylinder bores are having liners, made of wear resisting material. They are made and to be maintained as perfectly round and should always possess a smooth surface.

The liners may be mostly of two types: (*a*) Dry liner or (*b*) Wet liner. The liners which are surrounded by cooling water are the wet liners and the dry liners should have a good fit to have thorough surface contact with the cylinder bore. Water jackets are there in the cylinder block.

The common defects in the cylinder is the wearing, pitting or scoring of the liners, cracking of the liners etc.

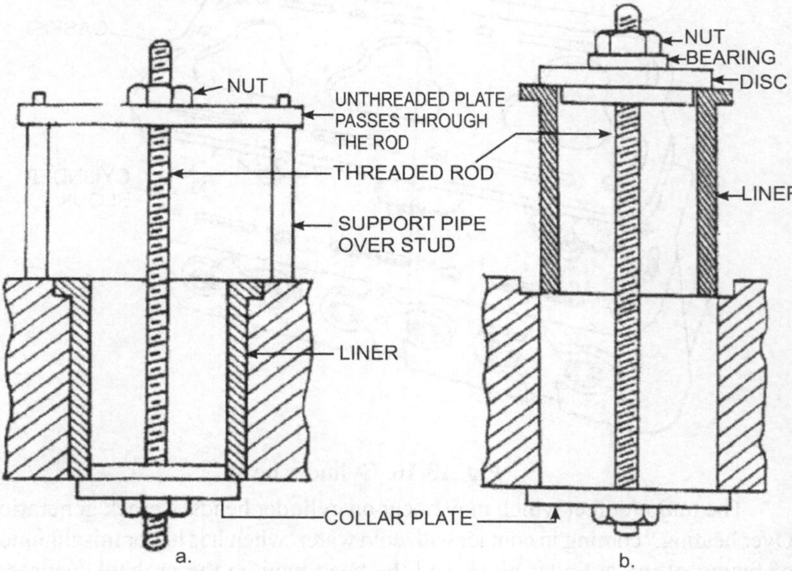

Fig. 13.17. (*a*) Cylinder liner pulling out, (*b*) Cylinder liner fixing.

If the wearing or scoring of the liners exceed the permissible limits, the engine will run hot. Smoke may come out from the crank case. The firing pressure and the effect of compression will be reduced. However, to restore the cylinder liner it is to be taken out. Generally, the pulling out of cylinder liner and the fixing of the liner as well, are done with the help of a puller as shown in Fig. 13.17. For pulling out, a step collar is used at the bottom of a long threaded rod to apply the pulling force only on the liner and the pulling force is applied by rotating the threaded rod tightening the nut placed on a bearing at the top [Fig. 13.1 7(*a*)].

But for fixing the liner, it is placed on the bore and the collar arrangements should be as shown in the Fig. 13.1 7(*b*).

After taking out the liner, it is to be thoroughly checked and cleaned. Metal spraying may be performed to restore its surface and dimension.

Reboring of liner may be done if the situation permits. It is rather better to replace a liner than to have a deep reboring.

The common cause of developing cracks on the liner may be overheating. The liner will also crack, if heated liner comes in contact with cold water. Furthermore, sometimes the gudgeon pin being loosened, strikes the liner, resulting crack on it.

Care should also be taken to add measured quantity of cold water to a hot engine to avoid localised heat stress which may initiate cracks and causes fracture as well.

The water jacket around the liner should be kept cleaned from deposited scales to avoid cracks or fracture in the liner.

During servicing, the adjustments of clearance between the liner and the piston ring, alignment of the bore with respect to the crankshaft axis etc. are performed carefully to achieve efficient running of the engine. A good compression is required to maintain the efficiency of an engine.

Piston and Piston Ring

Piston compresses the air or gas within the cylinder receiving motion from the crank shaft through the connecting rod. To prevent leakages from the clearance in between the liner and the piston, piston rings are fitted in the grooves designed for the purpose. Moreover, the lubricating oil is passed through the oil holes provided on the surface of the piston. So a piston is designed with such provisions like grooves for compression rings and scraper rings, oil holes and the gudgeon pin hole to connect the connecting rod as shown in Fig. 13.18. The piston may be of different types depending on the design of its head. It may be flat, concave, crowned or bevelled, being used for different purposes.

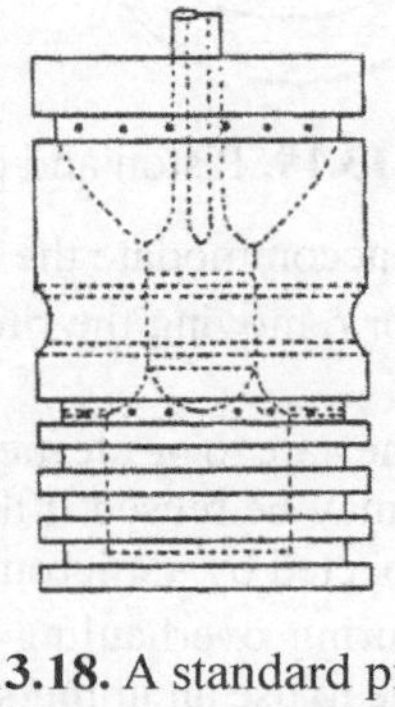

Fig. 13.18. A standard piston.

While fitting a piston, great care is taken to keep its axis aligned with the cylinder box and it should be at right angles to the crankshaft. The gudgeon pin is generally free to rotate in the piston hole. However a wooden drift may sometimes be used to remove it.

To take out the piston from the cylinder box, the cylinder head, cylinder block and crank case are to be taken out. The piston may then be lifted along with the connecting rod after dismantling the big end bearing of the connecting rod.

The common defect found in the piston is its wearing which is generally caused due to the presence of impurities in the lube oil, high temperature of the cooling water, overloading etc.

If significant wearing occurs, the clearance between the cylinder liner and the piston ring at the work region will increase allowing the piston to slap on the liner in each stroke. Thus, the slapping indicates the wearing either in the liner or on the piston rings.

A piston possesses three or four numbers of compression rings at the top side. The main function of compression rings is to seal the compressed gas preventing it from leakage as stated earlier. But there is a different ring termed as scraper ring fitted generally at the skirt of the piston. It is fitted to scrap excess oils from the cylinder walls, allowing it to return to the oil sump through the oil holes provided in it.

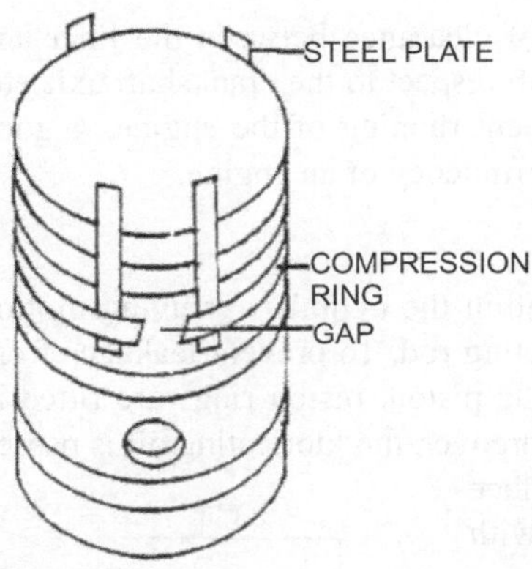

Fig. 13.19. Piston and piston ring.

The piston ring fitting or removing is to be done with proper care so that the rings are fitted with exact clearances at exact groove without being seized or loosened. While fitting a new ring, it may be rubbed over a fine emery cloth to have a freedom of movement of the ring in the groove. Then three or four fine steel plates are taken in contact with the piston surface upto the groove in which the ring will be fitted. The ring is placed through the plates (Fig. 13.19) and is allowed to move on them upto the groove. A slight manipulation of plates will place it into the groove. The ends should have a definite gap to accommodate the expansion of ring during rise of temperature.

For removing the rings the same technique is used with the help of steel plates.

They are to be cleaned thoroughly and carbon deposition is to be removed and it may be reused if the condition permits. The depth of the grooves may be inspected by a micrometer depth gauge.

During overhauling pistons are to be thoroughly cleaned. It is a good practice to use an immersion technique to clean the pistons. They are cleaned usually by some solvents forced through a hole on the piston. They may be finally cleaned with hot water and are dried by compressed air.

Crankshaft and Main bearing

Crankshaft is the key part of the engine connected with the piston through the connecting rod. It is placed in the crank case on a main bearing and it is fitted with the flywheel at the other end.

Crankshaft actually converts the reciprocating motion of the piston to a rotary motion.

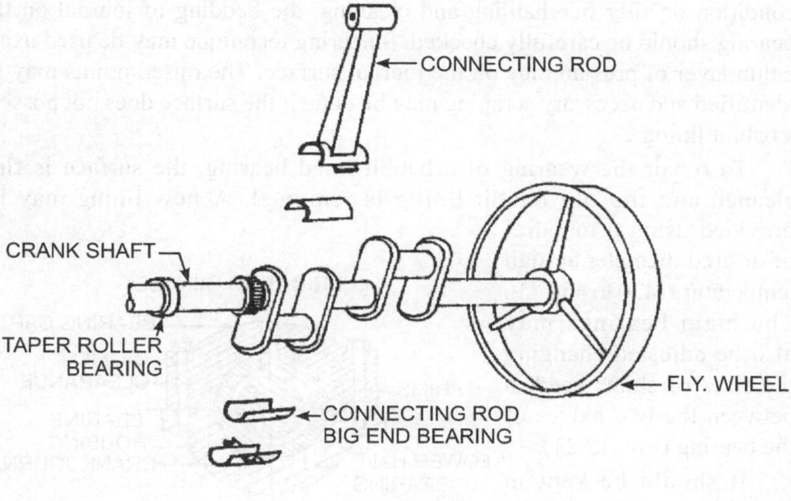

Fig. 13.20. Details of a crankshaft.

A section of a crankshaft is shown in the Fig. 13.20. It is important to note that it is connected with a number of reciprocating elements (connecting rods) for a multi-cylinder engine and so the resulted couples are to be balanced providing a suitable flywheel.

It may also be noted that big end bearings of connecting rods and thrust bearings are also mounted on the crankshaft alongwith the main bearing.

So, careful maintenance of the crankshaft at regular interval is an essential task. It involves inspection of bearing surfaces, checking of alignment of the crankshaft with the bearings, inspection of crankshaft surfaces etc.

Generally, overhauling of crankshaft is performed annually. The major defect found on the crankshaft is the formation of ridges on journal surfaces after a long run. The height of ridges from the journal surface indicates the extent of wearing of crankshaft journal. The height of ridge, if found, may be measured by feeler gauge or measuring the diameter of ridges and journal surfaces by vernier micrometer.

The ridges may be rubbed and removed using emery cloth, getting it wetted in oil.

The maintenance of bearing emphasises the maintenance of bearing clearances, proper bearing surfaces etc.

Whenever the bearing will be mounted on the crankshaft either in a new condition or after overhauling and cleaning, the bedding of journal on the bearing should be carefully checked. Smearing technique may be used using a thin layer of prusian blue on the journal surface. The raised points may be identified and necessary scraping may be done if the surface does not possess a babbit lining.

To repair the wearing of a babbit lined bearing, the surface is first cleaned and the old babbit lining is removed. A new lining may be provided using a mandrel of desired diameter at high temperature (430 to 450°C). The main bearings may also be adjusted changing the sizes of shims used in between the two halves of the bearing (Fig. 13.21).

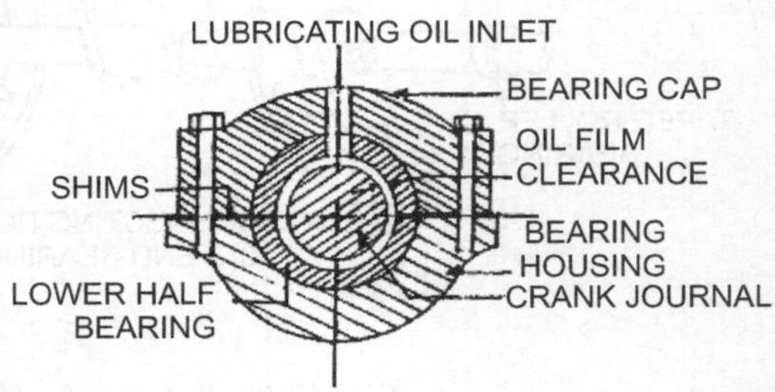

Fig. 13.21. Two halves of the main bearing.

It should be kept in mind that its alignment with the crankshaft, making provisions for specified clearance and proper lubricating material and procedure may enhance the life of the system.

A knocking sound may indicate the faults of a main bearing. But due to overloading of any particular cylinder, the temporary disbalance may also produce a similar knock. However, the temperature monitoring of individual cylinder will directly indicate the overloading of respective cylinder and the knocking near the crank case-door may indicate the bearing fault.

Fuel Oil Injector of Diesel Engine

The combination of diesel engine is of compression ignition type. The fuel is sprayed by the injector into the compressed gas. The injector is fitted at the recess of a cylinder head. A fuel pump is in control of delivering fuel oil and exert pressure to open the valve to pass the oil to the orifice for spraying.

There is a spring load on the nozzle valve spindle (Fig.13.22) which operates at a pre-fixed pressure. Fig. 13.22 shows only the outline of a typical injector.

The oil is pumped through the drilled inlet passage and is led to annular groove around the tapered portion of the needle valve at the bottom. An annular ring may be provided there. When the injection pressure attains the desired level, it lifts the valve off its seat against the spring load and the oil is injected through the orifice. The spray tip may be having a number of holes, the size of which are dependent on the size of the engine.

The injector deserves regular attention for cleaning but does not require repeated maintenance and repair as it is generally a precision member. An injector should always be handled with great care. If, at any time, it is to be dismantled, care should be taken to have a dust free atmosphere. The parts should not be rubbed off and should not come in contact with any abrasive material.

However, the deposition of carbon may be cleaned by soft wire brushes. The common failure of an injector may be a spring breakage, wearing of valve seat, jamming of nozzle valve, carbon deposition at spray tip holes.

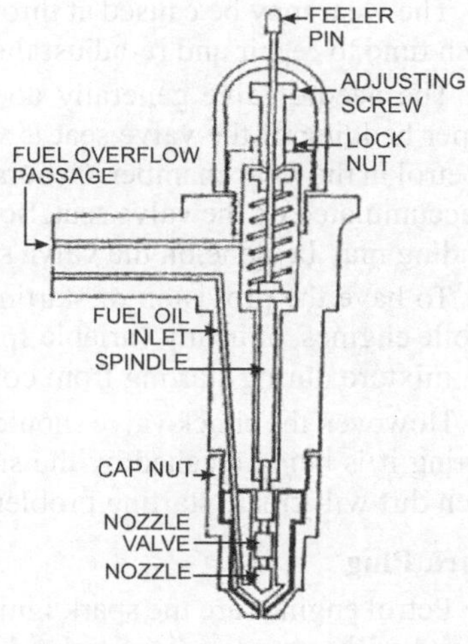

Fig. 13.22. Fuel oil injector.

The spring, if gets cracked or broken, should be replaced. The nozzle valve jamming may be cleared with a fine grinding.

The spray tips may be cleaned with steel needles. Repeated carbon deposition on the orifice hole may weaken its surface and the holes may be enlarged. The nozzle is to be replaced in such cases.

The common practice during overhauling of an injector is to dismantle its parts and to keep them dipped in a clean paraffin oil drum. They should be assembled without wipping out the oil from the surface of the parts.

Petrol Engine

A generalised discussion has been made so far regarding the maintenance of different parts of internal combustion engine. However, when the attention is given to the combustion factors, special attention should be given separately for a petrol and a diesel engine. Without going into the details of the combustion systems, special reference is hereby made regarding the curburettor and the spark plug.

Curburettor

The curburettor is the device which atomises and vaporises the fuel and mixes it with the air at required proportion, which may be varied under various situations.

Different adjustments at curburettor regarding avoidance of formation of too weak or too rich mixture, slow-running etc. are very much important during the running of the engine.

The wear may be caused at throttle valve stem or on the spindle. It takes much time to repair and re-adjust the valve.

The needle valve generally does not create any wearing problem. But proper bedding of the valve seat is very essential to restrict continuous flow of petrol in the float chamber. Generally the impurities of the petrol itself may be accumulated on the valve seat. So proper cleaning is required. If necessary grinding may be done on the valve seat.

To have the provision of starting and running jets, specially in the auto mobile engines, or in any variable speed engines, choke valves are used to get rich mixture during starting from cold condition

However, the chock valve should be always kept clean. Sometimes during starting it is being cleaned by the suction pressure, but accumulation of too much dirt will create starting problem.

Spark Plug

Petrol engines are the spark ignition engines. An electric spark, occurs at the end of the compression stroke, ignites the compressed charge.

Spark is flashed at the sparking gap of the plug (Fig. 13.23) as the spark plug is connected with an external electric system and is regulated accordingly.

The electrical system may be of two types :

(*i*) Coil ignition system and

(*ii*) Magneto ignition system.

The main part of a spark plug is the central electrode, covered all around by insulators, usually of porcelain. A screwed metal body is fitted at the bottom providing gas tight seals with the insulator. A small tongue, a projected metal strip is welded with the metal body and is made bent to form a gap with the electrode.

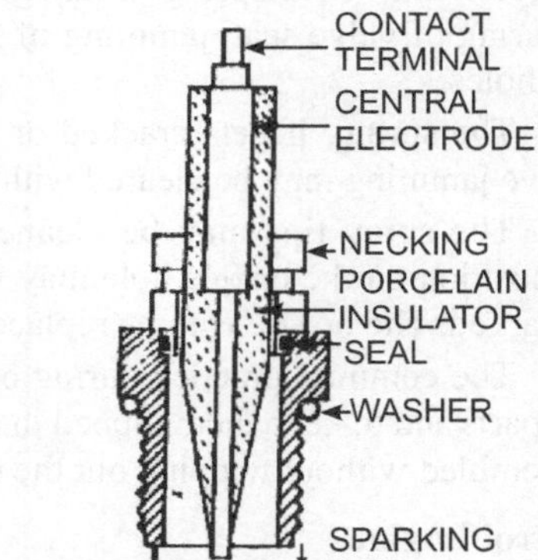

Fig. 13.23. Spark plug.

The spark plug is screwed into the cylinder head with copper asbestos washer.

Deposition of carbon on spark plug is a regular feature. It can be dis mantled holding the neck ring by a vice lightly and unscrewing the porcelain part. The central electrode is to be taken out and the insulator may be cleaned thoroughly to remove carbon deposits.

Spark gap may be adjusted if required by bending the metal tongue. But care should be taken to handle the electrode. It should never be bent, otherwise, it will lead to generate cracks at the insulator.

A spark plug may be burnt if it becomes too hot or if the mixture of gas becomes too weak.

A spark knock may indicate severe heating of spark plug, advance sparking, carbon deposition at the sparking gap. Of course, a weak mixture and hot engine may also generate spark knocks.

Proper cleaning of spark plug, maintenance of timing of sparking and a proper mixture of fuel will prevent early failure of a spark plug.

It may be noted that proper inspection and maintenance of fuel pump, and outer electrical system are also important for a petrol engine.

Operational troubles of a Diesel Engine

A few operational troubles are mentioned below. The engine should be stopped if such troubles occur. The probable causes are outlined. Specific remedial measure for the identified cause may then be undertaken.

Engine not getting started

Causes:

(*i*) Insufficiency of fuel in the reservoir.

(*ii*) Unsatisfactory quality of fuel.

(*iii*) Resulting of improper mixture.

(*iv*) Fuel valve timing is wrong.

(*v*) Incomplete compression, worn out valve seat or dirty and sticky valve may produce loss in compression. Cylinder liner, compression ring, other leakage may also be responsible.

Remedies:

(*i*) Fuel tank is checked.

(*ii*) Quality of fuel is assured.

(*iii*) Condition of valves, fuel valve timing may be inspected. Condition of liner, piston ring and cylinder head gaskets are also checked if necessary.

Engine is knocking

Causes:

(*i*) Errors in any bearing.

(*ii*) Overload may cause seizing of any piston.

(*iii*) Fuel injection occurs at faulty timing.

Remedies:

(*i*) Bearings may be checked.

(*ii*) Piston and piston rings are checked. The carbon deposition may be cleaned.

(*iii*) Timing of fuel injection is examined.

Engine runs irregularly

Causes:

(*i*) Insufficiency of fuel.
(*ii*) Presence of air or water in the fuel.
(*iii*) Problem in the injector nozzle; that is, failure of ignition.
(*iv*) Defects in valve seating.

Remedies:

(*i*) Fuel tank is checked.
(*ii*) Fuel oil system is inspected.
(*iii*) Adjustment as well as wearing of nozzle, if any is attended.
(*iv*) Valve timing, valve wearing and leakages are thoroughly checked.

Engine gets overheated

Causes:

(*i*) Insufficiency of cooling water.
(*ii*) Improper lube oil application.
(*iii*) Occurrence of late injection.
(*iv*) Accumulation of too much dirt or scale in the water jacket.

Remedies:

(*i*) Water level is checked. The performance of pump is also inspected. All the cocks or valves in the system are also checked to keep them open.
(*ii*) Lube oil should be as per the requirements.
(*iii*) Fuel injection timing is established.
(*iv*) Water jacket is thoroughly cleaned if necessary.

Engine stops on its own

Causes:

(*i*) Insufficiency of fuel.
(*ii*) Presence of water in the fuel.
(*iii*) Application of overload.

Remedies:

(*i*) Fuel tank is checked. Fuel filter is cleaned thoroughly, if it is full of foreign matters, dirts etc.
(*ii*) Oil is changed after cleaning the fuel system.
(*iii*) Load is reduced.

Engine exhaust is full of smoke

Causes:

(*i*) Accumulation of dirt on air cleaner, overloading of engine, too low compression pressure or thicker injection of fuel may cause black smoke.
(*ii*) Presence of water in the fuel or in the cylinder. Excessive compression of fuel pump pressure, early injection etc. may generate white smoke.
(*iii*) Presence of too much oil in the cylinder may produce bluish smoke.

Remedies:

(*i*) Air cleaner is thoroughly cleaned. Timing of fuel injection is checked. Nozzle valve seat wearing is examined and load in the engine is adjusted.

(*ii*) Water leakage in the cylinder and the fuel quality is checked. Fuel injection valve timing is inspected. Performance of fuel pump is verified.

(*iii*) The wearing of piston rings and liners are checked. The valve guides are also inspected whether there are any leakages or not due to the wear of them.

Trouble-shooting of a Petrol Engine

Engine not getting started

Causes:

(*i*) Improper mixture of fuel.

(*ii*) Problem in the spark plug.

(*iii*) Presence of water particles in the fuel or using of contaminated fuel.

(*iv*) Error in the valve timing or in the ignition timing.

Remedies:

(*i*) Air and petrol are to be mixed at required proportion.

(*ii*) Spark plug may be made cleaned, may be replaced if necessary.

(*iii*) Fuel should be of definite quality and filter may be checked.

(*iv*) Valve timing and ignition timing may be adjusted as necessary.

Engine gets overheated

Causes:

(*i*) Insufficiency of cooling water.

(*ii*) If the amount of petrol in the mixture is more than specified.

(*iii*) Ignition timing is wrong.

(*iv*) Jamming in the exhaust pipe.

Remedies:

(*i*) Water circulation is checked and re-established.

(*ii*) Mixture should never be rich or weak. Definite ratio of air and petrol is maintained.

(*iii*) Retarded ignition will make the engine heated. Ignition timing is adjusted.

(*iv*) Exhaust line is properly cleaned.

Engine is stopped suddenly

Causes:

(*i*) Non-availability of petrol in the curburettor.

(*ii*) Due to weak sparking.

(*iii*) Fault generation in the line of connection such as, high tension wire may be separated from carbon. Spring may become defective on wearing due to carbon deposition.

Remedies :

(*i*) Curbrettor is checked. Fuel jet is cleaned thoroughly.
(*ii*) Sparking plug is cleaned or replaced.
(*iii*) Thorough checking is made in the connections. Necessary replacement may have been done. Ignition switches may be checked to make proper contact.

Engine is knocking

Causes:

(*i*) Carbon deposition in the spark plug or in the combustion chamber.
(*ii*) Engine or the spark plug is overheated.
(*iii*) Occurrence of advance sparking.

Remedies:

(*i*) The spark plug and combustion chamber is thoroughly cleaned.
(*ii*) Sufficiency of cooling water is established to prevent overheating of the engine. The sparking plug clearance is checked to avoid over heating of spark plug.
(*iii*) Ignition timing is corrected.

REVIEW QUESTIONS

1. Write explanatory notes on maintenance services of cylinder, cylinder heads, piston and piston rings of a steam engine.
2. Write the procedure for starting up a steam engine.
3. State the procedural steps for the erection of a turbine.
4. Write a brief note on the servicing of a turbine.
5. Write a systematic approach for starting a steam turbine.
6. State the major steps for installation of an I.C. Engine.
7. Explain the maintenance service factors to be considered in different stages of C.I. and S.I. engines.
8. What possible troubles may come up in a CI. engine operation ? State the probable causes and measures for remedies for those troubles.
9. What action is to be taken and why------ if,
   (*a*) A momentary vibration of shaft is frequently felt in a turbine.
   (*b*) Admission valve in a turbine is jammed.
   (*c*) A diesel engine is not getting started.
   (*d*) A diesel engine gets overheated.
   (*e*) A petrol engine is stopped suddenly.
10. Identify the following sentences as correct or false and finally correct the wrong statements.
    (*a*) Drain taps are kept closed during starting a steam engine.
    (*b*) Start air pump at slow speed during starting of a turbine.
    (*c*) The major defect which occurs on cylinder head is wearing.
    (*d*) A piston of an I.C. engine possesses three or four number of scraper rings.
    (*e*) Carbon deposition in the spark plug may cause knocking of I.C. engine.

14

MISCELLANEOUS MAINTENANCE

PIPES

Pipes may be made of cast iron, wrought iron, steel, copper, brass, lead and they may also be made of cement, concrete and various plastic materials.

Use of pipes made of different materials are as follows :

CI. Pipe	:	To transmit cold water and in the underground works due to its corrosion resistant property. They are generally used at low pressure-low temperature purpose.
Wrought iron	:	Such pipes are stronger than C.I. pipes. For high pressure lines, wrought steel pipes are preferred.
Steel	:	General use is to transmit water and for general work. To convey corrosive liquids steel pipes may be lined with cement or lead.
Brass	:	Very much used when corrosive products are to be handled. Naturally use of brass pipe is very much seen in chemical industries.
Copper	:	They are mostly used where flexibility of pipe is a major criterion. Of course, they are expensive and not suitable for high pressure.
Lead pipes	:	Lead pipes are also highly corrosion resistant. They are also very much flexible. They are used in plumbing work.
Cement : Concrete	:	Such pipes are mainly used for water supply and for drainage purposes.

Plastic pipes : They arc flexible pipes and may easily form coils. Moreover, they are available in smaller diameter. General use is to transmit water at small quantity to some nearby distance.

Pipe Fittings

The figures 14. l (a to k) explain the shape of most common pipe fittings.

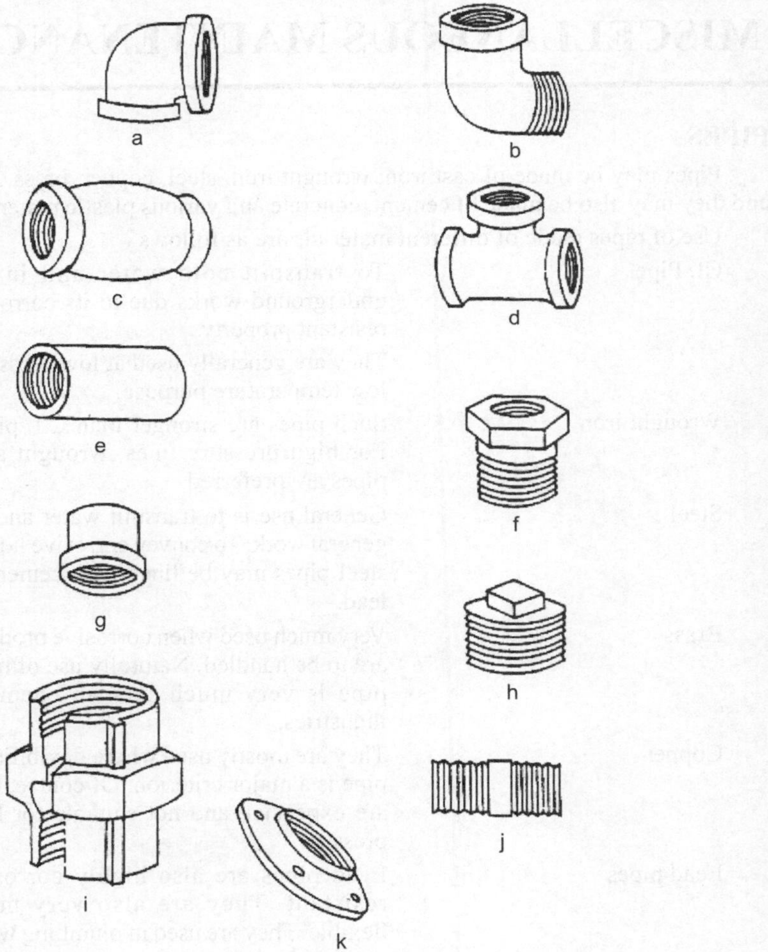

Fig. 14.1 Pipe fittings.

(*a*) Elbow is sometimes called 'ell'. It is used to make right angle turn of pipe line. 45° turn angle is also available.

(*b*)'Street ell' is also an 'ell' but its end serves the purpose of a nipple, as one of its end is a threaded end. A certain degree of flexibility is also available by fitting 'street ell' to allow a limited movement of parts without causing any strain on the pipe-joints.

(*c*) A 'reducer' means a reducing coupling which permits different sizes of pipes to be connected in the running line through it.

(*d*) A 'tee' is used when a main line is to be diverted in both sides.

(*e*) A 'coupling' is a hollow piece having threads at both ends. Two pieces; of pipes may be joined in a straight line through a coupling.

(*f*) A 'bushing' is having inner and outer threads. It is like a short sleeve to reduce the size of the threaded opening. The hexagonal end is provided to accommodate a wrench.

(*g*) A 'cap' is used to close any end of pipe. It can be fitted on a threaded end.

(*h*) A 'plug' also can be fitted on a threaded opening to close it.

(*i*) The 'union' joins two pipes. It is used when none of the pipes can be turned. It possesses three pieces of pipe. Two pieces may be screwed on two pipes and the third piece is a nut which draws the two parts closer together.

(*j*) A 'nipple' is a short piece of pipe having outer threads on both the sides.

(*k*) A 'floor flange' is fitted for fastening the end of a pipe to a wall or floor.

Major causes of faults

Faults in pipe line connections may be caused due to

(*a*) Leakages, (*b*) Swaying of pipes, (*c*) Water hammer and (*d*) Corrosion.

(*a*) Leakages: Leakages are often seen in pipe fittings, specially in flange joints. Besides corrosion, faulty assembly of fittings can create such problems. Say, in case of a flange, sequence of bolt tightening is to be maintained in a definite fashion, common practice is to tight the bolts, placed on diametrically opposite sides, then there may be a movement of 90° and similar procedure may be followed.

For ordinary fittings also, the threads may be coated well with the thread compounds like a mixture of graphite and heavy grease or cotton fibers dipped with zinc oxide to ensure tight joints.

(*b*) Swaying of pipes : Swaying is the leaning down of pipes unsteadily to a different direction. This may be caused by continuous oscillations due to vibration. Such defects may also be initiated from a faulty design or from non-availability of supports.

Generally for long piping installation, different brackets are used to arrest the swaying of pipes.

(*c*) Water hammer: When a pipeline experiences repeated shocks, it produces a sound like hammer blows. This ill phenomenon is known as water hammer.

It is generally caused due to the sudden reduction of flow of fluids.

The common cause may be to close a valve too fast suddenly. Such practice should always be avoided.

Moreover, in steam pipes, when the condensation is not being taken care of by adequate trapping, such recurring sound may be produced.

So attention should be given for proper drainage of condensate from a steam pipe. Checking should also be made to be sure that there is no sagging of any part of the pipe line.

(*d*) Corrosion: Corrosion is a common ill usually found in pipe line joints and should be taken care of properly. Much has been discussed regarding the corrosion and its prevention procedure in the relevant chapter before.

However, suitable coating, wrapping by hasian cloth, glass fibre etc. and sheathing (application of tough bitumen-asbestos compound) may be provided for external protection of water or gas pipes.

Lastly, it should be mentioned that maintenance work of pipe lines should be followed by a preset schedule. Preventive maintenance is always preferred, wherever possible.

Steam Piping and Fittings

A special mention may be made for steam pipe lines. Cast steel or cast iron pipes are used as steam pipes. Copper, bronze, monel metal etc. are the most common non-ferrous materials used for valves, fittings etc.

Brass may be used for handling salt waters, whereas lead lined cast iron, steel pipes may be used for handling acids, mineral water etc.

All pipes, fittings etc. should be installed as per the common standards.

Pipe Joints

Steam pipe joints may be either threaded or bolted. Calked and welded joints are also used. Flange fittings are very much common in pipe connections.

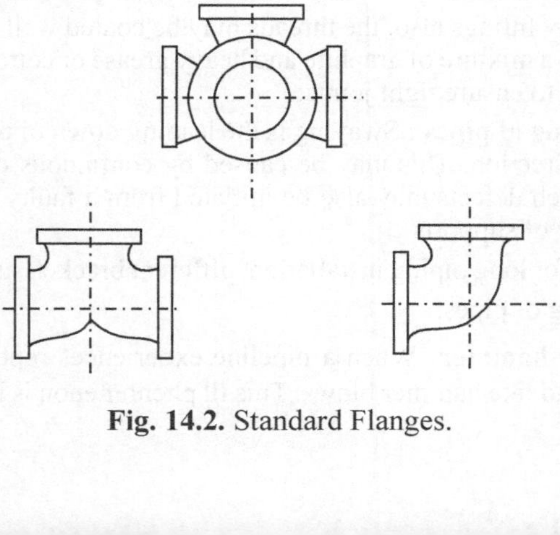

Fig. 14.2. Standard Flanges.

A few standard flanges are shown in the above Fig. 14.2. The types of flange joints to be used, depend on the pressure and the temperature. Such as, plane faced flanges (Fig. 14.3) are recommended for upto a medium pressure. Definite gasket is used in between the flanges, whereas raised faces flanges are used for high pressure joints.

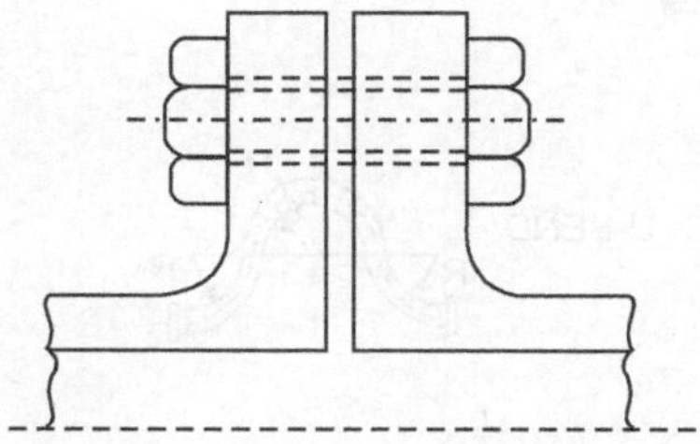

Fig. 14.3. Plane faced flange.

Again for low pressure, welded-on flange joints may be preferred.

Pipe Welding

Joining of pipes by welding is becoming very much popular now-a-days to reduce installation cost for better efficiency. To-day, welding is being followed widely for joining of pipes in the thermal and nuclear power generating stations, refineries, chemical plants etc.

Pipe welds are made under various requirements and in different welding situation. Most thinner walls may be welded as shown in Fig. 14.4(*a*). For a thicker wall, a square groove produces good weld [Fig. 14.4(*b*)]. As the thickness increases an angular grooving is provided [Fig. 14.4(*c*)].

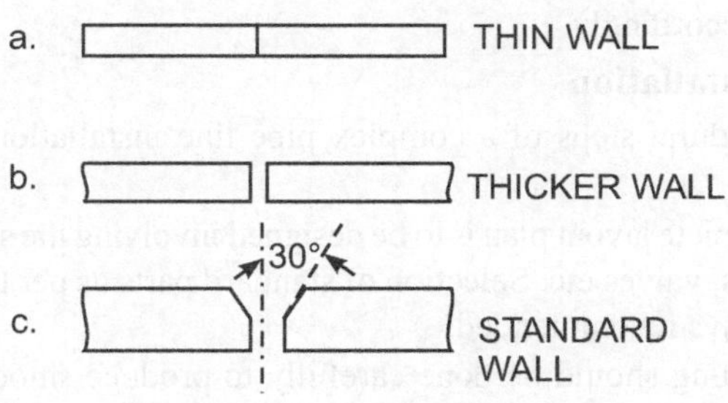

Fig. 14.4. Pipe Welding.

Pipe joint alignment is an important task before welding. When the pipe diameter is more, an internal line up checking may be provided.

Expansion bends

The common problem of steam pipe system design is to accommodate the expansion and contraction of pipes due to change of temperature.

Generally, long radius bends are provided to relieve stresses due to expansion. The practice of utilising long radius bends utilises the elasticity of the pipe. It is having some other advantages also. The uses of such bends facilitate to have clearance from other obstructions and eliminates unnecessary fittings.

Pipe Supports

Pipe lines should be well supported to avoid sag and to protect them from excessive stresses. The support-technique will obviously be guided by the diameter and the thickness of pipe. Fig. 14.6(*a*) shows a bridge method of pipe support. Light structure of angle iron of sutable section is very much used. Again when the movement in the axial line is to be restricted the pipe may be held between two rollers as shown in Fig. 14.6 (*b*).

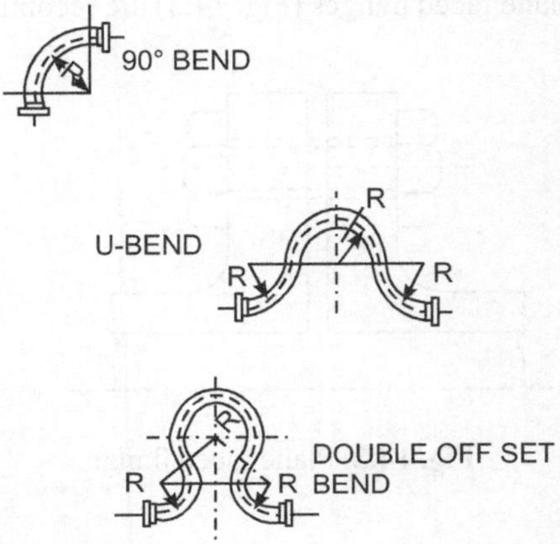

Fig. 14.5. Pipe Bends.

The sling type of support Fig. 14.6(*c*) suspension is also very popular and universally used.

However, selection of support technique is a job of pipe installation and erection and is designed as per requirement. The junction of support should be properly protected from chances of forming rust or corrosion and should be inspected accordingly.

Pipe lines installation

The procedural steps of a complex pipe line installation are described below:

(*i*) A complete layout plan is to be designed involving the size of the pipes, flanges, valves etc. Selection of standard parts as per the requirement is always recommended.

(*ii*) Threading should be done carefully to produce smooth threads and some kind of compounds (e.g. zinc oxide) along with waste cotton fibers etc. are to be used for effective joining.

(*iii*) All pipes, pipe holes, flanges, bolt holes etc. are to be thoroughly cleaned from dirt's, burrs etc. before fixing.

(*iv*) Gasket material is to be properly chosen and the gasket is to be cut carefully for any flange joints.

(*v*) Attention should be given to keep the flange surfaces parallel.

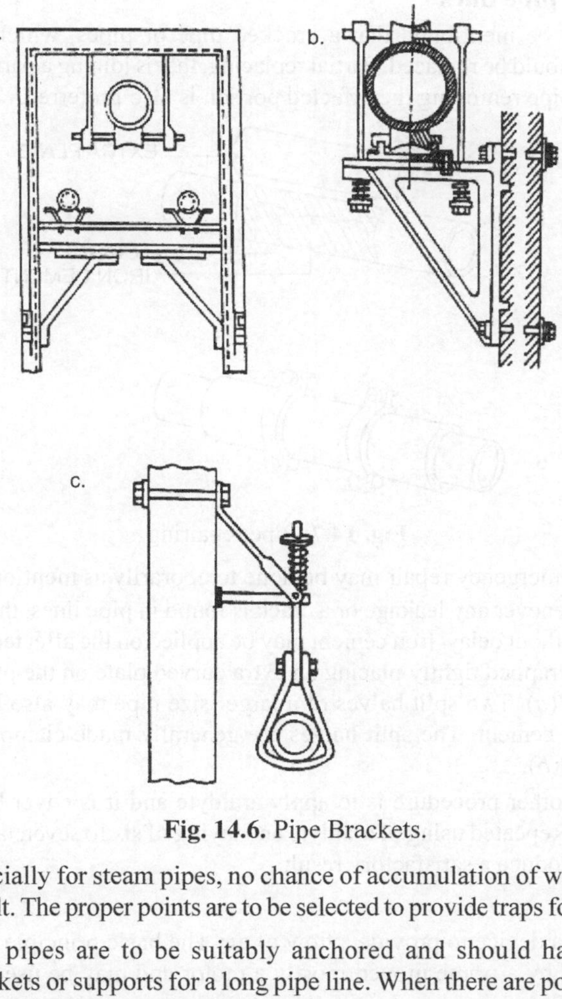

Fig. 14.6. Pipe Brackets.

(*vi*) Specially for steam pipes, no chance of accumulation of water should result. The proper points are to be selected to provide traps for drainage.

(*vii*) The pipes are to be suitably anchored and should have proper brackets or supports for a long pipe line. When there are possibility of movements of pipes in ups and downs, due to too much temperature change, springs of soft material is to be provided in the supports.

(*viii*) Insulation if necessary is to be provided. Calcium and asbestos are widely used for insulation. For too much temperature, double layer insulation may be provided.

(*ix*) Definite colour coating for identification of gas, water or any other pipe lines may be provided.

Repair of pipe lines

It may be mentioned that a cracked pipe or pipes, which are having leakages, should be replaced. Partial replacing, that is joining a particular length of similar pipe removing the affected portion is also preferred.

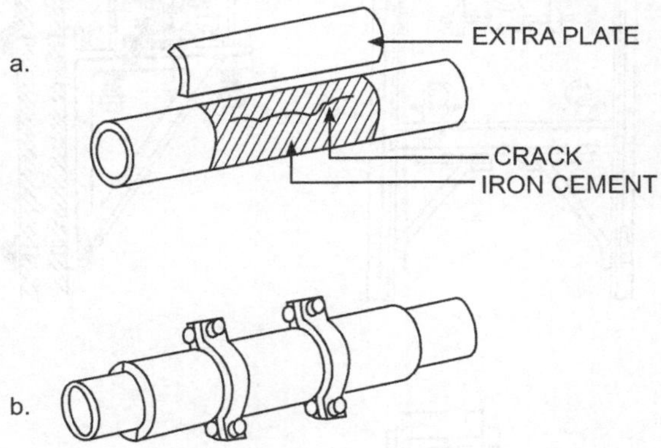

Fig. 14.7. Pipe repairing.

However, emergency repair may be done temporarily as mentioned below:

(*i*) Whenever any leakage or a crack is found in pipe lines, they should be attended without delay. Iron cement may be applied on the affected portion and it may be wrapped tightly placing an extra curved plate on the pipe as shown in Fig. 14.7(*a*). Two split halves of a larger size pipe may also be placed on the applied cement. The split halves are generally made clamped as shown in Fig. 14.7(*b*).

(*ii*) Another procedure is to apply araldyte and it is cover by a layer of glass tape. Repeated using of araldyte and laying of six to seven layers of glass tape will produce a satisfactory result.

FANS

Fan is a device to provide a flow of air. The basic principle is to revolve an impeller by a prime mover, mostly a motor and may be used for various purposes. Fans may be classified in different ways. However, fans may generally be of three types as per the types of flows through the impeller.

(*i*) Axial flow
(*ii*) Radial flow
(*iii*) Special type, like mixed or cross flow

The different types of fans have different capacities of the volume of air throwing and pressure range to overcome the resistance against flow.

Axial flow fans

The axial flow impeller is used for different types of fans.

Propeller fan: Propeller fans are having very small hubs. The housing and the blades may be made of cast steel or aluminium or plastic. They are properly coated.

They may be used for any required capacity but the flow pressure available is not high. They may generate sound if they are connected to the motor at high speeds. Both the vane axial fans and tube axial fans are placed in tubular casings. Vane axial fans can provide high pressure range along with high capacity. The flow of air can be made adjustable using diffusers between the fan and the discharge duct. They may be used where temperature is high or the corrosion and abrasion may occur.

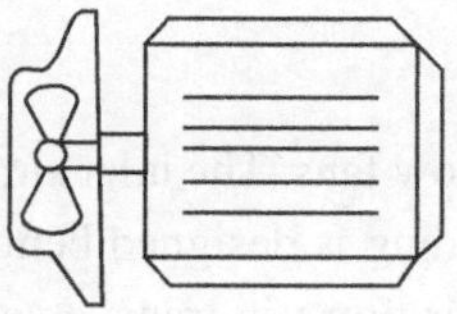

Fig. 14.8. Axial flow fan.

Radial flow fans

Centrifugal fans use radial flow impellers (Fig. 14.9*a*) and are mostly used to provide mechanical draft in the industries. Centrifugal fans are normally high capacity fans and are available in various types. The change of curvature of the blade tip will change the performance characteristics of a centrifugal fan.

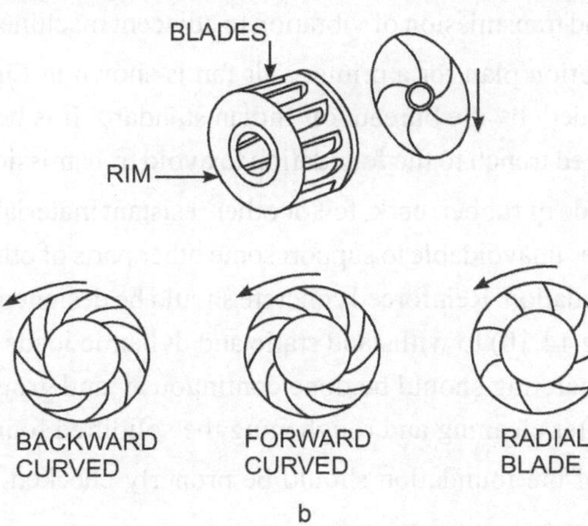

Fig. 14.9. Curvature of blade.

Fig. 14.9(*b*) shows different types of curvature of blades. The backward curved centrifugal pumps have got wide application due to its high capacity and pressure range. They are self adjustable, if resistance increases , the rated volume of discharge will automatically decrease. No problem of noise is there with such fans.

Centrifugal fan with forward curved blades is a velocity fan. They are used generally where high outlet velocity is required. But it has got problems of noise generation. It can also work at any range of pressure. But it should not be used in high dust areas.

Fans with radial blades are most rugged fans. Though the efficiency is less but it can maintain a constant pressure over a wide range of capacity. Major industrial application is seen as because it can handle heavy dust particles and capable to function at high temperature.

Special type flow

The special type may include mixed flow or cross flow fans. The inlet air, in case of a mixed flow, flows radially but the inlet blading is designed bent downwards over the hub and in a cross flow fan the air flows in transverse direction due to special design of the casing.

Installation of a fan

It is difficult to put forward a generalised installation technique for different kinds of fans on different uses .

However a few salient features may be discussed.

The fans usually are the rotary type devices of low frequency. Major care is needed to avoid transmission of vibration to adjacent machineries or devices. A typical foundation plan for a primary air fan is shown in Fig. 14.10 as has been recommended by the bureau of Indian standard. It is better to provide sand filled covered trench to the foundation to avoid transmission of vibration.

Gaskets made of rubber, cork, felt or other resistant materials may be used, in case it becomes unavoidable to support some other parts of other machineries on the same foundation. Reinforced concrete should be designed for foundation (as shown in Fig.14.10) to withstand static and dynamic loads and to provide rigidity. The concreting should be done continuously and grouting should be done as usual after cleaning and roughening the solidified foundation block.

The level of the foundation should be properly checked before placing the fan on it.

Selection of fans

Selection of a fan is obviously related to the purpose of use. It may be noted that improper selection of fans will undesirably increase the maintenance and repair work.

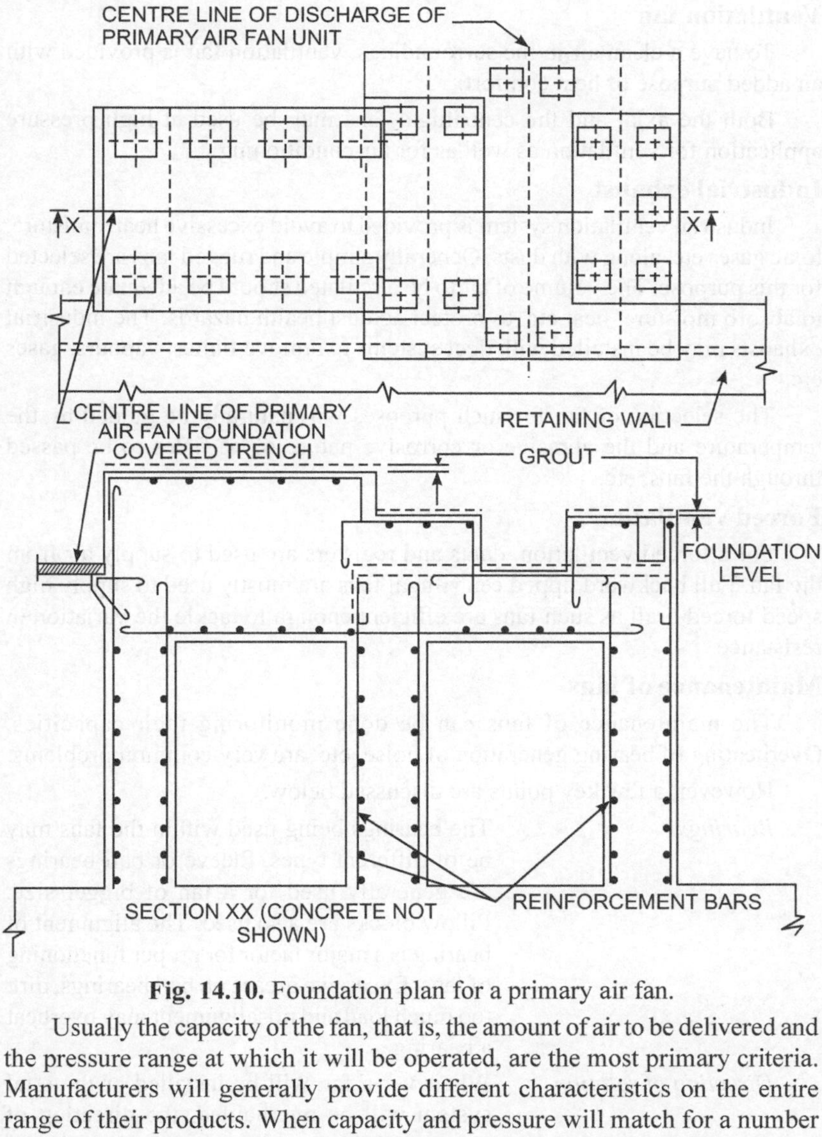

Fig. 14.10. Foundation plan for a primary air fan.

Usually the capacity of the fan, that is, the amount of air to be delivered and the pressure range at which it will be operated, are the most primary criteria. Manufacturers will generally provide different characteristics on the entire range of their products. When capacity and pressure will match for a number of fans, the available space and the speed of the fan may be the guiding factors of selection. It may be noted that a fan of lower speed will increase the rate of power consumption.

The purpose of installation of a fan may be of different types. Some of them are discussed below.

Ventilation fan

To have a clean air in the surroundings, ventilation fan is provided with an added purpose of heat comfort.

Both the axial and the centrifugal fans may be used at high pressure application for ventilation as well as for air-conditioning.

Industrial exhaust

Industrial ventilation system is provided to avoid excessive heat, moisture, toxic gases etc. along with dusts. Generally simple and rugged fans are selected for this purpose. The volume of air to be circulated should be adequate enough to absorb moisture, heat etc. to protect against health hazards. The industrial exhausts may be installed with duct systems to remove fumes, vapours, gases etc.

The selection of fan for such purposes will naturally be guided by the temperature and the abrasive or corrosive nature of particles to be passed through the fans, etc.

Forced ventilation

For a forced ventilation, ducts and registers are used to supply air from the fan. Full backward tipped centrifugal fans are mostly used to supply high speed forced draft as such fans are efficient enough to tackle the variation in resistance.

Maintenance of fans

The maintenance of fans can be done monitoring their capacities. Overheating of bearing generation of noise, etc. are very common problems.

However, a few key points are discussed below.

Bearings : The bearings being used within the fans may be of different types. Sleeve or ball bearings are generally used for a fan of bigger size. Pillow blocks are also used. The alignment of bearing is a major factor for proper functioning of fans. Excessive grease on ball bearings, dirt, too much load and misalignment may overheat a bearing.

Direction of rotation : Whenever a fan will be installed or the total system will be overhauled, the direction of rotation of fan is to be tested in the primary run. Generally a reversal occurs due to faults of electrical connections.

Lubrication : The bearing used in fans, does not require frequent lubrication. But one should be assured of the grade of lubricants before using. Generally a soda-base grease of high quality is applied at fixed intervals.

Other problems, those are generally encountered, are due to vibration or generation of noise. A bent shaft, lack of alignment of bearing, fan and motor or a worn coupling may generate noise and produce vibration.

A poor installation may also be responsible for noise and vibration. All the bolts whether in the foundation or in the system should be checked at regular interval.

Positioning of fans

The fans may be installed with the objective of either exhausting the used air from a room or shed or shop, or it may be used to supply air to a process or to a surrounded space. An efficient positioning will depend on the purpose of its use. Numerous types are described below.

(*a*) A forced draft fan will be positioned depending on the procedure to be followed either to draw air from the boiler room itself or from a air preheater.

(*b*) A forced draft fan may be placed near the boiler at a definite height, where air is delivered to the furnace. Again, in case of pulverized fuel burning boiler, the air carries the products of combustion through the unit.

(*c*) To remove smoke and heat, exhaust fans may be placed in the ceilings.

Generally, exhaust fans are placed at an height of 8 to 10 ft. from the floor. Care should be taken to select their position, so that fresh air entering into the space does not get an easy path to go out through the exhaust fans. Usually such installations are done on the opposite wall of the provisions, through which fresh air comes in.

DUST COLLECTORS

Dust collectors are mostly installed to resist the pollution of the atmosphere, to prevent damages of the surrounding environments and to maintain a clean working atmosphere in the plant. Dust separating or collecting equipments are, so, very commonly used in boiler houses to eliminate particles from the flue gas, to separate fumes, mists etc. from a gas flow in chemical industries and so on.

A large variety of dust separation devices are available now-a-days. A very few types are described here. They may be mostly of two categories :

(*i*) Mechanical separators

(*ii*) Electrostatic precipitators

Both the types may be run with dry or wet units.

Mechanical Separators

Gravity separator. The simplest type is gravity separator. The gas to be cleaned is passed through a large horizontal drum. The velocity of the gas is reduced within the drum to such an extent that the carried particles can drop down below. An hopper may be used at the bottom of the drum to collect the dusts. Such separators can only eliminate particles of larger sizes.

Cyclone separators. The smoke is allowed to flow at high velocity through a thick wall tube, having a conical bottom. The flow-path passage is designed as spirals, as shown in the Fig. 14.11. The upper involute cover restricts the flow at entry to move upwards and ultimately the smoke is to come out through the outlet tube crossing over the spiral path. Due to this cyclonic movement of the smoke, the dust particles will be forced to drop down and may be collected in a hopper placed at the bottom of the separator. Multiple number of small units in parallel are generally used in the flue path of a boiler before exhaust.

Cyclone separators can eliminate 60% of the particles at the maximum.

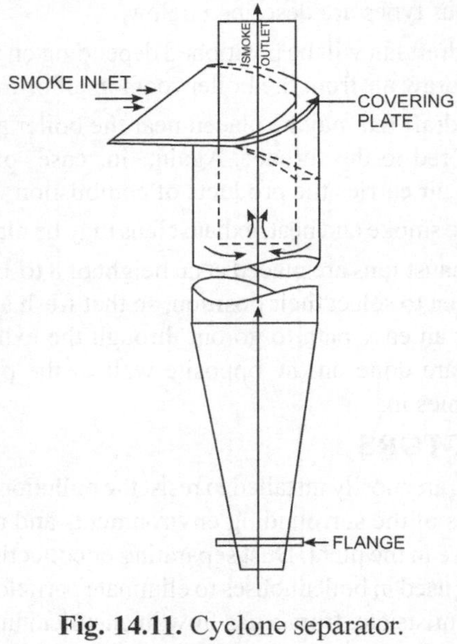

Fig. 14.11. Cyclone separator.

Packed tower separators. A high tower is installed along with a bed of packed materials, through which the gas is to be passed. It is generally used in chemical industries (Fig. 14.12). The gas inlet is at the bottom and the outlet is at the top end. The packing layers are commonly of randomly packed ceramic tiles, metal spirals etc., supported on a grillage. It works in wet condition. Water is sprayed from the topside and the gas mixes with the liquid within the packed layers and relieves the dust particles within the packing.

The general maintenance required for such a system is to clean the packing layers frequently or to change any layer if leakages are formed.

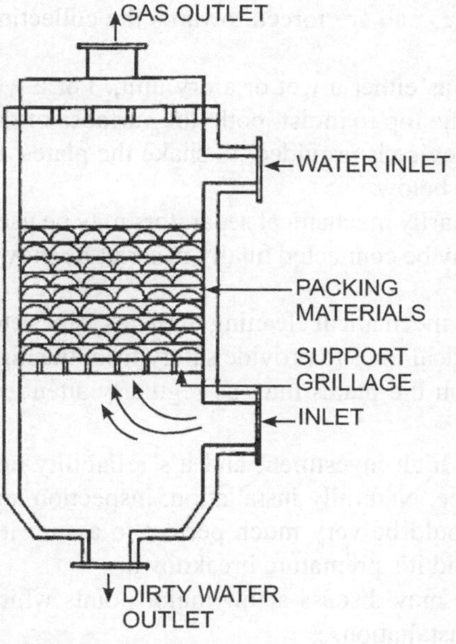

Fig. 14.12. Packed tower separator.

Here also the efficiency is not upto the satisfactory level and this system is generally used as a first stage dust separator chemical industries.

Electrostatic Precipitator

Electrostatic precipitator is having inherent advantages over all other devices. The technical supremacy lies in low pressure drop and high eliminating capacities of small particles. The procedure is very simple, though it involves complicated ionisation of gases and charging of the dust particles, migration and deposition of charged particles. There are two sets of electrodes, one in the form of wire (discharge electrodes) and the other set in the form of pipes or plates (collecting electrodes).

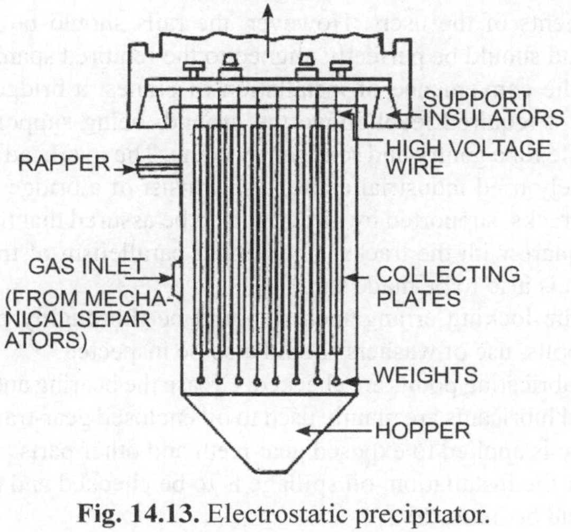

Fig. 14.13. Electrostatic precipitator.

When the suspended particles are trapped in between two high voltage electrodes, that is within a high electric field, they acquire negative charge, due to the ionisation of gas molecules and are forced towards the collecting plates where they get deposited.

This device can again function as either a wet or a dry unit,. For a wet unit, separate boxes are placed on the top to moist both the surfaces of the plates and for a dry unit a rapping device is provided to shake the plates to discharge the particles in the hopper below.

To achieve high efficiency, primarily mechanical separators may be used and the electrostatic precipitators may be connected finally in series as shown in Fig. 14.13.

The maintenance work involves mechanical cleaning of plates, pipes etc. along with the maintenance of electrical lines to provide safety from the high voltage line. Corrosion or erosion on the plates may be regularly attended.

Cranes

Installation of cranes require a high investment and it's reliability and safety aspects are of vital importance. Naturally installation, inspection and maintenance of such equipment should be very much perfect to assure it's trouble-free performance and to avoid it's premature breakdown.

However, regarding cranes, we may discuss a few major points which should be taken care of during it's installation.

(*a*) The manufacturing companies mostly provide match-marks to the elements, required to be assembled one with the other. So, before the installation of cranes, a thorough check should be made on all the assembled parts, in this regard.

(*b*) The runway rails of cranes has got certain span as per the specification and requirements of the users. However, the rails should be kept straight throughout and should be perfectly aligned to the required span.

(*c*) For the convenience of installation of cranes, a bridge structure is assembled at a required level above the ground, being supported by track girder or trestle for a gantry and semigantry crane. The overhead bridge ranes, the most widely used industrial crane, also consist of a bridge travelling on wheels over tracks, supported by girder. It is to be assured that the girders are assembled square with the tracks. The perfect parallelism of tracks through out the length is also to be made sure.

(*d*) All the locking arrangements should be thoroughly checked. The tightness of bolts, use of washers should also be inspected.

(*e*) The lubricating points are checked. Grease the bearing and the hoisting cables. Liquid lubricants are mainly used to oil enclosed gear-trains and worm drives. Grease is applied to exposed gear teeth and other parts.

Just after the installation, oil spillage is to be checked and all oil sports, if found should be made dry.

(*f*) The individual mechanisms, the brakes, the limit switches, limit stops and all other safety points are to be checked.

The crane should be operate slowly throughout the length in all directions to be assured on it's effective functioning, just before the crane is finally placed in service.

Inspection - to decide the degree of maintenance

Maintenance involves replacements, repairing, cleaning, lubricating etc. Some activities are followed as routine jobs. But the maintenance work is completely dependent on the report of technical inspections, which ascertains that the machine is in good repair and it is running in conformity with the standing regulations.

Hoisting equipment is liable to regular minor inspection, may be after six months, whereas regular major inspections are carried out annually or within two years, as per the instruction of the manufacturer.

The major points to be noted during minor inspections may be the 'freedom of structural parts' from cracks, the condition of the mouth of the hook (the wear in this part is generally allowed up to 10% to 12% of the hook cross-section), the swiveling performance of the hook, the condition of rope fastening, safety drives etc.

Such inspection also includes visual inspection of mechanical units to avoid mechanical defects. Not testing is required for minor inspection.

Static and dynamic tests are done to assure safety during the major inspection.

Static test ascertains the strength and capacity of the machine as a whole. The static load used may be 1.25 times the rated capacity. The trolley is taken at a point, where the deflection of the bridge may be the maximum, just over the crane track. After the application of load the book is kept lifted around 200 mm. above the floor, for a considerable time, say 10-12 minutes. A plumb with a wire may be suspended from the bridge. If the position of the plumb remains unaltered after the test, the performance of the bridge is proved to be satisfactory, as no permanent deflection is detected from the plumb.

For the dynamic test, a load more than rated load, say 1.2 times, is applied and the mechanisms are lowered and lifted for a number of cycles to check their performance. Brakes are also inspected through dynamic test.

The testing of controlling and limit switches and other motions are also inspected simultaneously. This may be termed as 'Functional test'.

Crane Hooks and Slings

So far as the hoisting installation is concerned, 'Hooks' are the most important handling attachment to pick up the load to be handled. Slings are used to suspend the load from the hooks. Slings, generally, may be of hemp ropes of chains.

Hook is always tested by the manufacturer, using a test load, say 1.25 the rated load. The time period of application of load is also considerably much more (may be about 8 to 10 times) while a hook is being tested.

The hook 'should be free from crack or distortion. Nicks and wears are always to be examined.

The hook require 'normalising' after a prolonged use, as because the work hardness is generally developed during continuous use. However the heat treatment may be done at an interval of one or one and half years. If the hook is seemed to be opening while being loaded, it is advisable to replace the hook. This may be noted that the defective hooks should never be used getting it welded or otherwise reconditioned.

Regarding slings, it is to be remembered that the slings, made of either steel wire rope or welded chain, should not have any knot or loop, while binding or securing the load to the hook of any hoisting arrangement.

It is always recommended to go for preventive maintenance for hoisting machinery's. The chains should always be kept free from dust. Inspection at reasonable interval is essential to avoid undue strain and rapid wear. The links, those have gone worn or stretched should readily be replaced.

WELDING EQUIPMENTS AND INSTRUMENTS

Welding is a process of joining two metals with the application of heat and with or without the application of pressure.

Welding may be classified in several types, such as, forge welding, gas welding, resistance welding, arc welding etc. They may again be classified into various categories.

Our aim is not to discuss the process or procedures but to identify the key points for servicing and maintenance of the equipments or instruments needed for performing welding.

Here, the gas welding and the electric arc welding have been taken into consideration.

Gas Welding

The combustion of gases in a gas welding process produces a high temperature flame and the flame is directed by a torch. The filler metal along with the definite region of the work are melted and the united molten metal is allowed to solidify on the work piece.

The gas welding may be performed with different combinations of gases.

Oxy-hydrogen, air-acetylene and oxy-acetylene welding are generally used. Of course, the most popular process is the oxy-acetylene welding.

Oxy-Acetylene Welding

The mixture of oxygen and acetylene is having the capability of producing a flame of about 3400°C.

A filler rod may or may not be used.

The equipments and apparatus needed for oxy-acetylene welding process are:

(*a*) Welding torch;
(*b*) Hose for oxygen and acetylene;
(*c*) Regulators for both oxygen and acetylene;
(*d*) Oxygen and acetylene cylinder;
(*e*) Spark lighter.

The system shown in the Fig. 14.14 does not require too much maintenance work. The fuel gas hoses for oxygen and acetylene are having specifically different threads to attach them with the welding torch so that they cannot be connected in a reverse way. Generally right hand threads and left hand threads are used in two connections.

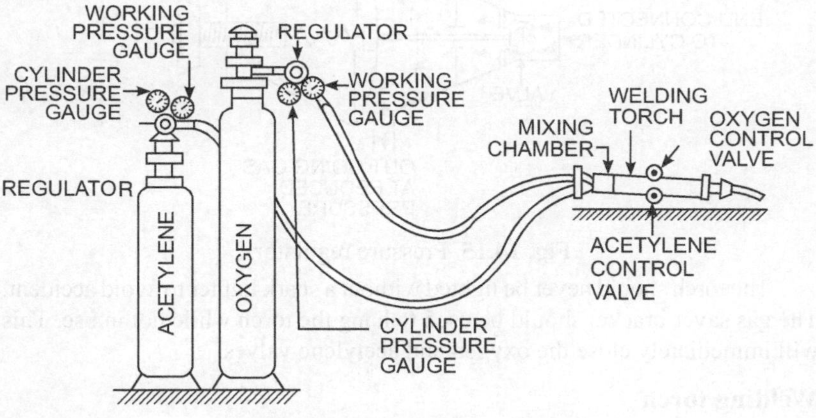

Fig. 14.14. Welding system.

Mixing takes place at the mixing chamber. But the volume of gases can be controlled separately by the controlling valves. The tips may also be of different sizes. Larger the orifice, more will be the flow of volume of fuel mixture and larger will be the flame size.

Selection of tip size is an important criterion for a specific job. The torch tip is to be maintained properly and should be stored in a careful manner so that the tip size does not get enlarged.

It may also be mentioned that pressure regulators must have to be used for both the cylinders to reduce the pressure. The pressure may be at the range of around 150 and 18 kg/cm^2 inside the oxygen and acetylene cylinders respectively. But the required working pressure for the respective fuels should be reduced to about 1.5 to 1.7 kg/cm^2 and 0.7 to 0.8 kg/cm^2. The pressure regulator (Fig.14.15) regulates the pressure by operating the knob. A clockwise turn of knob will press the flexible diaphragm through the compression spring, resulting wide clearance in the valve seat and allowing more gas to flow into the gas chamber. The regulating knob is to be adjusted in such a way that required pressure is obtained at the outlet as per the requirement of the working pressure.

While working the pressure should be maintained at specified level. The hoses are to be maintained in good order. Any leakage, if develops, should be immediately repaired or the hose is to be replaced. Definite colour code should be used to differentiate hoses. A hose, once used for one gas should never be used for the other gas.

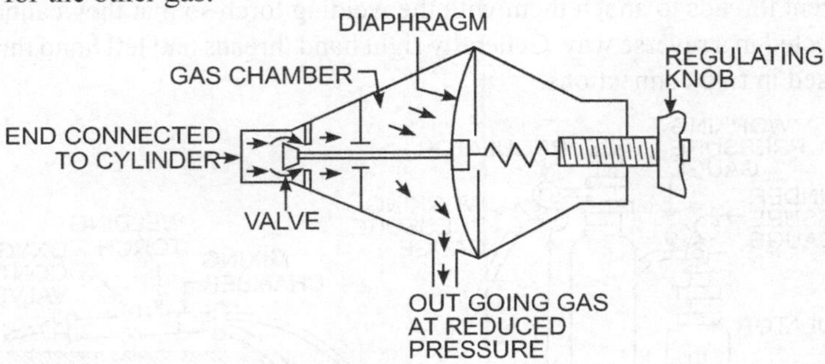

Fig. 14.15. Pressure regulator.

The torch should never be lighted without a spark lighter to avoid accident. The gas saver bracket should be used to hang the torch while not in use. This will immediately close the oxygen and acetylene valves.

Welding torch

A special mention is required for the maintenance of welding torch. A welding torch may accommodate various sizes of welding tips. While a welding tip will be changed, the threads and sealing surfaces should be properly cleaned. Nozzle should be cleaned with a specific tool, designed for this purpose only.

The flame may go off, if the nozzle is blocked with dirt etc. or if the nozzle becomes too hot. Too much heat at the nozzle will make the bore size narrower.

Again, on the other hand if the welding tip is not being connected properly, or, the gas mixture flows at a too slow speed due to considerable pressure reduction of working gases etc., the flame may strike back into the torch. The following procedure may be followed to establish a desired flame.

(*a*) Oxygen valve is opened slowly up to half a turn.

(*b*) Acetylene valve is opened similarly up to half a turn.

(*c*) The gas mixture is ignited.

(*d*) The flame is regulated to get the proper flame.

And while closing the system, the acetylene valve is closed first and then the oxygen valve.

Arc Welding

The following power sources are generally used for electric arc welding :

To supply direct current	:	(*a*) Generator
		(*b*) Rectifier
To supply alternating current	:	Transformer

Installation of Power Sources

Installation of welding power source should be performed strictly as per the instructions of the manufacturer. The location of installation should be at a proper ventilated place. Moisture, fume or excessive dirt should always be avoided.

The motor generators should be checked to be sure that the direction of rotation has been made proper and during the installation of a transformer, care is to be taken to balance it on a three phase line.

Lastly, it may be mentioned that installation procedures should be organised according to the state or national rules and regulations.

Maintenance

Generally preventive maintenance is employed for the maintenance of welding power sources. A few significant points are described below.

(*a*) All the machines should be maintained clean.

(*b*) Dirt may create several problems like the blockage of cooling passages, reduction of cooling efficiency of windings, if the dirt is collected inside.

(*c*) Inside cleaning is generally done by low pressure compressed air.

(*d*) All the internal and external connections should be checked at regular intervals. Valves, pipe lines, hoses etc. should be inspected periodically.

(*e*) Any internal or external damage or sign of corrosion should be attended without delay.

(*f*) If acetylene cylinder is found to be getting hot frequently, it is to be kept cool applying jet of water outside. It is better to return it to manufacturer, making it empty.

(*g*) For the cause of safety, the maintenance work should be undertaken by qualified and experienced personnel.

Common troubles, their causes and remedies

Troubles	Probable causes	Remedies
(*a*) Machine not getting started.	(*i*) Power line may be dead or damaged.	(*i*) Check power line.
	(*ii*) Line voltage is not proper.	(*ii*) Check line voltage.
(*b*) Welding current is stopped immediately after primary starting up.	(*i*) Improper overload relay.	(*i*) Replace the elements.
	(*ii*) Application of too much load.	(*ii*) Welding current may be reduced.
	(*iii*) Non-functioning of cooling elements.	(*iii*) Ventilation and fan if any, may be checked.
(*c*) Too much spattering and loud welding arc.	(*i*) Current flow may be too high.	(*i*) Amperage should be checked.
	(*ii*) Wrong electrode may be used.	(*ii*) Proper electrode may be used. Generally A.C. or A.C. and D.C. electrodes are used at A.C. welding.
(*d*) Sluggish welding arc.	(*i*) The flow of current may be too low.	(*i*) Check amperage.
	(*ii*) Connecting cables may be too long.	(*ii*) Cables are used as per specification.
	(*iii*) Loose connections.	(*iii*) All connections are properly checked.
(*e*) Getting shock at machine or at work.	(*i*) Faulty machine earthing	(*i*) Earthing is checked and rectified properly.
	(*ii*) Faulty earthing of work or work table.	

Safety Measure

To protect the operator and the viewer of operation from spark or hot metal particles or even from harmful radiation, filter, goggles, hand shield and helmet should be used. Of course, such equipments can only protect above the shoulder of the operator. A leather apron, leather gloves, arm protector, compact footwear etc. are also recommended for the operator.

However, the hazards involved in the operation of welding and cutting may be classified in ascending order of the degree of hazard involved as shown in the following table. The table indicates the protection equipments (for eye and face) permissible for each class of hazard.

Table -14.1

SELECTION OF EYE AND PACE PROTECTION EQUIPMENT

Hazard involved	*Permissible equipment*
Direct exposure to moderate injurious radiation of light and heat and to sparks, for example, in gas welding and cutting.	Goggles, hand shields and helmets may also be used.
Direct exposure to high intensity radiation of light and heat, sparks and particles of hot metal, for example, in electric arc welding and cutting.	Hand shields or helmets
Exposure by direct radiation and by reflection to large amounts of infra red radiation, for example in gas shielded metal arc welding and cutting.	Helmets but with provision for auxuliary heat absorbing filter.

ELECTRIC MOTORS

Electric Motor is the most common prime mover, suitable for various applications.

The reliability and life of the electric motor are mostly dependent on the selection, installation and maintenance of the system.

The different types of electric motors, commonly used in industries are described below in brief.

A.C. Motors

A.C. motors may be of two types : (*i*) Synchronus motor or (*ii*) Induction motor. The starter is almost same for both A.C. synchronus and induction motor [Fig. 14.16(*a*)]. But they differ in rotors [Fig. 1.16(*b*) & (*c*)]. In a synchronus motor A.C. is applied to the stator and D.C. is applied to the rotor. However, polyphase A.C. induction motors are very much popular and mostly used. They contain steel-frame stator with laminated iron core along with the winding of individual coil in the slots and a 'squirrel cage type' or 'wound type' rotor.

The rotor construction in a squirrel cage motor is of bars of copper or aluminium conductors. These bars are connected to each end of the rotor by circular copper rings.

Whereas the wound rotors are made of polyphase windings and are connected to three slip rings.

A.C. motors are always connected to the line through motor starters as because a direct connection will draw a huge current amounting three to six times the normal current, causing low voltage and consequently disturbances in the system.

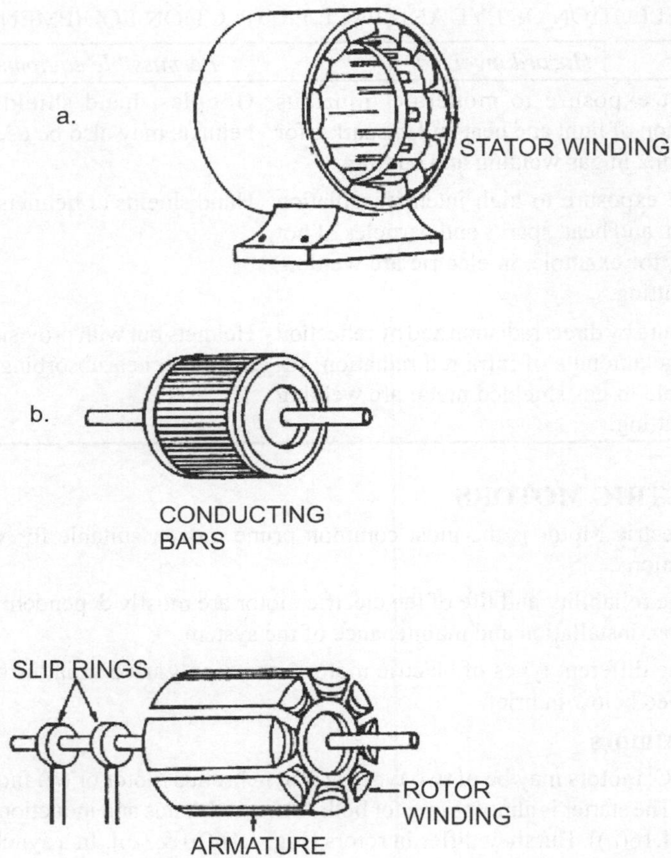

Fig. 14.16. (*a*) Stator (*b*) Rotor of induction motor (*c*) Rotor of synchronus motor.

D.C. Motors

The stationary part of the D.C. motor is the field which contains field poles or interpoles, either permanent magnets or laminated steel with field windings and the rotating part is the armature which is consisted of the laminated core, shaft, windings and the commutator. The commutators hold the carbon brushes to flow current from the armature.

The D.C. motors may be 'series wound' which provides high starting torque, 'shunt wound' which can provide minimum variation of speed or 'compound wound' motors.

Inspection on arrival

On receipt of the motors, the name plate particulars should be checked after removing the packing. The motor should be stored in a clean, dry place, if not erected at once. The insulation resistance will be reduced due to the presence of moisture. If it is kept in the store for a long period, the safe insulation resistance should be checked before putting it into service.

The low insulation always indicates the dampness and the motors are then to be dried out with suitable procedure.

Selection of motors

The selection of motor is generally guided by the following factors :

(*i*) **Supply available :** It is to be noted whether the supply is A.C. or D.C. Moreover, the supply voltage frequency, number of phases etc. are also to be taken into account.

(*ii*) **Requirement of speed :** The limits of ' speed' along with the variation of speed requirement, is to be considered.

(*iii*) **Starting torque or current :** Mostly to drive any equipment, a starting torque is necessary. So the torque requirement or the limitation of current during starting is a very important precondition.

(*iv*) **Type of power transmission :** The power transmission device is to be given due weightage. Power may be transmitted by belt or chain. The motor may be of direct coupled type or other type of coupling may also be engaged.

(*v*) **Mounting feature :** The mounting may be on the horizontal or vertical plane. It may also be mounted for an inclined shaft. However, requirement of special bed plate or mounting flanges are also the guiding factor of selection. Besides the above factors, the environment, the type of bearings or the altitude at which the motor will operate and of course, the cost etc. are also vital factors regarding the selection of electric motors.

(*vi*) **Load requirement:** The load of the motor is undoubtedly a primary criterion. The horse-power requirement is to be calculated with an anticipation of overloading .

Installation

Before the installation, one should be very much careful to finalise the location for setting up the motor. The location will be selected naturally considering the position of the machines or equipments to be driven. But a clean, dry, cool, well ventilated and well lighted area will be preferred. They are to be located away from heat radiation, from steam pipe lines etc.

The lay out for installation should provide adequate space for handling materials and maintenance work.

The foundation should be rigid. All possible measures are to be taken to avoid the transmission of vibration. Though the depth of foundation is dependent on the type of the soil but it is also dependent on the maximum output or horse power of the motor.

Provision for cable-ducts, required pipe lines etc. should be made along with the foundation.

The levelling and grouting of bedplates are done accordingly.

Maintenance work

Bearing : The ball or roller bearings are to be lubricated as per the maintenance schedule. Over lubrication may create problem. The vibration, wearing, under lubrication or excessive heat may cause bearing failure.

A regular check up of the temperature on the housing of the bearing, by touching with the hand is necessary.

On the other hand, the sleeve bearings should be inspected to check the height of the oil in the reservoir, to be assured that oil rings are supplying steady flow of oil.

A bearing may be taken as safe, if it operates below 70°C with minimum temperature variation.

The end play or the extent of wear of the bearing are to be inspected during servicing.

Slip rings and brushes

The slip rings should be maintained always clean and smooth. They should remain in glossy condition. The slip rings are taken for turning if any ridges etc. are formed on it, but no emery cloth is generally use.

Brushes should not be allowed to wear too far. They should safely ride on their holders and the brush holding spring should properly press the top of the brush for free functioning. Proper grade of quality brush is to be selected while replacing.

Commutator

The surface of the commutator should be kept always clean. On wearing, it may be made true in a lathe. The mica separators are usually recessed slightly below the commutator surface. Care should be taken, so that no carbon dust etc. accumulate in these grooves.

Commutator may be cleaned by wiping with a piece of cloth dampened with solvent. Emery cloth should never be used.

Operational problems for an A.C. Motor

A motor may have various problems while it is in operation. However, a chart for the trouble shooting of an A.C. motor is given below. The chart contains only a few of the major problems, their causes and remedies.

Motor is not getting started

Causes:

(*i*) Electrical connections may be loose, broken or wrong.
(*ii*) Availability of voltage is too low.
(*iii*) Power trouble.
(*iv*) Problem in the stator or rotor.

Remedies:

(*i*) Motor terminal connection may be checked and restored.
(*ii*) Proper voltage is to be made available.
(*iii*) Power at all lines are checked. A three phase motor will not start at one phase.
(*iv*) The stator is to be checked. Coils may have got opened or shorted. The rotor bars are checked. They are repaired or replaced if necessary.

Motor starts but does not take load

Causes:

(*i*) Voltage is not sufficient.
(*ii*) Starting load is too high.
(*iii*) Stator winding problem.
(*iv*) Problem in the rotor.

Remedies:

(*i*) Proper voltage is to be restored.
(*ii*) The load is checked and action may be taken accordingly.
(*iii*) The short circuiting of stator winding, if found, is repaired.
(*iv*) The rotor-bars are checked. Repairing is to be done if they are broken. They may also be replaced if necessary.

The motor is running noisy

Causes:

(*i*) Problem in the foundation or mounting.
(*ii*) Air gap between stator and rotor is non-uniform.
(*iii*) Unbalance of voltage in between the phases (for a three phase motor).
(*iv*) Ambient temperature may be too high.
(*v*) Rubbing on the stationary part.
(*vi*) Bearings are running dry.

Remedies:

(*i*) Foundation is checked. It may be redesigned (with the provision of vibration insulation) if necessary.
(*ii*) The gap is checked and restored. Bearing is also checked and aligned.
(*iii*) The voltage is to be balanced.
(*iv*) Adequate ventilation is established.

(*v*) Rubbing, if occurs, is identified and is repaired. The fan may specially be checked.

(*vi*) Rotor bearings are checked and lubricated properly.

Motor is getting overheated

Remedies:

(*i*) The load may be high.

(*ii*) Voltage is too high or too low.

(*iii*) Voltage variation among the phases.

(*iv*) Shorted winding.

(*v*) Insufficient ventilation.

Remedies:

(*i*) The load is checked. The load may be reduced or a larger motor may be used.

(*ii*) Specified voltage is to be restored.

(*iii*) Equal necessary voltage is established for all phases.

(*iv*) The winding is checked and rewinding is done if necessary.

(*v*) Proper ventilation is provided and blowers may be used if necessary to remove dirt and dust from the frame.

Operational problem for a D.C. Motor

A fault finding analysis for a D.C. motor is given below. The probable causes are mentioned. Remedial action may be taken according to the finding of fault.

Motor is not getting started

Causes:

(*i*) Non-availability of power.

(*ii*) Connections may be wrong.

(*iii*) Starter may be defective.

(*iv*) Applied load is excessive.

Commutator sparking of brushes

Causes:

(*i*) Applied load may be too high.

(*ii*) Brush contact is poor.

(*iii*) Armature winding may be shorted.

Motor is noisy or getting vibrated Causes:

(*i*) Lack of armature balancing.

(*ii*) Defective bearings. End play may be a factor.

(*iii*) Striking between armature and poles may occur.

Motor runs too fast

Causes:

(*i*) Improper connections.

(*ii*) Field current is either too low or too high. Open field may also be shorted.

(*iii*) Short circuiting in armature winding.

REVIEW QUESTIONS

1. What are the major causes of failure of pipes ? State the maintenance of pipes with special reference to the steam pipes.

2. State the major factors which affect life of a fan. What methods are to be adopted to minimise their pernicious effects ?

3. What are the ventilation and exhaust fans? State special features of positioning of exhaust fans.

4. State the purposes of installing dust collectors in a factory shed. How they are maintained? Describe with a neat sketch any type of dust collecting equipment to be used in chemical industry.

5. Write short notes on:

(*a*) Mechanical separators

(*b*) Electrostatic precipitator

6. State the major apparatus needed for oxy-acetylene welding. How are they maintained.

7. State the installation procedure of electric arc welding. Point out the safety measures to be adopted and identify the common troubles, their causes and remedies in case of electric arc welding.

8. State the important factors to be considered for selection of an electric motor. Discuss the operational problems of an A.C. motor.

9. What steps you will adopt, if

(*a*) An A.C. motor gets overheated.

(*b*) A D.C. motor runs noisy.

(*c*) Welding current is stopped immediately after starting.

(*d*) The bearing of a fan gets overheated.

(*e*) A hammer blow sound is produced in a steam pipe.

(*j*) A crack is found in a pipe line.

10. What special care is to be taken in the following cases ?

(*a*) The stresses in a steam pipe to its expansion is to be released.

(*b*) A pipe of smaller diameter is to be joined with a pipe of larger diameter.

(*c*) Smoke and heat is to be minimised from a laboratory.

(*d*) A motor at store will not be installed immediately.

(*e*) A high efficiency is to be achieved to collect dusts in boiler house.

15

ENGINEERING MATERIALS

Introduction

A large varieties of materials are used in the industries. The Engineers are expected to have a thorough understanding on the behaviour and characteristics of such materials along with their properties at different environments to use them fruitfully and purposefully.

Selection of materials for a particular situation may create a lot of problems to an engineer, who is supposed to recommend a simple solution readily.

The maintenance personnel should also have a thorough knowledge on the physical and chemical properties of various material along with their structural changes at different conditions.

Basically, the materials may be (*a*) Metallic or (*b*) Non-metallic. Furthermore, the metallic materials may be either ferrous or non-ferrous. Both ferrous and non-ferrous materials may again be classified as shown below:

Ferrous materials	*Non-ferrous materials*
(*i*) Pig Iron	(*i*) Copper alloys
(*ii*) Cast Iron	(*ii*) Zinc alloys
(*iii*) Wrought Iron	(*iii*) Nickel alloys
(*iv*) Different types of steel, e.g. Structural steel, Alloy steel, Tool steel etc.	(*iv*) Aluminum alloys
	(*v*) Magnesium alloys
	(vi) Miscellaneous alloys

We will discuss the ferrous and non-ferrous materials in brief. The non-metallic materials will also be discussed later in this chapter.

Ferrous Materials

The ferrous materials are largely used in the industries, as because they have got unique magnetic properties along with their availability in abundance and their cost per unit strength is very much economical.

Pig Iron

Pig iron is the first phase during the production of other ferrous materials. It is obtained by smelting iron ore. Pig iron contains 92 to 97 per cent iron and remaining portion contains silicon, manganese, sulphur and phosphorus along with carbon. The carbon remains in the pig iron either as iron carbide, being combined with iron or as graphite being free carbon. It may be stated that there is practically no direct use of pig iron in the industries.

Wrought Iron

Wrought iron is very much soft and ductile. In this phase, most of the impurities are removed from the molten pig and carbon content is less than 0.05 %. It may be termed as the purest form of ferrous materials. It is, so, very much malleable and can be forge-welded better. Wrought iron may be used for welded steam and water pipes, boiler tubes etc. Bar iron for stay bolts, rivets, chains may also be made of wrought iron for its corrosion and fatigue resistance. It is widely used for forging applications.

Cast Iron

It is basically the pig iron, which has been refined in structure and composition. Pig iron is melted and cast usually after addition of steel scrap.

Cast iron may be of five types, grey, ductile, malleable, high alloy and white. The form of excess carbon categories the types of cast iron.

Grey Cast Iron

Carbon remains mostly in graphite form and silicon content is very high in grey cast iron. Grey cast iron can be easily machined. It is having low tensile strength and it is comparatively soft. Grey cast iron is having very little tendency of shrinkage and that is why it can be used to have a sound casting effectively even in complex shapes. Air brakes, control valves etc. may be made of grey cast iron. Machine tool beds, pistons and cylinders for internal combustion engine, fly wheels etc. are generally made of grey cast iron.

Ductile Cast Iron

Ductile iron may be considered as a newly developed cast iron. Addition of a small amount of magnesium causes the graphite in the molten metal to condense into spheroids (nodular in form).

The fluidity, weldability, wear resistance, castability of such type of cast iron is very high along with its high strength and machinability.

Ductile iron is being largely used for heavy duty machineries, gears, dies, crankshafts, cams and other shock-resisting parts.

Malleable Cast Iron

The manufacturing of malleable cast iron is done through the heat treatment procedure 'annealing.'

Depending on the process of annealing the malleable cast iron may either be (*a*) white heart or (*b*) black heart cast iron. White cast iron is the basic material in both the cases. When the casting after being heated at about 1 000°C, are packed in oxidising medium, white heart (WM) cast iron is produced. But for black heart (BM) cast iron they are packed in an inert material and the rise of temperature remains in the range of 850°C to 900°C.

However, malleable cast iron is tougher and more resistant to bending. The machinability is also improved. The major uses are to manufacture Automobile parts, Gear blanks, pipe fittings etc.

High Alloy Cast Iron

To achieve better strength, more resistance to wear and corrosion, nickel, chromium, molybdenum, manganese etc. are added to cast iron. Such addition results a radical change in the micro structure. High alloy irons are more costly and can have special application. Mostly, cylinder heads, pistons etc. are made of nickel-molybdenum cast iron alloy.

White Cast Iron

Carbon is present in this type of cast iron as cementite. The colour of such iron is white as no graphite is present in it. This is also termed as a chilled cast iron. White cast iron is mostly manufactured by casting the grey iron and allowing it to cool rapidly. Naturally, it is hard and brittle. Their uses are also restricted to the part where no much machining is required and the requirement of abrasion resistance is more.

Steel

Steel is manufactured by adding carbon to pure iron. It may be roughly said as an alloy of iron and carbon. The carbon content may be in the range of almost zero *e.g.* in mild steel and upto 1.8 per cent in Tool Steel. Various chemical elements are added to achieve required physical properties of alloy steel.

It may be mentioned that as the carbon content in steel increases, the proportion of pearlite to ferrite increases in the micro-structure and the steel becomes stronger but its ductility will be reduced.

Plain Carbon Steel

Carbon steels e used for different purposes considering its carbon content.

(*a*) Mild Steel: This is a class of low carbon steel which contains carbon from 0.04 to 0.25-per cent and is used in the form of plates, angles, bars etc. It is also used for light utensil components. As it is having greater strength and elasticity than wrought iron and as it can be forged easily, Such steel is also used for boiler and tank plates, angle channels etc.

(*b*) Medium Carbon Steel: Such steels contain carbon from 0.25 to 0.65 per cent. They are popularly used for manufacturing rails, axles, steel tyres, drop forgings, castings etc.

(*c*) High Carbon Steel: The carbon content in this classification may be from 0.6 to 0.8 per cent. The steel becomes comparatively harder. They are considerably used for making springs clutch plates, die blocks etc.

Table 15.1

APPLICATION OF SOME HIGH CARBON STEEL

% of Carbon	*Weldability/Forge ability*	*Application*
0.75	Good	Hammers, dies, stone and coal drills.
0.85	Good	Dies, chisels, springs, Shear blades.
1.0	With care	Large taps, punches, chisels.
1.15	Difficult	Taps, dies, punches, reamers.
1.25	Difficult	Machine tools, drills.
1.5	Difficult	Machine tools, hard steel plates etc.

Structural Steels

Structural steels are basically low carbon steels. Gas cylinders, pressure vessels, sections and sheets of steel structure can be easily fabricated from such steels. Weldability of this steel is also very good. It is designated as St. 30, St 32, St 50 etc. The number associated with 'St' indicates the minimum tensile strength. As an example St 50 will have a minimum tensile strength of 50 kg/mm^2 or 500 N/mm^2.

Tool Steels

Tool steels are having the properties of very good shock and wear resistance, strength and toughness. This is a class of high carbon alloy steel. Tool steels obviously contain carbon (0.8" 1.3%), manganese (0.2 - 1.6%), silicon (0.5 - 2%), chromium (0.25 - 14%), tungsten (1.5 to 20%), vanadium (0.15 to 3%), molybdenum (0.8 - 5%) and cobalt (0.75 - 12%) in combination with iron.

Chemical composition and typical uses of some tool steels are given below (table 15.2) along with their designation.

Table 15.2

Types	*Designation*	*Uses*
Carbon Tool Steels	T 118, T70	Files, Drills, Shear blades, Chisels, Press tools etc.

	T60 and T85	Hand tools, Die blocks etc.
Carbon-Chromium Tool Steel	T 105 Cr 1 Mn 60 T 105 Cr 1	Knurling tools, Lathe centres, Cold forming tools
High Carbon, High Chromium Tool Steels	T215 Cr 12 T 160 Cr 12	Press tools, Shear blades etc.
Fast Finishing Tool Steels	T 140W_4 Cr 50	Marking tools, Finishing tools.
Shock resisting tools, steels	T 50 Cr I V 23	Heavy Duty Punches, chisels etc.
High speed and Tool steels	T 75 W 18 Co 6 Cr 4 V1 Mn 75	Reamers, broaches, drills, milling cutters, slitting saws etc.

Note: It may be noted that the associated digit with 'T' indicates average carbon percentage, *e.g.* T 60 means steel with 0.55% to 0.65% C.

Generally Tool Steels are selected considering their wear and abrasion resistance, strength of toughness. Hardenability and ability to perform work at elevated temperature.

Alloy Steels

A steel is called alloy steel when the elements like manganese, nickel, chromium, molybdenum, silicon, tungsten, cobalt etc. are present in sufficient quantity. The presence of these elements, other than carbon and iron will modify the properties of the steel. The alloy steels can provide better toughness, ductility or strength than those of carbon steels and as a result for high impact loading or to withstand high stresses, alloy steels can be recommended.

The effects of the alloying elements must have to be considered to select an alloy steel for a specific purpose. Effects of different alloying elements are listed below:

Manganese	:	Contributes to strength and hardness, decreases the critical cooling rate.
Nickel	:	Offers much toughness and corrosion resistance. Allows deep hardening .
Chromium	:	Improves hardenability, corrosion resistance.
Molybdenum	:	Increases tensile strength, improves hardenability.
Silicon	:	Promotes resistance to high temperature. Deoxidizes, increase the critical temperature for heat treatment.

Tungsten : Refines grain size, increases cutting hardness and magnetic retentivity.

Vanadium : Promotes fine grain sizes, deoxidizes.

Typical uses of a very few alloy steels are given in the following table (15.3).

Table 15.3

USES OF ALLOY STEELS

Alloy Steel	*Uses*
20 Mn 2/27 Mn 2	Levers, shafts, spindles, welded structure etc.
37 Mn 2	Shaft, axle, crank, connecting rods etc.
35 Mn 2 Mo 45	Crank shaft, bolts, levers etc.
40 Cr 1	Gears, Wear resistant plates etc.
40 Cr 1 Mn 28	Gears shafts, high tensile bolts/studs etc.
40 Ni 3	Cold tough steel – used in compressors, refrigerators, turbine blades, screws , bolts or nuts used under high stress etc.
35 Ni I Cr 60	Air craft, heavy vehicle components.
40 Ni I Cr I Mo 15	General machine parts like bolts, gears etc.
14 Mn 1S 14	Parts where good finish and machinability are required.
55 Si 2 Mn 90	Mostly used for springs in the hardened and tempered condition.
50 Cr 1V 23	

Note: The number ahead of each alloy steel indicates the average carbon content *e.g.* in 37 Mn 2 the carbon is 0.32 to 0.42%.

There are so many varieties of alloy steels, being developed and used commercially, that, it is simply impossible to discuss them even briefly with the scope of this book.

However, it may be mentioned that the costs of alloy steel vary widely.

Some may be a little more expensive than plain carbon steels, whereas some of them may be even twenty times costly.

Of course, the minimum changes of composition do not always restrict the selection of steel for a definite purpose. Because, a considerable range of properties can be developed in any steel (particularly the alloy steel) after the heat treatment.

Heat Treatment of Steels

The major importance of heat treatment is that, it changes the micro-structure of the materials and modifies physical properties. A few common processes of heat treatment for carbon and alloy steels are described in brief.

A. Hardening

It is the process to raise the temperature of a piece of steel upto 30 to 50 degree above the upper critical temperature and to cool it in a media like water or oil or brine. The usual temperature of hardening depends on the carbon content and it generally belongs in the range of 760 to 800°C. The time duration of maintaining the temperature is dependent on the required depth of hardening.

The degree of hardness can be varied by the rate of quenching. For a low and medium carbon steel, water may be the cooling media for rapid quenching, whereas for high carbon steel and alloy steel, oil is most suitable. Moreover, to avoid cracks, interrupted quenching may be followed. Such cooling may be done in oil bath, the temperature of which may be maintained in the range of 150° to 250°C.

Case Hardening

Case hardening is performed on a low carbon steel (0.04 to 0.25%) to enrich the surface artificially with carbon upto a certain depth. Filing or drilling becomes impossible after case hardening but it provides a hard bearing surface with a tough core. Journals or gudgeon pins etc. are mostly case hardened.

Carburising

Carburising is a common procedure of case hardening. The article is placed in an iron box, packed with carbonaceous material, such as charcoal, and is heated upto a temperature of 900° to 950°C. The temperature is to be maintained for several hours depending on the thickness of case hardened surface. Generally several pieces are packed together in a single box. The surfaces, which do not require hardening should be kept covered. Mixture of asbestos powder and fire clay may be used for such covering.

Cyaniding

A cyanide bath is used in this process to immerse the articles and to raise the temperature upto 850° to 950°C. This process produces a layer of iron nitrides on the surface. It provides considerable hardness and good resistance to wear. Generally smaller parts are selected for this treatment. Another advantage of this technique is that if required, the bright finish of machined parts can be maintained in this process.

Nitriding

Nitriding is another technique of case hardening. The ammonia gas is used in this process to introduce nitrogen to the surface at a temperature of 450° to 500°C. Actually, the formation of iron nitrides on the surface produce the required hardness. Nitriding is generally done for tools which take light cuts.

Flame Hardening

It is a form of localised hardening. An oxy-acetylene flame may be used for heating the surface and it is to be quenched immediately by means of a jet of water. Crank shaft bearing surfaces, cams, gear teeth etc. may be hardened with this process. In this process the distortion comes to a minimum due to localised heating and consequently little machining is required for finishing. For a high carbon steel, quenching rate may be adjusted and reduced or otherwise cracks may be developed.

Induction Hardening

Induction hardening is widely used in the industries. The induction heating circuit is a transformer when a copper conductor may be used for heating purpose. This conductor acts as a primary coil and the materials to be heated act as a secondary coil. A current of a bout 1000 to 10000 cycles per second is allowed to pass through the conductor. The heating of the material is performed by the thermal effect caused due to the induction of high frequency current. Thus the material is heated by alternating current, though the material is not directly connected to the power source.

The current flowing through the inductor will set up a magnetic field and the alternating magnetic lines will pass through the surface of the material. The induced flow of energy will cause to develop an internal molecular friction and this will ultimately dissipitate in the form of heat.

It is a rapid process. Moreover, by changing the frequency of voltage or by controlling current the depth of hardness can be controlled more easily.

B. Annealing

The internal stresses, developed by the operations like forging, welding or rolling may be relieved by this heat treatment. In this process the material is heated to a high temperature and is allowed to cool down slowly.

This process not only increases the ductility through refinement of grain size, or relieves the internal stresses but it also assists the material to undergo other required heat treatments.

Generally the cast iron is heated upto a temperature of 400 to 800°C depending on the degree of softening required and the Mild Steel is heated around 850°C depending on the carbon content. The cooling is performed generally keeping the material within the furnace and sometimes the work may be kept within packed charcoal or sand to have gradual cooling.

Steel is annealed mainly to improve the machinability characteristic.

C. Normalising

It is to some extent similar to annealing. It produces a uniformity in the structure of the material and improves mechanical properties. In this process the cooling is done in still air at ordinary temperature. Normalised steel will have better finish when machined with edge cutting tools.

Gears, bolts, nuts etc. may be heat treated to avoid major distortion. The sequence of operation may be followed as given below:

ROUGH MACHINING

↓

NORMALISING

↓

FINISHING ⟶ OTHER HEAT TREATMENT ⟶ GRINDING

Tempering

After hardening, when a steel is quenched, it becomes brittle and hard and contains internal stresses for the sudden cooling. Tempering is a reheating process to relieve this internal stress and to achieve a definite combination of mechanical properties.

Generally, tempering is done below the lower critical temperature. For cutting and measuring tools of low alloy steel, the temperature may be about 200°C, whereas for laminated springs or coils it may be upto 350°C. For structural steel the reheating may be done at about a temperature upto 500°C.

Different types of tempering *e.g.* austempering or martempering may be employed for different purposes. In austempering, distortion and quenching cracks may be avoided achieving an uniform microstructure. The interrupted process martempering may produce some residual stresses. A minimum distortion of material may also occur.

NON-FERROUS MATERIALS

Non-ferrous materials are the metals, in which iron is not present as a main constituent. These materials have got wide applications in the industry. Mostly the non-ferrous materials used in the industries are the alloys. However, copper, aluminum, nickel etc. are used extensively in unalloyed forms also.

Actually for specific requirement of combinations of some particular properties, the non-ferrous alloys are selected. Of course, before using such alloys the justification of the cost factor should be considered.

A number of useful non-ferrous materials are discussed below with reference to their properties and applications.

Copper

Copper is one of the most popular non-ferrous metal. The significant causes of its popularity are because of (*i*) its high conductivity and (*ii*) the alloys which it can form.

It is a yellowish red metal having a specific gravity of 8.93 and a melting point of 1083°C.

Pure copper is very much malleable and weak and naturally copper is not mostly used in its pure form.

Probably the most extensive use of copper is done in the electrical industries and in the chemical plants. Large quantities of cables, bus-bars, pipes etc. are made of copper.

Copper Alloys

Brass

It is one of the oldest alloys being used. It is an alloy of copper and zinc. Lead and tin may also be added.

A wide range of physical characteristics, *e.g.* corrosion resistance, machinability, workability in hot and cold form, may be manipulated by changing the proportions of copper and zinc and by adding lead, tin, nickel, aluminum etc. in different proportions.

Moreover, the physical properties are also dependent on the processes undergone, for example whether they are extruded or rolled or cast etc.

However, composition and characteristics of some special brasses are shown in the following table.

Cu	Zn	Sn	AI	Characteristics
60%	40%	–	–	Ductility is less, tensile strength is high. Mostly used for manufacturing valves and other fittings.
70%	30%	–	–	Ductility is more, comparatively strong. Mostly used as sheets, tubes are also made of it.
70%	28 -29%	1-2%	–	The resistance to corrosion increases.
76%	22%	–	2%	Corrosion resistance is increased for this alloy also.

Bronze

Bronze is a copper alloy being alloyed with tin. For various types of bronzes, the copper content may be some 85 to 95%. The remaining portion is tin. Sometimes a small amount of zinc (say upto 3%) is also added. It is mostly used for pipe fitting, valves and as a bearing metal. Gun metal with more percentage of tin yields more strength of the gun metal. The most common composition of gun metal is Cu - 88% Sn - 10% and Zn - 2%.

Small amount of phosphorus may sometimes be added to improve its mechanical properties. Such bronzes are called Phosphorus bronze. It can produce greater ductility and ability to withstand shock besides strength.

Bearings, slide valves etc. may be made of phosphor bronze. Other uses are worms, gears, pump rods, propellers etc.

Aluminum bronze has got wide field of application and is popularly used for its excellent resistance to corrosion and more strength. Copper-aluminum alloys having 7% aluminium possess excellent cold working properties for its enhanced malleability. When aluminium content is within 8% to 12%, the material may be worked even at an elevated temperature upto 950°C. They may also be used for casting.

Nickel bronzes are called as Monel Metal. It is having an average composition of 67% nickel and 28% copper. Tin, zinc, silicon, manganese along with about 5% iron may also be present in such nickel alloys. They are mostly used in chemical plants for pipes, vats, retorts etc.

Aluminium

Aluminium is a material of low tensile strength and is too soft in its pure state. Pure aluminium is of course quite satisfactory for household utensils etc. It is also used for motor-coach body panels and architectural purposes.

However, in its alloyed form, it can improve its mechanical properties upto such an extent that regarding even strength, they can be compared with alloy steels. Aluminium alloys are widely used to-day not only for domestic purposes but also for industrial and engineering purpose. Some of them are discussed below.

Aluminium Alloys

Aluminium - copper - magnesium alloys are very much strong. The magnesium content lies in between 3% to 7%. More often manganese or chromium (about I%) is also added.

The aluminium silicon alloys have got the best castability but its strength is not that much more.

Nickel is also sometimes used to maintain its mechanical properties at high temperature.

Duralumin is a widely used aluminium alloy having Cu – 3.5% to 4.5%, Mn - 0.4% to 0.7%, Silicon above 0.3% to 0.4%. A certain percentage of magnesium (about 0.6%) is also sometimes used. Duralumin is used mostly in form of forgings, pressings and rolled sections. It is not suitable for castings. The wing structure of aeroplane, connecting rods of petrol engine etc. are generally made of duralumin.

Nickel and its alloys

Nickel is a tough and silvery white magnetic material. It is mostly used as an alloying element but direct uses are also there for electrodeposition.

Monel metal is the most popular nickel-copper alloy. It contains nickel about 65% to 70%. Iron (3%), manganese (upto 2%) and silicon (upto 4%) are also there. The rest is the amount of copper. It is having high resistance to corrosion and is strong and tough. It is extensively used for manufacturing tubes, strainers, pumps, valves etc.

Nickel chromium alloys are widely used in electrically heated furnaces. They may be used as sheet or cast plates to place them at the bottom of the furnace. Nichrome is also a nickel-chromium alloy popularly used as heating coils.

Nickel - molybdenum alloys are having the quality of high corrosion resistance even in the presence of acids or alkalis. They are mostly used for the components required in the chemical plants.

The other alloy, known as 'Inconel' (Ni – 80%, Cr – 14%, Fe – 6%) is also popularly used for food processing industries. It is corrosion and heat resistant. This alloy is available in the form of rod, sheet, wire and in cast form.

Some other alloys

Zinc alloys : Zinc base alloys are also widely used in the industries. Carburetors, fuel pumps etc. in auto industries, machine parts like gears, blower rotors etc. and even household appliances may be made of zinc alloys.

The alloy, having Zn – about 95%, Al. – about 5% with a little amount of copper (say about 0.25%) is very much used for casting purpose. Of course all the zinc base alloys can be easily cast.

Tin base alloys : The main use of tin base alloys may be as 'solder' or as 'bearing materials' etc.

Tin and lead (say, Sn – 40%, Pb – 60%) produces soft solders. Hard solders may be produced adding copper and zinc, reducing the tin content at a minimum level.

More commonly used alloy is 'Babbit metal', principally used for bearing. An example of babbit metal may be,

Tin	–	91%
Copper	–	4.5%
and Antimony	–	4.5%

To make the alloy harder and stronger antimony content may be increased to a range of 7% to 7.5%, reducing the percentage of tin to 79% and also reducing amount of copper about 1%.

For higher speeds or for fluctuating loads, bearings made of babbitt metal is preferred more than that of bronze.

Non-Metallic Materials

The non-metallic substances mostly used in the industries may be classified in three categories, as follows:

A. Ceramic Materials B. Polymers and

C. Others

A. Ceramic Materials

The important ceramic materials used for industrial purposes may be of following types:

(*i*) Oxides (SiO_2, Al_2O_3, MgO)
(*ii*) Refractories (Fire clay)
(*iii*) Silicates (Asbestos)
(*iv*) Glasses
(*v*) Enamel (ceramic type surface coating)
(*vi*) Abrasives (Corundums, emery, SIC)
(*vii*) Cement and concrete

Oxides like SiO_2 may be used as refractory bricks, specially in the open hearth furnace. .Al_2O_3 is used widely in spark plug bodies, grinding wheels etc. other than bricks. Refractories are used for fire bricks, or to manufacture crucibles etc.

Silicates are mainly used to have electrical insulators. Asbestos itself is a complex magnesium silicate and mica is a potassium-alumina silicate. They may be used as thermal insulators also.

The amorphous substance glass has a wide ranging applications due to its transparency and high chemical resistance. Bottles, tubes, phials, window panels etc. are some of the examples of its extensive uses. They are widely used in chemical laboratories.

Enamel coating on steel or iron parts offer a finished and glossy surface. Diamond or Corundum (Al_2O_3 mineral) are the natural abrasives. Various synthetic abrasives are also now-adays used, specially for grinding wheels. The silicon carbide may be referred as the most important synthetic abrasive. They are also called carborundum.

Lastly, the cement is a mixture of silica, alumina and lime stone, or dolomite and sand. Its use does not require any special mention.

B. Polymers

It is now-a-days seen that the light weight and low cost polymers are replacing a good number of metals and ceramics. They are long chain 'organic molecules or macromolecules.' The important use is in the form of plastics. Plastic is becoming popular for its chemical resistance, appearance, strength, formability and machinability. They are used in huge quantity in electrical industries as insulators. Plastic may be classified into two groups :

(*i*) Thermoplastic and (*ii*) Thermosets

The thermoplastic can be remelted and softened by heat and can be made to any shape through injection moulding and extrusion. PVC (Polyvinyl Chloride) is type of Thermoplastic. They are very much used for hose pipes, electric wire insulation etc.

Thermosets cannot be remelted after solidification. They are available in the form of powder or liquid. They are formed by setting them with resins. Backelite is one of the important thermosets. Thermosets are generally harder and stronger than thermoplastic, Natural and synthetic rubbers are also plastics. Such plastics may also be extruded to form films, filaments etc. Vulcanised rubber is used mostly in tyre industries. The roll forming process 'Calendaring' is mostly performed to produce sheets from heated plastic materials. The fiber reinforced plastics are also becoming popular for its high mechanical strength.

C. Others

There are materials like leather, corks etc. being used in the industries.

Leather would have been mostly used for belts. But synthetic rubber is replacing it due to its non-availability and high cost.

Cork sheets, cork tiles etc. are industrially used specially as isolating pads for machineries. Corks are also used for bottle stoppers etc. They may be used in the polishing wheel of glass also.

REVIEW QUESTIONS

1. Classify the materials generally used in the industries. What are the different types of ferrous, materials ? State some of their applications.

2. What are various types of cast irons ? Write a brief note on each of them.

3. What are different types of steel ? Specify some of the applications of high carbon steel.

4. What are the effects of alloying elements? Mention their effects and some specific uses.

5. What is the importance of 'heat-treatment'? Write short notes on (*a*) Hardening and (*b*) Annealing.

6. Write a brief note on the compositions and applications of Brass and Bronze.

7. How may the non-metallic materials, used for the industrial purposes, be classified. Describe some of them.

8. Write short notes on:

(*i*) Tool Steel

(*ii*) Cyaniding and Nitriding.

(*iii*) Flame hardening and Induction hardening.

(*iv*) Thermoplastic and Thermoset

9. Identify the following sentences as true or false and rewrite the false statements correctly.

(*i*) Wrought iron is very much hard and brittle.

(*ii*) Duetile iron is being largely used for heavy duty machineries.

(*iii*) Malleable cast iron is termed as chilled cast iron.

(*iv*) Tool Steels are very much shock and wear resistant.

(*v*) Case hardening is performed in high carbon steel.

(*vi*) Inconel is very much used in food-processing industries.

10. Suggest suitable materials to manufacture following products :

(*a*) Heavy vehicle component (*b*) Springs (*c*) Gears (*d*) Spindle (*e*) Taps and dies (*f*) Reamers and (*g*) Grinding wheels.

INDEX